Systemic Risk and Complex Networks in Modern Financial Systems

New Economic Windows

Vincenzo Pacelli

Editor

Systemic Risk and Complex Networks in Modern Financial Systems

 Springer

Editor
Vincenzo Pacelli
University of Bari Aldo Moro
Bari, Italy

ISSN 2039-411X   ISSN 2039-4128 (electronic)
New Economic Windows
ISBN 978-3-031-64915-8   ISBN 978-3-031-64916-5 (eBook)
https://doi.org/10.1007/978-3-031-64916-5

The project was supported by University of Bari Aldo Moro with the co-financing of the European Union—Next Generation EU under the program "MUR—Fondo Promozione e Sviluppo—DM737/2021" managed by the University of Bari Aldo Moro (reference: S47–"Verso una gestione integrata e prospettica del rischio sistemico attraverso modelli di 'explainable artificial intelligence' (XAI)"—CUP H99J21017600006—Principal Investigator: Prof. Vincenzo Pacelli).

This Springer imprint is published by the registered company Springer Nature Switzerland AG
The registered company address is: Gewerbestrasse 11, 6330 Cham, Switzerland

If disposing of this product, please recycle the paper.

To my beloved children
Christian Francesco and Perla

Preface

In modern financial systems, complexity is intrinsic, inbred, inevitable, multifaceted, and systemic but this does not mean it cannot be understood, revealed, or explained and therefore hopefully managed. As researchers, we need to ask the right questions and search for the most pertinent answers with patience, critical spirit of observation, method, and self-sacrifice. The road is rough but fascinating to travel. In this book, we try to start the journey—or perhaps it is more correct to say that we try to continue that journey already started by other research—asking ourselves some questions and looking for some answers, in the awareness that many other questions will arise from our work. But that's what we hope for to continue the journey.

How is it possible to reveal the complexity of financial systems and how does complexity influence systemic risk? What is systemic risk? How does it originate? What are its determinants and how does it propagate? How can it be analyzed, measured, and predicted? What are its impacts on the economic and financial systems? What are spillover effects and how do interconnections influence systemic risk and contagion mechanisms? Can network science help unravel complexity and therefore better understand, analyze, and predict systemic risk and contagion mechanism? These are all relevant questions that this book aims to answer, analyzing the topic both with the eyes of economists and also with those of physicists, mathematicians, and jurists. And this is to delve deeper into the topic through the different sensitivities and experiences of scholars who study systemic risk from different angles.

So, this research goes in the direction of bringing together different worlds, which are still significantly distant although less than a few decades ago, so long as economists, physicists, mathematicians, and jurists often still speak different languages and struggle to fully understand each other, due to different studies, experiences, and sensitivities. This book can therefore also be understood as an attempt to bring together different worlds, from whose contamination of knowledge and more intense scientific collaboration, I believe promising prospects can arise.

The crises of the last years have underlined how much the modern financial systems are today more exposed and vulnerable to systemic risk, defined as the risk of uncontrolled propagation of a crisis of a single player or area of an economic system to a wider system through contagion mechanisms. Systemic risk is more relevant today than in the past due to the increasing interconnection between the players in economic and financial systems and the increasing speed of flows of goods, money, and people. Furthermore, we live in an era of systemic (we could say structural) instability, as the succession of economic cycles is much faster today than in the past and this inevitably causes greater turbulence and instability in the financial markets. All this prompts us to reflect on the need to analyze, predict, and manage systemic risk holistically and through the logical-conceptual schemes that can be borrowed from the network science. It is now obvious to everyone that modern financial systems are indeed complex systems with a large number of parts which interact in a not simple way. Furthermore, modern financial systems are characterized by complex and multi-layered dependencies that can only be discovered through the use of complex techniques borrowed also from the network science and the collaboration of researchers from different scientific areas. In fact, many dependencies can only be revealed through the appropriate and efficient analysis of big data and their understanding and decoding. While the efficient analysis of big data through the most effective methodologies is a subject of study of physicists, mathematicians, or statisticians, fully understanding what lies behind the data requires the knowledge and experience of economists and social scientists. Therefore, it appears extremely evident how today the study of economic and financial phenomena is necessarily interdisciplinary and this is the vision that inspired the idea of this volume. The hurdle often lies in the difficulty of meeting and communication (or understanding) between researchers coming from different and often distant areas and we hope that this book will be a fruitful opportunity for meeting and collaboration.

Starting from this consideration, I built an interconnected and contaminated index of essays which address the topic of systemic risk through different visual angles and different methodological approaches, but all sharing the awareness that the paradigms offered by network science allow a broader and more holistic view of the phenomenon.

In this volume, we host 20 essays, written by 47 different authors, coming from 19 universities from different countries.

Considering the nature of the object of investigation, a book on systemic risk needs to have a contaminated and interconnected index and so this book is organized as follows.

The first contaminated part of the book is dedicated to the issues of theory, policy, and methodology with 11 essays (chapters).

My first introductory chapter, deliberately titled like the volume "Systemic Risk and Complex Networks in Modern Financial Systems", embarks on a journey through the complex world of systemic risk, adopting a logical-conceptual framework that draws upon the strengths of network science. Recognizing the interdisciplinary nature of this challenge, the chapter promotes collaborative efforts that draw insights from diverse domains—economics, mathematics, statistics, physics, and computer science—to unravel the complex dependencies within financial systems. Shifting the focus from a siloed, micro-prudential perspective to a holistic, macro-prudential lens becomes paramount. This change acknowledges the inherent interconnectedness of financial entities, conceptualizing them as nodes within a dynamic network. This chapter underscores the centrality of network science in comprehending intricate shadow links, treating economic and financial systems as complex networks susceptible to cascading failures. These multilayer network approaches can be valuable tools for predicting crisis propagation and identifying early-warning signals of system-wide instabilities. From a macroperspective, the chapter acknowledges global cooperation's key role as a cornerstone for building sustainable economic and social systems in an increasingly interconnected world. The chapter extends an invitation to engage with the intricate dynamics of systemic risk and underscores the urgent need for collaborative solutions to safeguard financial and economic stability.

Our Chap. 2 "Systemic Risk and Network Science: A Bibliometric and Systematic Review", by Pacelli, Panetta, and Povia, using bibliometric analysis, explores research on systemic risk issues, revealing a decade of rapid growth and new research frontiers. Network science emerges as a powerful tool, providing deep insights into systemic risk mechanisms. Climate-related financial risks take center stage, highlighting the pivotal role of network models in stress testing and navigating the low-carbon transition. Moreover, digitalization and cryptocurrencies demand heightened attention due to their unique systemic risk profiles.

Pacelli, Canana, Chakraborti, Di Tommaso, and Foglia, in our Chap. 3 "A Holistic Journey into Systemic Risk: Theoretical Background, Transmission Channels and Policy Implications", undertake a meticulous examination of the multifaceted nature of systemic risk. Our analysis delves into various aspects, including its measurement, models, determinants, interconnections, and key variables. The chapter serves as an insightful guide, unraveling the intricate transmission channels through which systemic risk propagates, casting a critical eye on diverse conduits such as contagion, credit, liquidity, market, and macro-economic risks. Crucial to their discourse is the acknowledgment of the interconnected nature of these channels, which serves to amplify systemic risk and its widespread consequences. We explore the interconnections between financial institutions and markets, underscoring their pivotal role as catalysts of systemic risk and emphasizing the vital importance of network analysis in elucidating these complex relationships. Furthermore, the chapter highlights the need to consider both visible and hidden (shadow) interconnections when assessing systemic risk, advocating for a holistic approach that encompasses the full spectrum of vulnerabilities within the financial system.

Our Chap. 4 "Macro-Prudential Policies to Mitigate Systemic Risk: An International Overview", by Pacelli and Povia, delves into the global response to the intricate threat of systemic risk in financial systems. We analyze the reforms undertaken by diverse economies like Europe, USA, China, Islamic nations, and Japan. Through a comparative analysis, the chapter dissects the interconnected threads of these reforms, evaluating their effectiveness in mitigating systemic risks. This insightful analysis paints a picture of the international financial panorama undergoing a universal shift toward macro-prudential regulation. Despite diverse contexts and structures, a clear convergence toward macro-stability and systemic risk mitigation is shown across geographies.

Chapter 5 titled "Systemic Risks and Multilayer Financial Networks: From Contagion to Mitigation", by Quirici and Moro Visconti, examines the role of complex networks in the financial sector and how they can lead to the rapid spread of systemic risk through various transmission mechanisms. Understanding these mechanisms is essential for risk management and designing effective measures to mitigate the spread of systemic risk.

Sánchez-Garcìa and Cruz-Rambaud in their Chap. 6 titled "The Impact of Inflation and Financial Stability on the European Financial System: A Network Approach" build an econometric design to estimate a network of volatility connectedness, and an Exponential Random Graph Model (ERGM) is proposed to analyze the structure, capturing both endogenous and exogenous effects on the network.

Focusing on the financial system and insurance sector, Chap. 7 of Moliterni "Credit Risk Transfer and Systemic Risk" delves into the interplay between the banking and insurance sector, specifically systemic risk. The practice of credit risk transfer arises from banks' desire to divest themselves of credit risks. Insurance companies, especially those dedicated to risk transfer services, naturally become the recipients of these risks. Notably, credit insurance firms possess specialized capabilities in risk evaluation and selection. Banks opt to transfer their credit risks to insurance firms to mitigate their exposure. This transfer is facilitated by the insurance companies' adeptness at assessing and effectively managing such risks.

Chapter 8 titled "Systemic Cyber Risk in the Financial Sector: Can Network Analysis Assist in Identifying Vulnerabilities and Improving Resilience?", by Panetta and Leo, aims to theoretically deepen the different methodologies to forecast and measure systemic risk and crises, by applying various network analysis techniques to identify key nodes, evaluate their centrality and criticality, and simulate cascading effects under different cyber-attack scenarios.

Chapter 9 titled "Time Sensitive and Oversampling Learning for Systemic Crisis Forecasting", by De Nicolò, La Rocca, Marrone, Monaco, Tangaro, Amoroso, and Bellotti, examines the development of early-warning systems for systemic crises. This chapter proposes the combination use of the following techniques: (*1*) Temporal Cross Validation, aiming at extending the classical cross-validation framework considering the chronological ordering of time series data; (*2*) the SMOTE (Synthetic Minority Oversampling TEchnique) algorithm, used to augment the minority class in datasets, to balance crisis and non-crisis time points.

Chapter 10 of Biswas and Chakrabarti "A Fiber Bundle Model of Systemic Risk in Financial Networks" presents a review of developments for catastrophic failures using the fiber bundle model. This chapter analyzes the failure data of banks in terms of the inequality indices and studies a simple variant of the fiber bundle model to analyze the same. It appears, both from the data and the model, that the coincidence of these two indices signals a systemic risk in the network.

Part I of the volume concludes with an overview of the main approaches of measuring systemic risk offered in Chap. 11 "Measuring Systemic Risk: A Review of the Main Approaches" by Pampurini and Quaranta. The chapter reviews five approaches: (*i*) the Probability Distribution Measures, as the Delta Conditional Value at Risk (Delta CoVaR), the Marginal Expected Shortfall (MES), the Systemic

Expected Shortfall (SES), and the Systemic Risk Measure (SRisk); (*ii*) the Network Analysis Measures; (*iii*) the Illiquidity Measures; (*iv*) the Contingent Claims and Default Measures (CCA); and, finally, (*v*) the Macro-Economic Measures.

Part II of the book explores different and varied empirical insights through nine other essays (chapters).

Chapter 12 "Systemic Risk and the Insurance Sector: A Network Perspective" by Sylos Labini, D'Apolito, and Nyenno examines both the primary and latest regulatory and supervisory interventions implemented by European authorities regarding systemic risk management within the insurance sector. It also explores potential macro- and micro-systemic supervisory tools envisioned for regulating the insurance industry.

In our Chap. 13 "Damping Systemic Risk. The Role of Cooperative Banks" by Pacelli, Pampurini, and Quaranta, we investigate the potential countercyclical effect of cooperative banks within the context of systemic risk. Focusing on Cooperative Italian banks, the authors find the crucial role of these banks in mitigating systemic risk due to their unique business model, emphasizing their useful commitment to local development. The empirical evidence supports the argument that cooperative banks act as a counterbalance to systemic risk, prompting a shift in supervisory paradigms toward recognizing them as *"Too Useful to Fail"*.

Continuing on the topic of local banks, Chap. 14 "Shocks at Local Banks, EU GDP Growth, and Banking Sector Stability" by Arca, Carosi, and Moro delves into the potential impact of shocks originating from local banks on both the economic growth of the European Union (EU) and the stability of its banking system. The authors identify locally influential banks, which can have substantial influence over their respective regional economic landscapes, and investigate the relationship between idiosyncratic shocks occurring at these locally dominant banks and the stability of the EU banking system. This chapter demonstrates that idiosyncratic shocks experienced by the locally dominant banks have nationwide ramifications, contributing significantly to fluctuations in the aggregate macro-economic landscape of the EU. Furthermore, the authors underscore the significance of local banks within the EU.

Our Chap. 15 "How Does NPLs Securitization Affect EU Banks' Systemic Risk?" by Dell'Atti, Di Tommaso, Onorato, and Pacelli investigates the relationship between NPL securitization and systemic risk in EU banks. Using unique data (2012–2020), the authors analyze 35 European banks, considering G-SIB designation and country risk. The findings reveal a quadratic relationship: excessive NPL securitization can be a risk transmission mechanism. In fact, the authors find a threshold level beyond which securitization harms EU banks' financial stability. The authors underscore that to design systemic risk indicators better, it is necessary to incorporate the bank's securitization exposure and NPL resolution plans.

In Chap. 16 "The Systemic Importance of Cyber Risk in Banks" by Birindelli and Iannuzzi, the focus is on evaluating cyber systemic risk with specific regard to the banking and financial sector by highlighting the progress made in academic studies, the systemic impacts of this risk as well as the point of view of supervisory authorities. Using some data from the ORBIS database, the chapter examines the current exposure of banks (and other financial institutions) to this new and sophisticated risk.

Exploring the dynamics of risk contagion in cryptocurrency markets over the past decade, Chap. 17 "The Dynamics of Crypto Markets and the Fear of Risk Contagion" by Aliano, Ferrara, and Ragni investigates the dynamics of risk contagion in cryptocurrency markets over the past decade and the authors introduce a Susceptible-Infected-Recovered model incorporating a time delay. They focus on the governance token prices of major cryptocurrency exchange platforms, along with their spillover effects, crash risks, and indicators of market participants' attention. The parameters derived from this assessment are employed in the Susceptible-Infected-Recovered model to replicate the intricate dynamics of risk contagion observed in the analyzed cryptocurrency markets.

Similarly, our Chap. 18 "Cryptocurrencies and Systemic Risk. The Spillover Effects Between Cryptocurrency and Financial Markets", by Pacelli, Di Tommaso, Foglia, and Ingannamorte, delves into the tail risk spillover effect between cryptocurrencies, capital markets, and systemic risk. By a VAR for VaR model, we investigate the cryptocurrency market's influence on global equity indexes under both bearish and bullish conditions. The analysis reveals complex dynamics in risk-return trade-offs, highlighting the risk interconnectedness of these financial (assets) markets and emphasizing the international importance of the U.S. market as a risk hub market.

Shifting the focus on the circular economy, Chap. 19 "Financial Challenges and Threats of Circular Economy Logistics" by Capozza, Mokiy, Zvarych, Ilyash, and Vankevych focuses on the financial challenges and potential threats that could impede the successful implementation of the circular economy, particularly examining the logistics involved. To conduct a thorough analysis of financial risks, the authors propose a step-by-step approach. Initially, the attention is focused on a key logistics-related indicator within the European Union countries. Subsequently, the authors identify financial risks that have the potential to disrupt the efficacy of circular logistics. Subsequently, the authors employ the circular material waste (CMW) indicator to evaluate how these risks impact the reuse of materials. Finally, building on the outcomes from the first phase, an additional exploration was undertaken to gather information on startups contributing to the efficient functioning of circular economy logistics.

Finally, assessing the systemic risks linked to capital investments in a conflict-afflicted economy such as Ukraine, the last Chap. 20 "Systemic Risks to Capital Investment Flows in the Post-crisis Economy of Ukraine" by Rubino, Mokiy, Fleychuk, Khaustova, and Salashenko analyzes the systemic risks associated with capital investments in the economy of a conflict-affected country. To have a comprehensive examination of it, the authors employ the Vector Autoregression (VAR) methodology and the Kalman filter.

Therefore, addressing the issue of systemic risk from different and varied angles, this volume "Systemic Risk and Complex Networks in Modern Financial Systems" aims to offer insights into understanding, measuring, and mitigating systemic risk within financial systems, emphasizing the interdisciplinary nature of this challenge and the importance of collaborative efforts in addressing it effectively.

Bari, Italy   Prof. Vincenzo Pacelli
April 2024

Acknowledgements

I am grateful to all the authors of the essays for their valuable contributions and I am also grateful to the Editorial Board and the Editorial Team of the New Economic Windows series of the publisher Springer for their support in getting this volume published in their esteemed series. I also address my thanks to Caterina Di Tommaso, Matteo Foglia, and Maria Melania Povia for providing valuable help during the preparation of the manuscript and for the continued commitment to research. I would also like to thank Stefano Dell'Atti and all my colleagues of the banking and finance research group of academic departments of Bari, Taranto, and Foggia for the continuous and faithful scientific discussion and cooperation.

In order to guarantee maximum dissemination of the results of our research, we have financed the open access availability of all the essays in this book. This book in open access was produced with the co-financing of the European Union—Next Generation EU under the program "MUR—Fondo Promozione e Sviluppo—DM 737/2021" managed by the University of Bari Aldo Moro (reference: S47—"Verso una gestione integrata e prospettica del rischio sistemico attraverso modelli di 'explainable artificial intelligence' (XAI)"—CUP H99J21017600006—Principal Investigator: Prof. Vincenzo Pacelli). I therefore wish to express my gratitude to the University of Bari Aldo Moro and to my Academic Department.

Finally, allow me a final digression into the sphere of more intimate feelings. Primarily, I sincerely hope that my deepest feeling of gratitude can reach my parents wherever they are. And then, I would like to thank my wife and my children Christian Francesco and Perla for their patience and dedicate this book to them.

Bari, Italy   Prof. Vincenzo Pacelli
April 2024

Contents

Editor and Contributors

About the Editor

Vincenzo Pacelli, qualified as Full Professor in Economics of Financial Markets and Institutions, currently is Associate Professor at University of Bari Aldo Moro, where he holds the courses of "Economics of Financial Markets and Institutions" and "Banking Strategy and Management" and where he is President of two degree courses. He holds a Ph.D. in Banking and Finance from the University of Rome "Sapienza" and he is author of more than 80 scientific publications, speaker in International Conferences, and lecturer in training days in University Master and in Ph.D. seminars.

Contributors

Mauro Aliano Associate Professor, University of Ferrara, Ferrara, Italy

Nicola Amoroso Associate Professor, University of Bari Aldo Moro, Bari, Italy

Pasqualina Arca Assistant Professor, University of Sassari, Sassari, Italy

Roberto Bellotti Full Professor, University of Bari Aldo Moro, Bari, Italy

Giuliana Birindelli Full Professor, University of Pisa, Pisa, Italy

Soumyajyoti Biswas Assistant Professor, SRM University-AP, Amaravati, Andhra Pradesh, India

Lucianna Cananà Senior Researcher, University of Bari Aldo Moro, Bari, Italy

Claudia Capozza Associate Professor, University of Bari Aldo Moro, Bari, Italy

Andrea Carosi Full Professor, University of Sassari, Sassari, Italy

Bikas K. Chakrabarti INSA Senior Scientist and Emeritus Professor, Saha Institute of Nuclear Physics, Kolkata, India;
Economic Research Unit, Indian Statistical Institute, Kolkata, India

Anirban Chakraborti Professor, School of Computational & Integrative Sciences, Jawaharlal Nehru University, New Delhi, India

Salvador Cruz-Rambaud Full Professor, Mediterranean Research Center on Economics and Sustainable Development, CIMEDES, Almería, Spain;
Full Professor, Department of Economics and Business, University of Almería, Almería, Spain

Elisabetta D'Apolito Associate Professor, University of Foggia, Foggia, Italy

Stefano Dell'Atti Full Professor, University of Bari Aldo Moro, Bari, Italy

Francesco De Nicolò Ph.D. Student, University of Bari Aldo Moro, Bari, Italy

Caterina Di Tommaso Senior Researcher, University of Bari Aldo Moro, Bari, Italy

Massimiliano Ferrara Full Professor, Mediterranea University of Reggio Calabria, Reggio Calabria, Italy

Mariya Fleychuk Professor, Stepan Gzhytskyi National University of Veterinary Medicine and Biotechnologies, Lviv, Ukraine

Matteo Foglia Senior Researcher, University of Bari Aldo Moro, Bari, Italy

Antonia Patrizia Iannuzzi Associate Professor, University of Bari Aldo Moro, Bari, Italy

Olha Ilyash Professor, National Technical University of Ukraine "Igor Sikorsky Kyiv Polytechnic Institute", Kyiv, Ukraine

Stefania Ingannamorte Ph.D. Student, University of Bari Aldo Moro, Bari, Italy

Viktoriia Khaustova Professor, Research Center for Industrial Problems of Development of the National Academy of Sciences of Ukraine, Kharkiv, Ukraine

Marianna La Rocca Research-track Assistant Professor, University of Bari Aldo Moro, Bari, Italy

Sabrina Leo Associate Professor, Sapienza University of Rome, Rome, Italy

Antonio Marrone Full Professor, University of Bari Aldo Moro, Bari, Italy

Anatoliy Mokiy Professor, Institute of Regional Studies Named After M.I. Dolishny, National Academy of Sciences of Ukraine, Lviv, Ukraine

Francesco Moliterni Full Professor, University of Bari Aldo Moro, Bari, Italy

Alfonso Monaco Senior Researcher, University of Bari Aldo Moro, Bari, Italy

Ornella Moro Full Professor, University of Sassari, Sassari, Italy

Roberto Moro-Visconti Associate Professor, Catholic University of the Sacred Hearth of Milan, Milan, Italy

Iryna Nyenno Full Professor, KU Leuven, Leuven, Belgium

Grazia Onorato Research fellow, University of Foggia, Foggia, Italy

Vincenzo Pacelli Associate Professor, University of Bari Aldo Moro, Bari, Italy

Francesca Pampurini Associate Professor, Catholic University of the Sacred Hearth of Milan, Milan, Italy

Ida Claudia Panetta Full Professor, Sapienza University of Rome, Rome, Italy

Maria Melania Povia Ph.D. Student, University of Bari Aldo Moro, Bari, Italy

Anna Grazia Quaranta Associate Professor, University of Macerata, Macerata, Italy

Maria Cristina Quirici Associate Professor, University of Pisa, Pisa, Italy

Stefania Ragni Associate Professor, University of Ferrara, Ferrara, Italy

Alessandro Rubino Associate Professor, University of Bari Aldo Moro, Bari, Italy

Tetiana Salashenko Senior Research Fellow, Research Center for Industrial Problems of Development of the National Academy of Sciences of Ukraine, Kharkiv, Ukraine

Javier Sánchez-García Assistant Professor, Mediterranean Research Center on Economics and Sustainable Development, CIMEDES, Almería, Spain; Department of Economics and Business, University of Almería, Almería, Spain

Stefania Sylos Labini Full Professor, University of Foggia, Foggia, Italy

Sabina Tangaro Associate Professor, University of Bari Aldo Moro, Bari, Italy

Ivan Vankevych Ph.D. Student, West Ukrainian National University, Ternopil, Ukraine

Iryna Zvarych Professor, West Ukrainian National University, Ternopil, Ukraine

Part I
Theory, Policy, and Methodology

Chapter 1
Systemic Risk and Complex Networks in Modern Financial Systems

Vincenzo Pacelli

Abstract The crises of the last years have underlined how much the modern financial systems are today more exposed and vulnerable to systemic risk, defined as the risk of uncontrolled propagation of a crisis of a single player or area of an economic system to a wider system through contagion mechanisms. Systemic risk is more relevant today than in the past due to the increasing interconnection between the players in the economic system and the increasing speed of flows of goods, money and people. All this prompts us to reflect on the need to analyse, predict and manage systemic risk holistically and through the logical-conceptual schemes that can be borrowed from network science.

Keywords Systemic risk · Network science · Financial crisis · Financial systems · Complex networks

Introduction

The crises of the last years have underlined how much modern financial systems are today more exposed and vulnerable to systemic risk, defined as the risk of uncontrolled propagation of a crisis of a single player or area of the economic system to a wider system through contagion mechanisms. Systemic risk is also much more relevant today than in the past, due to the increasing interconnection between the players in the economic system and the increasing speed of flows of goods, money and people. Just as, from a health point of view, the transmission of a virus is closely linked to the relationships between individuals and the speed of their flow, the same can be said for the transmission of a financial crisis. We could, in other words, note how the genesis and evolution of the crises of the last years have highlighted the relevance and danger of systemic risk and the need to analyse and manage it

V. Pacelli (✉)
University of Bari Aldo Moro, Bari, Italy
e-mail: vincenzo.pacelli@uniba.it

© The Author(s) 2025   3
V. Pacelli (ed.), *Systemic Risk and Complex Networks in Modern Financial Systems*,
New Economic Windows, https://doi.org/10.1007/978-3-031-64916-5_1

through logical-conceptual schemes borrowed from network science. A lot of relevant phenomena in modern financial systems can be understood only in terms of interactions between financial actors, as I will try to explain later.

Furthermore, modern financial systems are characterised by complex and multi-layered dependencies that can only be discovered through the use of complex techniques borrowed also from network science and the collaboration of researchers from different areas (economics, mathematics, statistics, physics and computer science). In fact, many dependencies can only be revealed through the appropriate and efficient analysis of big data and their understanding and decoding. While the efficient analysis of big data through the most effective methodologies is a subject of study by physicists, mathematicians or statisticians, fully understanding what lies behind the data requires the knowledge and experience of economists. Therefore, it appears extremely evident how today the study of economic and financial phenomena is necessarily interdisciplinary and this is the vision that inspired the idea of this volume. The hurdle often lies in the difficulty of meeting and communicating (or understanding) between researchers coming from different and often distant areas, and we hope that this book will be a fruitful opportunity for meeting and collaboration.

Besides the evolution and rapid global spread of the 2007–2008 financial crisis have also highlighted *inter alia* the limits of the architecture of regulation and control in the financial sector, suggesting a broadening of the spectrum of analysis from a merely micro-prudential perspective (focusing on banks or intermediaries individually) to a macro-prudential one, which adequately takes into account the complex networks of relationships that characterise financial systems. Until the outbreak of the great financial crisis, international financial regulation and supervision were disproportionately oriented toward individual intermediaries and players in the financial system and underestimated the interconnections between them. As mentioned, the events of the last years have highlighted the relevance and the danger of systemic risk, reflexively inducing the international supervisory authorities to a sudden, radical and opportune widening of their spectrum of analysis, or from the conditions of solidity and solvency of the single actor to the overall stability of the entire economic system.

In this introductory essay of this volume, we investigate the need to analyse the complex phenomenon of systemic risk in a holistic way and through logical-conceptual schemes borrowed from network science. To support this thesis, in the next paragraph, I will outline the causes and propagation mechanisms of systemic risk, while in the third paragraph, I will deepen the logical-conceptual paradigms that make network science a fundamental tool for understanding and analysing systemic risk, and then conclude in the fourth paragraph with some reflections on the current economic situation.[1]

---

[1] This chapter is a revised and updated version of an essay previously published in the volume edited by V. Pacelli and I. C. Panetta, *Systemic Risk, Monetary Policy and Portfolio Diversification in the Great Crises' Era*, Quaderni del Dipartimento Jonico n. 19/2022, Edizioni DJSGE, Università degli Studi di Bari Aldo Moro, 2022 (ISBN: 9788894503098).

Systemic Risk: Definition, Causes and Propagation Mechanisms

Although it is not easy to attribute an unambiguous and exhaustive definition to "systemic risk" due to its complexity and multidimensionality as well as its many causes and varied facets, nevertheless, wishing to circumscribe the perimeter of the discussion, it is possible to affirm that systemic risk is defined as the risk that the crisis, bankruptcy or the mere perception by the market of the risk of insolvency of one or more relevant players in an economic system—essentially large companies, financial intermediaries or sovereign States—may lead to generalised crises, insolvency or chain failures of other players in the economic system. Therefore systemic risk is the risk associated with the manifestation of an event capable of causing, through propagation and contagion mechanisms, structural effects and a situation of systemic and generalised instability. Hence, systemic risk appears to be multidimensional and extremely complex as its origin lies in multiple phenomena and dynamics that are significantly interconnected, while its propagation is influenced by the various forms of interaction and interconnection between the players in an economic system.

As well illustrated by the International Monetary Fund (Blancher et al. 2013), systemic risk tends to manifest itself through sequential events that start from one or more shocks of various nature which then propagate in the economic system with a chain effect generating a crisis of systemic nature. Systemic risk must be defined as typically endogenous to the financial system, not only because it derives from the collective behaviour of economic agents whose choices may be rational at an individual level, but harmful to the financial system as a whole, but above all because it tends to feed on itself, to accumulate over time and then spread, in an uncontrolled way, throughout the financial system when a shock occurs.

To deepen and understand the complex nature of this risk, it is, therefore, necessary to examine the possible triggers and mechanisms of propagation of systemic risk.

First of all, it is important to clarify that when talking about triggers, it is necessary to distinguish between the initial shocks that give rise to the mechanisms of propagating the crisis and the causes that feed and favour these propagation mechanisms.

A shock can be defined as any event that can significantly or structurally modify or transform a financial system, limiting or even inhibiting its ability to carry out its specific functions. There are many shocks capable of generating mechanisms for the systemic propagation of a financial crisis, but they can be classified, at the cost of some simplification of a taxonomic nature and some overlapping, in four closely linked categories:

- sudden increase—not justified by a phase of economic expansion—of the cost of money, which incentivises the phenomena of moral hazard and adverse selection (Delli Gatti 2008; Rötheli 2010);
- incorrect or short-sighted economic, social and fiscal policies that lead to imbalances in public accounts, trade and balance of payments, or that inhibit a country's

growth, lending itself to speculative attacks on sovereign debt (Eser and Schwaab 2016; Pacelli 2014; Pagano and Sedunov 2016);

- crisis, bankruptcy, or mere market perception of the risk of default of a financial institution, sovereign state, or industrial firm characterised by significant size and significant economic and financial correlations (Allen and Gale 2000; Beirne and Fratzscher 2013; Cronin et al. 2016; Engler and Große Steffen 2016; Freixas et al. 2000; Michaelides et al. 2015; Nelson and Katzenstein 2014; Pagano and Sedunov 2016);
- idiosyncratic, exogenous events of varying nature and origin that lead to an abrupt contraction of demand or supply in certain markets, reflexively causing the sudden reduction of prices of real and financial assets and, in particular, the collapse of the value of the residential or commercial real estate (Reinhart and Rogoff 2013; Rötheli 2010).

Listed the main shocks that can give rise to the systemic propagation process of a crisis, it is then necessary to dwell upon the causes that favour and feed these propagation mechanisms. So, how does a shock become systemic? The causes that determine the propagation of an initial shock, through the mechanisms that we will examine later, to an entire economic system are essentially the following:

- crisis of confidence, uncertainty and information asymmetries in financial markets (Cottarelli et al. 2010; Duarte and Eisenbach 2021; Flannery et al. 2013);
- high indebtedness of players in the economic system and thus high financial dependence of debtors on creditors and vice versa, which makes the whole system vulnerable in times of crisis (Brunnermeier et al. 2016);
- high interconnectedness between the main players in the economic and financial systems, in particular high exposure of banks to sovereign debt and interbank markets (Blundell-Wignall 2012; Brutti and Sauré 2015; Flannery et al. 2013; Böhm and Eichler 2020; Hoque et al. 2015);
- microeconomic causes such as, for example, gaps in financial regulation in some countries, limitations in prediction and risk management techniques, deficiencies in asset valuation models or international accounting rules (Beirne and Fratzscher 2013; Eser and Schwaab 2016; Gonçalves and Guimaraes 2015; Pacelli 2014).

In particular, the empirical evidence of the main international financial crises confirms how it is precisely the high level of interconnection between the main players in the economic and financial system that is the main factor that feeds the mechanisms for the propagation of a systemic crisis. The genesis and subsequent evolution of the U.S. financial crisis characterised by over-indebtedness, initial low-interest rates, uncontrolled development of financial engineering, ineffective controls by the competent authorities, opportunistic behaviour of various players and speculative bubbles in various markets are too well known to dwell further. The US subprime mortgage crisis of 2007 thus became systemic due to the interconnection between international economic and financial systems and spread throughout the world and in particular to weaker areas, just like a virus that attacks and does greater damage to weaker subjects with lower immune defences. This is what has happened in Europe since 2008 that is international financial speculation has taken advantage of imbalances in public finances and the balance of trade and payments of some European countries (PIIGS) (Portugal, Italy, Ireland, Greece and Spain), of the absence of a European political union capable of supporting monetary union and therefore of a weak political and economic governance at the community level with limited powers, which has therefore detected inadequate to prevent and deal with the crisis. In recent months, there has been great concern about the sustainability of the Chinese economy's debt, which risks producing significant systemic effects in the event of deflagration due to the abnormal size of China's public and private debt[2] and due to speculative bubbles in the real estate sector, which expose the main Chinese real estate companies (Country Garden, Poly, Evergrande, Vanke and Sunac) to the risk of default. The current crisis in the Chinese real estate sector or the crisis of sovereign debt in Europe in 2009 can certainly be considered, therefore, as consequences of the systemic nature of the US crisis of 2007, but it is equally true that all these crises were favoured by evident errors in economic policy.

After describing the possible initial shocks and the causes that can feed the systemic propagation of these shocks to an entire economic system, it is now necessary to deepen the mechanisms and methods of transmission and propagation of a systemic crisis. In this regard, Aharony et al. (1986) underline the existence of two channels for the propagation of shocks: the "direct exposure channel" and the "information channel", which can act independently but also jointly. While the direct exposure channel refers to the domino effects that can occur due to the significant interconnections present within the interbank market or due to the sovereign exposures held by the banking system; the information channel reconnects to information asymmetries or errors in the interpretation of signals by market participants and savers, who are imperfectly informed respect to the nature of the shock (Di Clemente 2016). As argued above, following a shock to the economy, first of all, the ability of borrowers to repay loans is reduced. Thus, the first effect of the shock is represented by a reduction in the value of banks' assets, and this reduction will be more pronounced the greater the percentage of assets recorded at market value on banks' balance sheets. To secure their balance sheets and preserve their capital and financial solidity, banks react to these shocks by rationing credit and activating strategies of deleveraging and asset sales, which fuel the process of falling prices of financial and real assets in the markets and thus the deflationary spiral that, also due to the reduction in the values of bank assets and assets, in general, available in the markets as collateral, further fuels the vicious circle of credit rationing, the collapse of investment and industrial production, the reduction in tax revenues and the increase in

---

[2] As of March 2021, China's aggregate household, corporate and public sector debt exceeded $46 trillion, or 287% of China's annual GDP. While at the beginning of the century China accounted for a fairly small share (less than 3%) of global non-financial sector debt (public and private), its weight is currently 21%, second only to that of the United States (28%). To understand, in particular, the enormous leverage of the Chinese construction sector, it is sufficient to consider that at the end of 2020 the top 5 Chinese real estate companies (Country Garden, Poly, Evergrande, Vanke and Sunac) had total liabilities (excluding "off-balance sheet" liabilities) of over $1 trillion, which is about 10 times more than 2011 levels.

public debt, with consequent difficulties in refinancing on the markets and increased speculative pressures on the cost of debt issued by a government, whose solvency may even be compromised in the most serious cases, leading to chain effects on the balance sheets of banks holding sovereign debt. It can therefore be seen how, in case of the occurrence of an event capable of fuelling the propagation mechanisms of a systemic crisis, a bank's defence strategies can determine, in the absence of concerted and efficient monetary and fiscal policies, perverse effects which causes exacerbation of the deflationary spiral and further undermining the economic and capital balances of the economic system concerned. All these risks feed through both the channel of direct exposure of banks' balance sheets and through the information channel, since the investment or disinvestment choices of institutional investors.

The Utility of Network Science for Systemic Risk Analysis

Today we live in a globalised, complex and significantly interconnected world, in which complex systems and phenomena, apparently distant from each other, in reality, mutually influence each other. This reality exposes societies to systemic risks and requires analysis through a holistic approach to complex phenomena in order to investigate, through the help of network science, hidden relationships between economic, financial, political, health and environmental phenomena. Hence, today, nothing happens independently. Phenomena and events are connected with countless others in a huge and complex universal puzzle, in which these phenomena and events mutually cause and interact with each other (Barabasi 2004). We, therefore, live in a "small world", in which everything is connected and often in ways that are difficult for human rationality to understand, especially when inadequate methods of analysis are used. Moreover, the total volume in value of financial transactions in international financial markets has greatly exceeded the volume in value of transactions in the real economy, thus generating a significantly wide, complex, and interconnected network of financial systems that exposes international financial systems to systemic risks. So, only by analysing economic and financial systems as networks, it is possible to discern complex relationships between phenomena of different origins and nature that can generate the propagation of systemic crises.

The study of financial systems as networks thus allows us to discern the way and direction through which a shock could propagate in the system. We can consider financial systems as networks of interconnected economic agents (nodes), whose relations between them are regulated through financial contracts (financial instruments) (links). A network is a set of nodes connected by edges (links). But the relationships (links) between economic agents (nodes) are not only formal and explicit (i.e., regulated by financial contracts between two or more operators), they can also be informal and implicit, through—for example—the common investment of two economic agents in the same asset. In this case, it speaks of indirect contagion through portfolio overlapping.

The relationships between the economic agents (nodes) of a network in the economic-financial sector can therefore be of different kinds and tend to change over time. This variety and complexity (and often lack of transparency) of the financial relationships between the various nodes of a network increases the complexity of financial systems, which produces information asymmetries, risks of moral hazard and, therefore, consequences in the propagation of risks of a systemic nature. As we have already argued above, therefore, the financial crisis born in the USA in 2007–2008, the subsequent—and in some ways consequent—sovereign debt crisis in Europe in 2010–2011 and then the crisis arising from the pandemic, the Chinese debt crisis and energy and inflation crisis are clear examples of how it is necessary to approach the study of economic and financial systems through the principles of network science, to aspire to interpret the hidden relationships that bind together complex phenomena.

The origin of network science dates back to 1736 when the Swiss mathematician Euler (1707–1783) inaugurated (perhaps unconsciously) a prolific branch of mathematics known as "graph theory", which is today the basis of modern network science.[3] Over the years, several other mathematicians contributed to the development of this science, but among them who deserve special mention are the Hungarian mathematicians Paul Erdos (1913–1996) and Alfred Renyi (1921–1970), founders of the theory of random networks.

Unlike the random networks of Paul Erdos and Alfred Renyi, which are purely static (i.e., the number of nodes and links tends to remain unchanged over time), the "real" networks (i.e., those with which we are confronted every day in economic systems) are dynamic, because they evolve over time or tend to grow. In other words, real networks are characterised by several nodes and connections among them that grow and evolve over time. In particular, in the real world, each network originates from a small nucleus and expands with the addition of new nodes and links that tend to be created according to a mechanism of "preferential connection", i.e., the new nodes, when they have to decide where to connect, tend to prefer the nodes that have more links.

Real networks are also characterised by a hierarchy of hubs, where a larger and therefore highly connected node is followed by many other less connected nodes, followed in turn by even smaller nodes. These characteristics of real-world networks are empirically confirmed in interbank markets by In't Veld et al. (2020), who show that financial networks in inter-banking have a configuration similar to a core-periphery structure, where the core forms a network of fully connected nodes, while the peripheral banks are connected only to the core.

In the real world, moreover, networks tend to be self-organizing, i.e., they are an example of how the independent actions of millions of nodes give rise to emergent behaviours, and all these characteristics of real networks must be taken into account when analysing economic phenomena.

---

[3] For a comprehensive discussion of network science, refer to Barabasi (2004); Caldarelli and Catanzaro (2016); Parisi (2021), among others.

In the economic-financial field, the basis of the use of network science to understand the phenomenon of systemic risk is represented by the awareness that individuals, companies, institutions and sovereign states are not independent or are not a collection of actors in isolation (as assumed instead by classical economic theory) but rather influence each other. Their operation, far from being completely rational (another fallacious assumption of classical economic theory), is significantly dependent on mutual influence. More simply, as we have already underlined, we live today in a complex and interconnected world. In particular, the work of Acemoglu et al. (2015) highlights that it is the highly interconnected nature of the financial system that contributes to its fragility, as it facilitates the spread of financial distress and solvency problems from one institution to others in an epidemic way. Acemoglu et al. (2015) find that until the volume of negative shocks is below a certain critical threshold, a more diversified pattern of interbank liabilities leads to less fragility, however when shocks exceed that certain critical threshold the high interconnectedness between nodes in the network becomes a source of systemic instability. When negative shocks are larger than a certain threshold, therefore, financial networks in which banks are only weakly interconnected are less prone to the propagation of systemic crises.

Studying economic phenomena through network science thus presupposes considering financial systems as complex networks in which the "nodes" are the economic agents (individuals, firms, financial intermediaries, central banks, supervisors, sovereign states, etc.) and the "links" are the economic and financial ties (edges) that connect them. Moreover, following the mechanisms of "growth" and "preferential connection", characteristic of "real networks", the more a network expands, the more the "hubs" (larger nodes) must expand, swallowing up the smaller ones. In economics, therefore, globalisation pushes the nodes to expand and therefore, mergers and acquisitions are the natural consequence of an expanding economy. The "Achilles heel" of a "small world" and, therefore, of the real networks are represented by the vulnerability due to the interconnection. An isolated shock can create chain effects that destabilise an entire economic system, and the probability that an isolated shock would undermine an entire system is higher if the nodes affected are the most interconnected. An interesting study in this regard is that of Battiston et al. (2012), which, through the DebtRank index, calculates for each node (bank) of an interbank network how much of the economic value of the network would be lost if that node (bank) failed. Thus, confirming that the greatest systemic effects are generated in the case of a crisis of financial institutions most connected to the other nodes of the system.

Wanting to list the main direct and explicit drivers (channels) of interconnection (interaction) in economic and financial systems, it is possible to consider the following:

– loans (and in general financial relations) between banks, companies, institutions and countries that feed the channel of direct exposure (illiquidity contagion);

- various financial contracts between the different economic agents (nodes) (individuals, firms, financial intermediaries, central banks, supervisors, sovereign states, etc.) of the financial network;
- commercial relationships and supply of goods and services between economic agents that create the conditions for financial interactions and dependencies;
- direct and indirect shareholdings of companies, businesses and institutions in other companies; as through chains of ownership, shareholders can influence the operation and business of firms owned directly or indirectly;
- the sharing of directors between different companies, or the presence of the same directors on the Boards of Directors of different companies (interlocking);
- interconnections between economic sectors and geographical areas which translate into correlations between the share prices of different companies;
- the network of international trade that generates imports and exports between countries and therefore very close links between the trade and payments balances of different sovereign states;
- the information channel, which is atavistically polluted by information asymmetries and the lack of rationality of market participants, who are imperfectly informed and tend to make their choices in an often irrational way, especially in particular conditions of systemic uncertainty.

As has already been highlighted in detail in the previous paragraph, through the various channels of direct exposure, the network effects (contagion) produced by the financial relationships between the various nodes of a financial network (banks, companies, states, etc.) can generally be explained. In particular, through the channel of direct exposure, the driver of "illiquidity contagion" explains how a bank tends to withdraw its loans granted to a counterparty as soon as it hears of the risk of the latter's insolvency, inducing this counterparty, in turn, to withdraw its loans from its respective counterparts to recover liquidity. In this way, the crisis spreads throughout the system (financial network) through the links between the various nodes of the network, generating liquidity crises and chain insolvencies and also a general reduction in prices fed by the consequent recessionary and deflationary cycle. Furthermore, if the price of an asset collapses, this event affects not only those who have invested directly in that asset (banks or companies) but also those who have acquired bonds or shares of companies that have invested in that asset (Bardoscia et al. 2021).

A second direct channel is represented by the deterioration of interbank assets and the subsequent write-down in the balance sheet of the non-performing assets carried out by banks with reference—for example—to the loans granted to companies in difficulty (i.e., the assets held by companies in difficulty). This devaluation mechanism impacts banks' balance sheets, reducing the value of assets and forcing greater provisions to reserve, and—as a consequence—generates pro-cyclical recessionary effects due to the contraction of credit and a general reduction in prices in the economic system. So, with the existence of complex chains of interactions, contracts or feedback mechanisms between the different actors of a financial system, the final effects can be much larger than the initial shocks.

Furthermore, modern financial markets are systems with a large number of elements (nodes as corporations, securities or stocks, economic agents, sovereign states), which can be differently affected by one another. It is certainly possible and useful to infer the network of most relevant links by studying the mutual dependencies between these elements-nodes (for example securities or stocks) based on an analysis of correlation measures. The elements-nodes (corporations, sovereign states, securities or stocks) that have highly correlated returns can in fact be linked in the network by links (edges), while the elements that are not correlated to each other, so that behave independently, are not linked by links (edges) in the network (Raddant and Di Matteo 2023). The relevant construction of such a network contains several critical issues and complexities that only the collaboration of economists and physicists (or mathematicians) can solve. The main critical factors relating to the construction of a network in the financial sector are the following: (i) identification of the most suitable method of data analysis for the phenomenon analysed; (ii) identification of nodes and relevant data-variables for each node; (iii) identification of the relevant variables to identify the significant relationships (interconnections or interactions) between the different nodes; (iv) identification of the most suitable network representation for the different financial relationship; (v) complete dataset availability and retrieval of all the necessary data (often private and sensitive and so not available); (vi) reconstruction of the missing data and information useful to build the network (network reconstruction); (vii) more granular timescale of the data necessary to an efficient network reconstruction as regulatory data are often reported quarterly or annually, allowing only for the analysis of temporal snapshots that could be too far apart to detect rapid build-ups of risk (Bardoscia et al. 2021).

But if a research group can solve the problems highlighted above and once the network is built, such a network can provide useful information to economic and financial operators and policy makers about the propagation of a shock in financial networks and therefore about the propagation of a systemic crisis. Wishing to broaden the field of analysis, it is clear that it is possible to build different networks of interconnected economic agents (nodes) and then also build links between these different networks. In this way, it is possible to predict the possible propagation of a crisis even between different networks. In this way, the correct network could provide an excellent tool for the identification of early-warning signals of system-wide instabilities. Evidently, everything depends on the correct and efficient construction of the network, which requires knowledge of both the method (relevant data analysis models) and the field of analysis (economic and financial systems).

As mentioned, moreover, relations between economic agents are not only formal and explicit and therefore give rise to the so-called direct (and explicit) drivers of interconnection. The relationships (links) between economic agents in a system can also be informal and implicit and thus give rise to indirect links, through—for example—the common investment of two economic agents in the same asset, thus giving rise to an indirect contagion via overlapping portfolios. This could happen if different economic agents (banks or other investment companies) are indirectly connected through some co-occurrence relationship as they invest in common assets. In this case, if a bank (or other investment companies) needs to sell some of its assets, this sale (if significant in amount as is often the case for larger financial intermediaries) could cause the devaluation of those assets and, therefore, losses for the other banks (or other economic agents) that have invested in them. This devaluation may cause these banks to sell their assets in turn, and so on (Bardoscia et al. 2021). In this regard, Jiang and Fan (2019) find that shocks propagate more rapidly when there is more overlap among the portfolios of banks in a system, i.e., when there are more common investments (assets) in the portfolios of multiple banks, while the propagation of a crisis slows down in the case of more heterogeneous and less overlapping portfolios. In this sense, in the coming years, a source of risk to be carefully monitored will have to be represented by the transition to a low-carbon economy, which may have implications for financial stability and determine a radical change in the allocation of resources. A devaluation of high carbon emission assets could have a significant impact on the balance sheets of the institutions that hold these assets in the coming years, thus generating indirect contagion phenomena through portfolio overlapping.

As already highlighted above, the relationships between the economic agents (nodes) of a network can be of various kinds and tend to change over time. This variety and complexity (and often lack of transparency) of the financial relationships between the various nodes of a network tends to increase the complexity of the financial systems and this produces information asymmetries, risks of moral hazard and, therefore, opacity and consequences in the processes of propagation of risks of a systemic nature. This complexity in the relationships between economic agents in a financial system has also been fed since the early years of the new millennium by the evolution (often uncontrolled) of financial engineering, which has made economic-financial systems more interconnected and therefore more complex, linking operators to one another in multiple ways and often unconsciously. Moreover, as already argued, the science of networks teaches us that the mechanisms of "growth" and "preferential connection" lead the "hubs" (larger nodes) to expand in phases of network expansion and thus incorporate smaller nodes. This phenomenon, otherwise known as globalisation, however, leads to the risk of extinction of smaller economic operators, such as local banks, whose disappearance or even simple competitive downsizing would, over time, lead to the loss of the extraordinary intangible and relational assets in dowry to these intermediaries, reflexively impoverishing the financial system and also exposing it to greater risks of a systemic nature (Pacelli et al. 2020).

Moreover, we cannot overlook the fact that globalisation, by making our world even smaller and thus exposed to the risk of contagion, has also highlighted the limits of the purely micro-prudential approach to the supervision of the financial system. The analysis of financial systems through the networks science highlights the need for a more macro-prudential supervisory approach, which considers the multiple and varied interconnections between the various economic agents and, therefore, the potential systemic effects produced by the crisis of a significantly interconnected agent as well as the mitigating role of systemic risk played by certain other smaller intermediaries, such as, for example, cooperative credit banks (Pacelli et al. 2020). As Masera (2021) has already authoritatively pointed out in the literature, analysing a complex system such as the financial system through a simplified approach (micro-prudential supervision), i.e., that considers separately the parts of a whole, neglecting instead the interconnections between these parts, inevitably leads to inappropriate, superficial and pro-cyclical regulatory prescriptions. Hence, there is a need for an evolution of supervisory approaches in a holistic and macro-prudential key, that is, in the direction of a more incisive enhancement of the quality of relationships, inter-actions and interconnections between operators in the financial system and so in the direction of an effective proportionality of regulation in banking systems. In this sense, it is believed that network science can provide an indispensable theoretical-conceptual paradigm, useful for allowing the financial supervisory authorities to perceive phenomena, dependencies, and interactions that are probably underesti-mated and to limit any competitive distortions. This is to ensure the conditions for a fair competitive comparison which takes into correct consideration the peculiarities of each category of financial intermediaries, also concerning its different systemic imprint and therefore its different aptitude to determine an acceleration or mitigation in the processes of propagation of a systemic crisis and therefore to the different role played in the preservation of financial stability.

Concluding Remarks

As already highlighted previously, today we live in a globalised, complex and significantly interconnected world, in which complex systems and phenomena, apparently distant from each other, mutually influence each other. This reality exposes societies to systemic risks and requires analysis through a holistic approach to complex phenomena in order to investigate, through the help of network science, hidden relationships between economic, financial, political, health and environmental phenomena. In other words, today nothing happens independently. Studying economic phenomena through network science thus presupposes consid-ering economic and financial systems as complex networks in which the "nodes" are the economic agents (corporations, securities or stocks, individuals, sovereign states) and the links (edges) are the various economic and financial ties that connect them. Furthermore, modern financial markets are "complex systems" with a large number of parts which interact in a not simple way. The relevant construction of such a network contains several critical issues and complexities that only the collab-oration of economists and physicists (or mathematicians) can solve. But once built effectively, such a network can provide useful information to economic and financial operators and policy makers about the propagation of a shock in the financial network and therefore about the propagation of a systemic crisis. Wishing to broaden the field of analysis, it is clear that it is possible to build different networks of interconnected economic agents (nodes) and then also build links between these different networks. In this way, it is possible to predict the possible propagation of a crisis even between different networks. In this way, the correct network could provide an excellent tool for the identification of early-warning signals of system-wide instabilities. Evidently everything depends on the correct and efficient construction of the networks, which requires both knowledge of method (relevant data analysis models) and of the field of analysis (economic and financial systems).

The world today is globalised due to the high level of interconnection of production and distribution, as well as of finance with the real economy, which are interconnected macrocosms capable of significantly influencing each other. By citing a recent example, an abrupt contraction in economic activity, such as that caused by the various restrictions due to the Covid-19 pandemic, initially generates liquidity tensions for companies and their lending banks and, later, the liquidity problem risks turning into a solvency problem for companies and, consequently, for banks, which could see their capital ratios deteriorate due to the flow of new non-performing loans, with systemic effects on credit, savings and private investment. Furthermore, in the first months of significant contraction in production, trade and economic activity in general following the pandemic, many national governments promptly intervened, supporting the sectors most affected with relief and compensation, but this led to an exponential growth in public debt. Banks have also been given various incentives, essentially through public guarantees, to finance companies in difficulty or to grant moratoria following the contraction of production and trade, but this threatens to lead in the coming months, when public protection will necessarily have to be reduced, to a worsening of the quality of the banks' credit portfolio, with a possible increase in non-performing loans and a consequent worsening of capital ratios. The possible insolvencies of lower quality entrusted companies could increase the NPLs of banks, deteriorating the quality of the credit portfolio and the capital ratios, with consequent potential recessionary effects only partially mitigated by public guarantees, which are—in any case—destined to further increase the interconnections between banks and States with an exacerbation of exposure to systemic risk. In other words, we are living in a complex historical phase in which it would be necessary to find a difficult balance between the need to mitigate the harmful effects of the crisis and the need to avoid the risk of "drugging" the markets with an excess of liquidity. And the inflation of recent years demonstrates that achieving this balance was indeed complicated. This unstable balance should also be found at both public and private levels, possibly avoiding confusing the various economic players with schizophrenic economic and monetary policies. In the path towards this desirable equilibrium, the role of the banking system will be fundamental, as banks will have to prove themselves capable of financing those firms that will prove themselves capable of creating sustainable value over time in a profoundly changed market and therefore evaluating with an entrepreneurial spirit, that is, critically and not supine, any clumsy public attempts to orient credit policies.

Otherwise, the risk would be that of artificially diverting or slowing down the free use of entrepreneurial energies, generating an inefficient allocation of capital in the short term with harmful consequences in the medium-long term.

In the hoped evolutionary process of the entrepreneurial system, the role of the banking system must not, therefore, be debased by too many regulatory constraints and excessive dirigisme of economic policies, which otherwise risk generating procyclical and recessive effects. Due to their intrinsic rigidity, those constraints also risk gagging the autonomy of banks and slow down the necessary process of selection of the most resilient companies in the credit market. This evolutionary process will have to bear the cost—including the social cost—of the disappearance of some companies and some productive segments (those that will prove to be anachronistic in the new post-pandemic world) and it will therefore be necessary to guarantee flexibility in the labour market and investments in personnel training to accompany the process of productive and entrepreneurial evolution. To do this, it will be necessary to have courage, "good debt" (citing Mario Draghi), and therefore investments that increase productivity and competitiveness. But also, greater cohesion and political and social solidarity (at all levels) to reduce inequalities and generate a global social renaissance.

In the medium term, therefore, the solution cannot be to persevere in short-sighted and unconditional welfarism, which risks generating laziness and immobility in the business world and creates the conditions for opportunistic or moral hazard behaviour. Instead, credit policy must be based on a careful analysis of creditworthiness and the entrepreneurial formula that pushes companies to innovate and compete and that, as a consequence, favours the generation of a more modern and competitive entrepreneurial system. If we wanted to give a sense, or rather a positive meaning to the tragic pandemic, we should interpret it as an opportunity for the business world to evolve. An opportunity, however, that we can't allow to miss, or else the survival of our economic system as we have understood it up to now. As mentioned above, the multiple government measures to support the liquidity of businesses and the public guarantees granted to banks in favour of loans to the private sector, on the one hand, temporarily sterilise the harmful effects of the blockage of economic activity following the pandemic, on the other hand, create the basis for a future exacerbation of systemic risk, because they indissolubly bind the banks with the sovereign States that are guarantors of the credit provided by them.

In a systemic crisis, the potential chain of contagion does not stop at banks and firms but also involves the sovereign States, which collect lower tax revenues due to the contraction of the economy, increase their public debt to support current spending and will see their reputation in the markets diminished as soon as the exceptional systems of public aid to the economy necessarily downsizing, with negative effects on the rates paid on the debt issued (Altinbas et al. 2018). This is especially true for those States characterised by high public debt and therefore endowed with little autonomous spending capacity. Moreover, the close link between the sovereign States and the banks, which are the main holders of the issued sovereign debt, ends up—as highlighted above—tying public fates with private ones, and this further fuels the risk of contagion with systemic effects.

It was also easy to foresee that the economic crisis due to the pandemic and its economic and financial consequences would increase inequality in income and wealth and, therefore, social inequalities, thus exacerbating the effects of a systemic nature. Let us not forget that the crisis that broke out in 2007 originated in the USA also because of evident inequalities and disparities of economic, social and regulatory nature.

A further aspect that should not be underestimated from a systemic perspective is certainly represented by the impact of the ecological transition on financial systems. In this dimension, the phenomenon of systemic risk intrinsic to the pandemic and that of environmental and climate risk have evidence in common that no one can say they are safe. This is an immediate indication of their common nature as "totalitarian" systemic risks of the same enormous natural ecosystem: the world ecosystem. Thus, we could recall the risk of a general breakdown of the ship: if the ship is on fire, no one is saved from fire or death at sea. The only factual rule is the same as always: if you want to save yourself, you must save the ship or help save the ship, obeying rules of necessary solidarity.

Scientific research has long highlighted the need to reduce global warming and promote green investments and this awareness is now widely spread at the social and political levels. The scientific community has begun to measure the impacts that the ecological transition can produce on real and financial markets and international political authorities have launched multiple initiatives aimed at making the ecological transition a reality. In such a scenario, the interconnection between finance and the environment appears central, since the need to rethink a model of sustainable development with a transition to a green economy cannot disregard the key role played by financial institutions, which are called to promote sustainable investments compatible with the desired green transition. But we must not underestimate the risks associated with a transition to a low carbon emission economy, due to the drastic and sudden loss of value of all those assets linked to industries with high carbon emissions. The desired green transition will therefore determine the crisis of many companies, of various economic sectors, the loss of jobs and the reduction in the value of many assets in the balance sheets of banks, which will therefore see the value of their assets reduced with obvious consequences on credit policies and economic growth.

So, to understand the complex and holistic nature of the phenomenon of systemic risk, today we cannot disregard the integrated and prospective analysis, through the logical-conceptual schemes borrowed from the network science, of the multiple and often obscure relationships that bind the various economic, political, social, health and environmental phenomena.

Starting from this conviction and wishing to offer a guideline for reflection in extraordinary historical periods such as the one we are living and are preparing to live in the coming years, the central and enlightened role of sovereign States in a strategic and operational transnational horizon appears indispensable. Disrupting financial relations and creating small, independent and disconnected autarchic hubs is not a viable solution in today's world. In other words, systemic risk today cannot be eliminated or fragmented; it must be lived with and managed in a forward-looking, proactive and integrated manner. A small and interconnected world, and therefore significantly exposed to contagion risks, requires risk sharing, social solidarity, "intelligent" forecasting and control tools as well as forms of public guarantees, especially transnational ones, which—possibly without interfering with the free deployment of entrepreneurial energies—protect and safeguard the market in the event of a crisis of individual operators. In this sense, great expectations should be placed on the improvement and, hopefully, enlargement of the economic and banking union in Europe, both from a financial and, above all, a political perspective. It is clear to all that in the coming years, we will live in a more indebted, interconnected and "small" world, where forms of community public guarantees, risk-sharing, and social solidarity will prove increasingly indispensable to preserve the competitiveness and sustainability of our economic and social systems.

References

Acemoglu, D., Ozdaglar, A., Tahbaz-Salehi, A.: Systemic risk and stability in financial networks. Am. Econ. Rev. **105**, 564–608 (2015). https://doi.org/10.1257/aer.20130456

Aharony, J., Saunders, A., Swary, I.: The effects of a shift in monetary policy regime on the profitability and risk of commercial banks. J. Monet. Econ. **17**, 363–377 (1986). https://doi.org/10.1016/0304-3932(86)90063-2

Allen, F., Gale, D.: Financial contagion. J. Polit. Econ. **108**(1), 1–33 (2000)

Altinbas, H., Pacelli, V., Sica, E.: The determinants of sovereign bond yields in the EMU: new empirical evidence. Int. J. Econ. Financ. **10**, 41–56 (2018)

Barabasi, A.L.: La nuova scienza delle reti. Einaudi, Torino (2004)

Bardoscia, M., Barucca, P., Battiston, S., Caccioli, F., Cimini, G., Garlaschelli, D., Saracco, F., Squartini, T., Caldarelli, G.: The physics of financial networks. Nat Rev Phys. **3**, 490–507 (2021). https://doi.org/10.1038/s42254-021-00322-5

Battiston, S., Puliga, M., Kaushik, R., Tasca, P., Caldarelli, G.: DebtRank: too central to fail? Financial networks, the FED and systemic risk. Sci. Rep. **2**, 541 (2012). https://doi.org/10.1038/srep00541

Beirne, J., Fratzscher, M.: The pricing of sovereign risk and contagion during the European sovereign debt crisis. J. Int. Money Financ. **34**, 60–82 (2013). https://doi.org/10.1016/j.jimonfin.2012.11.004

Blancher, M.N.R., Mitra, M.S., Morsy, M.H., Otani, M.A., Severo, T., Valderrama, M.L.: Systemic risk monitoring ('SysMo') toolkit—a user guide. International Monetary Fund 13/68 (2013)

Blundell-Wignall, A.: Solving the financial and sovereign debt crisis in Europe. OECD J. Financ. Market Trends. **2011**, 201–224 (2012)

Böhm, H., Eichler, S.: Avoiding the fall into the loop: isolating the transmission of bank-to-sovereign distress in the Euro Area. J. Financ. Stab. **51**(100763), (2020)

Brunnermeier, M.K., Garicano, L., Lane, P.R., Pagano, M., Reis, R., Santos, T., Thesmar, D., Van Nieuwerburgh, S., Vayanos, D.: The sovereign-bank diabolic loop and ESBies. Am. Econ. Rev. **106**, 508–512 (2016). https://doi.org/10.1257/aer.p20161107

Brutti, F., Sauré, P.: Transmission of sovereign risk in the Euro crisis. J. Int. Econ. **97**, 231–248 (2015)

Caldarelli, G., Catanzaro, M.: Scienza delle reti. Egea (2016)

Cottarelli, C., Forni, L., Gottschalk, J., Mauro, P.: Default in Today's Advanced Economies: Unnecessary, Undesirable, and Unlikely. IMF Staff Position Notes (2010)

Cronin, D., Flavin, T.J., Sheenan, L.: Contagion in Eurozone sovereign bond markets? The good, the bad and the ugly. Econ. Lett. **143**, 5–8 (2016). https://doi.org/10.1016/j.econlet.2016.02.031

Delli Gatti, D.: La crisi dei mutui subprime. In: Osservatorio bancario, pp. 28–58 (2008)

Di Clemente, A.: Stabilità finanziaria e rischio sistemico. In: Rischio sistemico e intermediari bancari. Aracne editrice (2016)

Duarte, F., Eisenbach, T.: Fire-sale spillovers and systemic risk. J. Financ. **76**, 1251–1294 (2021)

Engler, P., Große Steffen, C.: Sovereign risk, interbank freezes, and aggregate fluctuations. Eur. Econ. Rev. **87**, 34–61 (2016). https://doi.org/10.1016/j.euroecorev.2016.02.012

Eser, F., Schwaab, B.: Evaluating the impact of unconventional monetary policy measures: empirical evidence from the ECB's securities markets programme. J. Financ. Econ. **119**, 147–167 (2016). https://doi.org/10.1016/j.jfineco.2015.06.003

Flannery, M., Kwan, S., Nimalendran, M.: The 2007–2009 financial crisis and bank opaqueness. J. Financ. Intermed. **22**, 55–84 (2013)

Freixas, X., Parigi, B.M., Rochet, J.-C.: Systemic risk, interbank relations, and liquidity provision by the central bank. J. Money, Credit, Bank. **32**, 611–638 (2000). https://doi.org/10.2307/260 1198

Gonçalves, C.E., Guimaraes, B.: Sovereign default risk and commitment for fiscal adjustment. J. Int. Econ. **95**, 68–82 (2015)

Hoque, H., Andriosopoulos, D., Andriosopoulos, K., Douady, R.: Bank regulation, risk and return: evidence from the credit and sovereign debt crises. J. Bank. Finance **50**, 455–474 (2015)

In't Veld, D., van der Leij, M., Hommes, C.: The formation of a core-periphery structure in heterogeneous financial networks. J. Econ. Dyn. Control **119** (2020)

Jiang, S., Fan, H.: Systemic risk in the interbank market with overlapping portfolios. Complexity **2019**, e5317819 (2019). https://doi.org/10.1155/2019/5317819

Masera R.: Per una vera proporzionalità nella regolamentazione bancaria dell'Unione Eu-ropea. Le sfide del Coronavirus e di Basilea IV. Ecra, Roma (2021)

Michaelides, A., Milidonis, A., Nishiotis, G., Papakyriakou, P.: The adverse effects of systematic leakage ahead of official sovereign debt rating announcements. J. Financ. Econ. **116**, 526–547 (2015)

Nelson, S.C., Katzenstein, P.J.: Uncertainty, risk, and the financial crisis of 2008. Int. Organ. **68**, 361–392 (2014)

Pacelli, V., Pampurini, F., Quaranta, A.G.: Co-operative banks and financial stability. Int. J. Bus. Soc. Sci. **11**(11) (2020)

Pacelli, V.: Consulenza finanziaria e ottimizzazione di portafoglio. Come gestire la relazio-ne con l'investitore in tempo di crisi. Bancaria editrice (2014)

Pagano, M.S., Sedunov, J.: A comprehensive approach to measuring the relation between systemic risk exposure and sovereign debt. J. Financ. Stab. **23**, 62–78 (2016)

Parisi, G.: In un volo di storni. Le meraviglie dei sistemi complessi. Rizzoli (2021)

Raddant, M., Di Matteo, T.: A look at financial dependencies by means of econophysics and financial economics. J. Econ. Interact. Coord. (2023). https://doi.org/10.1007/s11403-023-00389-6

Reinhart, C., Rogoff, K.: Banking crises: an equal opportunity menace. J. Bank. Financ. **37**, 4557–4573 (2013)

Rötheli, T.F.: Causes of the financial crisis: risk misperception, policy mistakes, and banks' bounded rationality. J. Socio-Econ. **39**, 119–126 (2010). https://doi.org/10.1016/j.socec.2010.02.016

Chapter 2
Systemic Risk and Network Science: A Bibliometric and Systematic Review

Vincenzo Pacelli, Ida Claudia Panetta, and Maria Melania Povia

Abstract Estimating systemic risk in networks of financial institutions is increasingly a challenge in policymaking. The complexity of financial networks may increase the difficulty of mitigating systemic risk and how the topology of connections can propagate the failure of an individual entity through the network in the system. Our study's primary purpose is to apply the bibliometric techniques and the systematic review method to understand the evolution of research on systemic risk and interconnectedness among financial markets and institutions and highlight the literature's progress during the period from 2008 to 2023. Results suggest that systemic risk and financial networks have experienced rapid growth during the last decade, and this can contribute to a future research agenda on the topic.

Keywords Systemic risk · Network science · Financial crisis

Introduction

Understanding and controlling systemic risk has become a crucial social and economic topic in the literature. The 2007 crisis of financial markets, possibly the worst economic disaster since the Great Depression of the 1930s, the 2010 sovereign

The chapter is a revised and updated version of an essay previously published in the volume edited by V. Pacelli and I. C. Panetta, *Systemic Risk, Monetary Policy and Portfolio Diversification in the Great Crises' Era*, Quaderni del Dipartimento Jonico n. 19/2022, Edizioni DJSGE, Università degli Studi di Bari Aldo Moro, 2022 (ISBN: 9788894503098).

V. Pacelli (✉) · M. M. Povia
University of Bari Aldo Moro, Bari, Italy
e-mail: vincenzo.pacelli@uniba.it

M. M. Povia
e-mail: maria.povia@uniba.it

I. C. Panetta
University of Rome La Sapienza, Rome, Italy
e-mail: ida.panetta@uniroma1.it

© The Author(s) 2025   21
V. Pacelli (ed.), *Systemic Risk and Complex Networks in Modern Financial Systems*,
New Economic Windows, https://doi.org/10.1007/978-3-031-64916-5_2

debt crisis, the COVID-19 crisis, and the war in Ukraine have brought to the fore how crises, extraordinary events with a wide potential scope impact markets linkages and financial integration. During the past two decades, the succession of these crises has given increased attention to the study of the financial system's architecture in creating systemic risk and the relationship between the structure of the financial network and the extent of financial contagion.

The study of financial systems as networks thus makes it possible to discern the way and direction through which a shock might propagate through the system. Indeed, we can consider financial systems as networks of interconnected economic agents (nodes), whose relationships are regulated through financial contracts (links) and which grow and evolve.

In financial economics, the basis for using network science to understand systemic risk is the realisation that individuals, firms, institutions and sovereign states are not independent but influence each other. Globalisation pushes the nodes to become more significant; therefore, mergers and acquisitions are the natural consequence of an expanding economy.

In light of the above, it is believed that only by analysing economic and financial systems as networks, it is possible to discern complex relationships between phenomena of different origins and nature that can generate the propagation of systemic crises (Pacelli 2021).

Systemic risk has long been identified as a potential contagion mechanism or impact that starts from the failure of a financial institution and propagates through the financial system and to the real economy itself (Poledna et al. 2021). A constant concern of bank regulators is that the collapse of a single bank could bring down the financial system as a whole. Economic systems are increasingly built on interdependencies and the dynamic interaction of many different agents, creating complex networks.

Dense interconnections could represent a dangerous mechanism for propagating shocks, leading to a more fragile financial system.

For this reason, estimating systemic risk in network science represents an important challenge for regulators to anticipate systemic events and reduce systemic risk, ensuring the objective of financial stability.

Synthesising the literature, we define systemic risk as "the risk of a systemic default, i.e. the default of a large portion of the financial system that starts from the failure of a financial institution and propagates through the financial system".

The failure of a financial institution, or a group of institutions, depends on the network of financial exposures among institutions, and their complexity and interconnectedness can increase vulnerabilities. Indeed, systemic risk can be quantified from the network structure, the dynamical evolution analysis, and the nodes' interaction.

Moreover, Battiston et al. (2016) introduce a novel measure of systemic impact, DebtRank, defined as the "number measuring the fraction of the total economic value in the network that is potentially affected by the distress or the default of node". Mezei and Sarlin (2018) present a new approach to measuring systemic risk in networks and suggest using RiskRank.

Network science can contribute to a quantitative assessment of systemic risk and estimate systemic events in a network, improving the financial system's stability.

The worldwide economic crisis of 2007–09 has turned attention to the need to analyse systemic risk in complex financial networks. Different contributions have developed the study of the relationship between the financial networks and systemic risk, suggesting how the structure of the financial networks can mitigate or amplify systemic risk along various channels. Therefore, in this chapter, we use the bibliometric method to understand the evolution of research on systemic risk and interconnectedness among financial markets and institutions and highlight the literature's progress during the period from 2008 to 2023.

Our chapter is structured as follows. Section "Background and Research Question Development" summarises the previous literature review on systemic risk and network science, providing the basis for our research questions. Section "Research Design and Methodology" outlines the research design, describing the metrics used to conduct the literature review and the data sample construction process. Sections "Descriptive Sample Analysis" and "Discussion of Results" are dedicated to presenting the most influential aspects of the literature reviewed and the results of the bibliometric analysis and systematic review. Finally, section "Conclusions" offers a future research question and proposes some considerations for the evolution of research.

Background and Research Question Development

The recent troubled years for the global economy have led to increased interest in the role of the financial network in systemic risk literature. Many studies on systemic financial risk have shown that interconnectedness can facilitate risk sharing, which can help minimise the uncertainty faced by individual agents: diversification reduces risk and improves stability. However, more numerous and complex interlinkages among financial markets can serve as a channel for propagating shocks and amplifying existing information asymmetries or other externalities and market frictions (Acemoglu et al. 2015).

Financial institutions create multilayer networks characterised by holding exposures to joint assets, a network of trading relationships and exposures between financial institutions (Battiston et al. 2016). Thus, topological features of financial networks influence how easily distress can propagate within the system. Market integration and diversification are processes that can stabilise the financial system. Still, these factors can contribute to instability and amplify financial distress, making significant crises more likely to happen (Bardoscia et al. 2017).

Roukny et al. (2018) and Benoit et al. (2015) analyse the importance of financial networks in understanding systemic events. The structure of those networks can facilitate the capacity of regulators to estimate systemic risk in terms of expected losses, and it can decrease the difficulty of mitigating systemic risk and the social cost of financial crises.

Recent research on financial networks offers essential insights into systemic risk measurement methods by studying contagious links and fragile network structures. However, methods that focus only on the investments and relationships of a few large institutions can ignore several potential crises (Neveu 2018). Thus, this analysis can incentivise banks to choose investments and partners that maximise the financial system's overall value and understand that systemic risk depends on complex interdependencies (Jackson and Pernoud 2020).

According to Silva et al. (2017), many articles analysed the systemic importance of specific institutions in the literature, but an essential gap in comparative research on systemic risk measurement is evident. As summarised in Table 2.1, all previous literature reviews use a *narrative* approach. These reviews encompass a variety of topics, including relatively new fields of research based on the importance of financial networks in understanding systemic events; despite this variety of subjects and approaches, a clear focus on the relationship between network science research and systemic risk is still not yet evident. Therefore, we would like to follow a bibliometric and systematic analysis to complement with quantitative and qualitative analysis of previous narrative reviews focused on the interplay of financial networks and systemic risk, trying to picture the evolution of systemic risk and networks research, also considering influential aspects of the literature such as authors, themes, and articles. Following this aim, we try to answer to one central research question: "How is the literature on systemic risk issues and financial networks evolving in recent years?".

Research Design and Methodology

After having defined in previous parts the aim of the chapter, in this section, we describe all the steps followed in our analysis, namely:

1. The definition of techniques to conduct the literature review.
2. The sample selection identifies articles related to systemic risk and financial networks to be processed.
3. The run of the quantitative analysis and reporting of main findings through the systematic review.

To answer the research question regarding analysing the evolution of systemic risk and financial networks literature, we adopted bibliometric measures (step 1) as an essential vehicle for highlighting and motivating emerging scholarship (Khan et al. 2021). Bibliometric reviews analyse and classify bibliographic material by framing representative summaries of the extant literature (Donthu et al. 2020) from a more objective, quantitative perspective (Albort-Morant and Ribeiro-Soriano 2016). Bibliometric approaches are increasingly used in the literature reviews across different disciplines due to several factors, including the introduction of software tools, cross-disciplinary, and the capacity to synthesise a large volume of data (Donthu et al. 2021). Indeed, this methodology has been applied in different fields of

Table 2.1 Summary of literature review on systemic risk and financial network[1]

Title	Authors	Topic	Methodology
Pathways towards instability in financial networks	Bardoscia et al. (2017)	Analyse how market integration and diversification can amplify financial distress	Narrative
Interconnectedness as a source of uncertainty in systemic risk	Roukny et al. (2018)	Focus on the importance of financial networks in understanding systemic events	Narrative
Where the Risks lie: a survey on Systemic risk*	Benoit et al. (2015)	Focus on the extensive literature on systemic risk	Narrative
The price of complexity in financial networks	Battiston et al. (2016)	Analyse the complexity of financial networks	Narrative
Systemic risk in financial networks: a survey*	Jackson and Pernoud (2020)	Analyse an overview of the relationship between financial networks and systemic risk	Narrative
Modelling systemic risk to the financial system: a review of additional literature	Markellof et al. (2012)	Focus on rigorous assessment of the performance of various systemic risk models	Narrative
A survey of Network-based analysis and systemic risk measurement*	Neveu (2018)	Focus on systemic risk and the network approach	Narrative
Application of systemic risk measurement methods: a systematic review and meta-analysis using a network approach*	Dičpinigaitienė and Novickytė (2018)	Analyse systemic risk measurement methods	Narrative
An analysis of the literature on systemic financial risk: A survey	Silva et al. (2017)	Focus on systemic financial risk	Narrative

business research, including business strategy (Kumar et al. 2021b), human resources (Andersen 2021), marketing (Backhaus et al. 2011; Hu et al. 2019; Samiee and Chabowski 2012), management (Ellegaard and Wallin 2015; Zupic and Čater 2015), electronic commerce (Kumar et al. 2021a) and finance (Durisin and Puzone 2009; Linnenluecke et al. 2017).

There are two main technique categories for bibliometric analysis: performance analysis and science mapping (Donthu et al. 2021). Performance analysis recognises the importance of contributions of research constituents to a given field (Cobo et al. 2011; Donthu et al. 2021; Ramos-Rodrígue and Ruíz-Navarro 2004). Two of the most important measures are the number of publications and citations per year or research

---

[1] The asterisk at the end of a paper stands for all the studies included in review.

constituents (Baker et al. 2021; Cobo et al. 2011; Ramos-Rodrígue and Ruíz-Navarro 2004).

Moreover, science mapping represents the relationships between research constituents. The analysis is based on the intellectual interactions and structural connections among research constituents (Donthu et al. 2021). Science mapping includes relevant techniques: citation analysis, co-citation analysis, bibliographic coupling, co-word analysis and co-authorship.

According to the main objective of this chapter, we decide to analyse the evolution of systemic risk and financial networks literature using the following techniques: (i) citation analysis, (ii) co-citation analysis, (iii) co-word analysis and (iv) content analysis.

In particular, the citation analysis allows us to identify the influential aspects of systemic risk literature and network science.

Co-citation analysis is essential to explain research clusters by examining the co-citation pairs and network, which can provide significant and objective insights into the intellectual structure of the selected research discipline (Calabretta et al. 2011).

The co-word analysis is also called the semantic network and refers to the relationships among the keywords considered as the unit of analysis. Thus, the assessment of the keywords of scientific documents allows for establishing the research trends in a specific field.

Finally, the content analysis is a systematic analysis that aims to discover relationships, themes, and concepts about the data to produce a complete examination (Krippendorff 2004).

Bibliometric analysis is conducted with two software, Biblioshiny and VOSviewer. The Biblioshiny software is a shiny app providing a web interface for the Bibliometrix package of R (Aria and Cuccurullo 2017). VOSviewer is a powerful visualisation software tool for creating maps based on network data and visualising and exploring these maps (Baker et al. 2020; Khan et al. 2020). VOSviewer is particularly helpful in running bibliometric mapping and visualising bibliometric networks and identity clusters.

The second step regards the building up process of a significant sample of chapter in the field to be analysed through bibliometric techniques and map the scientific production on systemic risk and network science. We selected the Scopus database as the source of the bibliographic data since as it is the largest database of peer-reviewed literature in social science research (Baker et al. 2020), covering scientific journals, books and conference proceedings (Singh et al. 2021), widely used in literature reviews. In performing our selection process, we followed the so-called PRISMA method (Preferred Reporting Items for Systematic reviews and Meta-Analyses) displayed in Fig. 2.1. The PRISMA statement allows for a better understanding of the selection process and improves reporting quality and transparency (Knobloch et al. 2011).

In the second analysis, we applied the Systematic Review method, which mainly focuses on the quantitative study and volume of previous research. The Systematic Review, in particular, aims to evaluate the best available literature according to quality criteria with a rigorous analysis approach (Pati and Lorusso 2018).

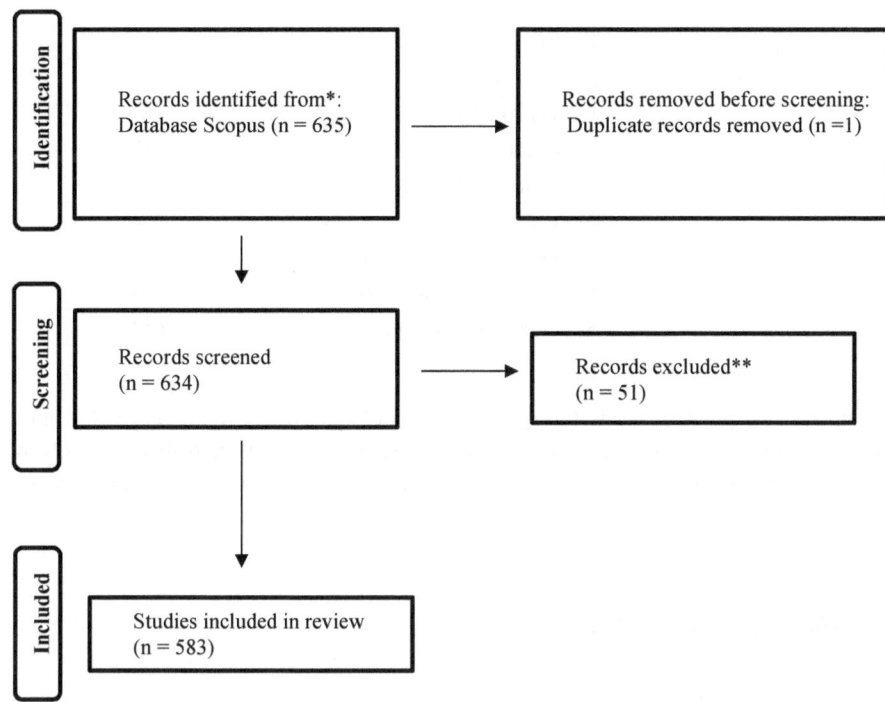

Fig. 2.1 PRISMA 2020 flow diagram for the included studies

After several simulation tests, we determined the most appropriate selection of keywords to include the most significant number of papers on systemic risk in application to networks science (Table 2.2). The keywords combination selected is: "systemic risk" and "network*" and its derivatives with "and" Boolean operator. This result comes from a keyword combination analysis. After an initial screening, we found the combination "systemic risk", "contagion" and "financial network": the word "financial" was eliminated because 'systemic risk' encompasses the whole group. Afterwards, we detected the keyword combination "systemic risk" and "network*", which allows the inclusion of the different suffixes through the asterisk at the end of the searched word. The exact keyword search is presented below: TITLE-ABS-KEY (systemic AND risk AND network*).

Then the sample was assessed using the following inclusion criteria: time range, subject area, selection of scientific articles and English language. In particular, the selection criteria contributed to the first screening of the sample, limiting the thematic areas of reference in: "economics" and "finance", business and management (ECO-BUSI), period from 2008 to 2023, excluding Conference Paper, Book chapter, Books, Editorial, Retracted, Conference Review and Letters.

Thus, we reorganised the dataset extracted (n = 635), removing duplicate records (n = 1) and eliminating those not compliant with the filter chosen (n = 51).

Table 2.2 Sample of articles on systemic risk

Keywords	Scopus
Systemic risk	3469
Financial networks	6181
Contagion	3515
Systemic risk measurement	136
Systemic risk; contagion; network	236
Systemic risk and contagion	464
Systemic risk and network*	635
Systemic risk and stability	611

The last stage (3) regards the quantitative and qualitative analysis of the main sample results of multiple tests; the main results, presented in the following section, allowed us to outline the evolution of research streams on systemic risk and network science.

Descriptive Sample Analysis

As summarised in Fig. 2.1, our final sample consists of 583 scientific articles, which results big enough to run a bibliometric analysis (Donthu et al. 2021). Looking at the sample distribution by year (Fig. 2.2), we observed that few articles in the sample were published in or before 2008, confirming that the topics surged greater attention only after the financial crises of 2007–8.

After the financial crises, we recorded steady scholarly growth in the literature and its citation impact, with a clear upward trend after 2011 suggesting increased academic interest in systemic risk and network science. From 2009 onwards, articles on systemic financial risk are regularly published yearly, but with low frequency until 2011. From 2012 onwards, with the worsening of the financial crisis and the pandemic crisis in 2020, the number of articles on systemic financial risk increased significantly. Further, the top 10 influential/authors in systemic risk and network science are recognised. Our findings show that the top three authors/researchers are Battiston with 11 articles and Caccioli and Silva with eight products each (Table 2.3).

We conduct bibliometric co-citation analysis with VOSviewer software, which identifies the research streams or clusters in the literature in the form of networks (Kim and Mcmillan 2008).

Co-citation analysis, as mentioned, is essential to explain research clusters by examining the co-citation pairs and network, which can provide significant and objective insights into the intellectual structure of the selected research discipline (Calabretta et al. 2011).

We created a map based on bibliographic data, applying the counting method (full counting), a unit of analysis (cited authors) and a minimum number of citations of

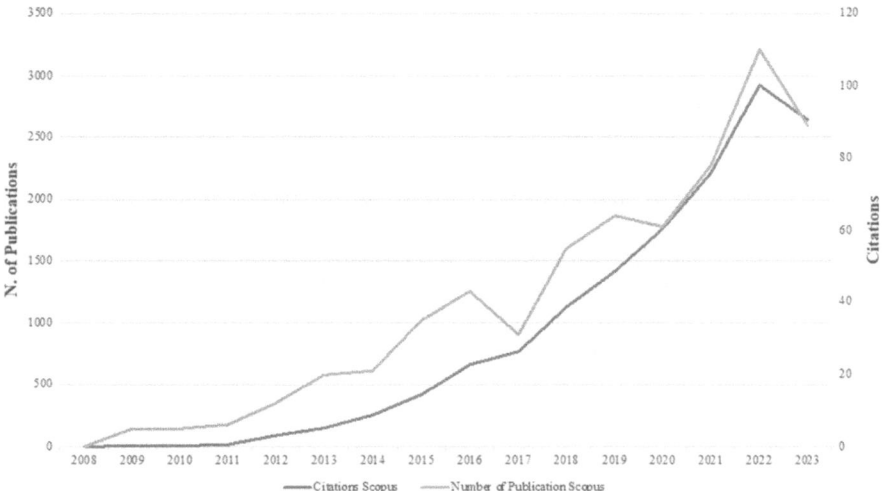

Fig. 2.2 Sample distribution by year (number of papers and number of citations)

Table 2.3 Most relevant authors: fractional authorship quantifies an individual author's contributions to a published set of papers (following the hypothesis of uniform contribution of all co-authors at each document)

Authors	Articles	Articles fractionalized
Battiston	11	3.16
Caccioli	8	1.93
Silva	8	2.42
Steinbacher	8	3
Thurner	8	2.78
Feinstein	7	3.67
Li	7	2.43
Tabak	7	2.08
Zhang	7	1.73
Gallegati	6	1.9

an author (20). As a result, of the 18,341 authors, only 494 meet the threshold. For each of the 494 authors, the total strength of the co-citation links with other authors has been calculated.

Figure 2.3 defines five clusters for the co-citation analysis: 494 items, 95,219 links and 1,131,595 total link strength. In particular, Adrian with 491 links and 243 citations, Kapadia with 490 links and 306 citations, Battiston with 487 links and 572 citations, Thurner with 487 links and 286 citations and Gallegati with 471 links and 237 citations.

Co-word analysis refers to the relationships among the keywords considered as the analysis unit. Thus, the assessment of the keywords of scientific documents allows for establishing the research trends in a specific field.

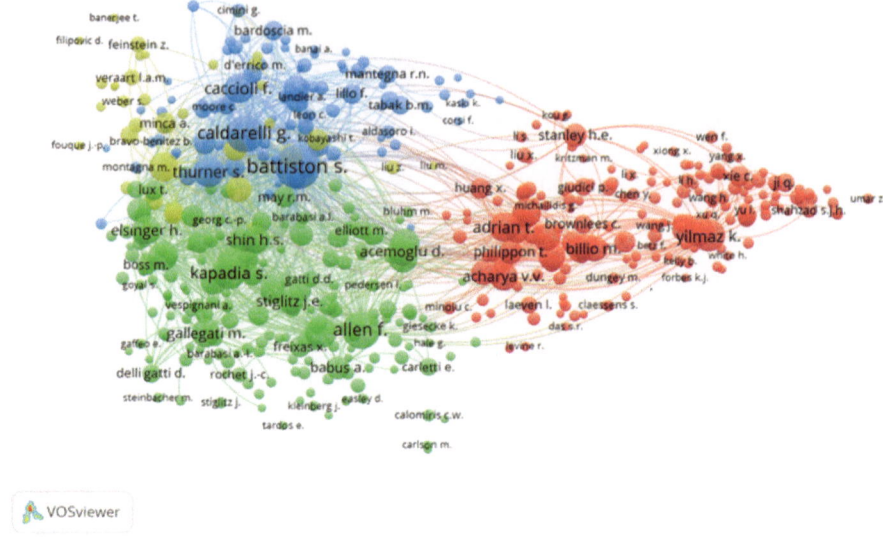

Fig. 2.3 Co-citation analysis (VOSviewer)

VOSviewer software was used to generate a keyword co-occurrence clustering view in systemic risk and network science. A total of 86 have been selected from 1768 keywords, and a co-occurrence analysis was performed on these 86 keywords, as shown in Fig. 2.4. For each of the 86 keywords, the total strength of the co-occurrence links with other keywords is calculated. The keywords with the greatest total link strength are selected. In Fig. 2.4, the node area and font size depending on the keyword's weight value. The more significant the weight value, the more times the keyword appears; the line between nodes indicates that a keyword appears in common with another. The thickness of the connection line indicates the co-occurrence strength between the two keywords (Tamala et al. 2022).

We identified an important limitation of these analyses associated with the evidence that the VOSviewer method does not reduce an ample space of related terms that are easier to understand but are also indicative of the actual partitions of interrelated concepts in the literature under consideration. Indeed, we find an essential association between terms, but there is no selection of semantic relationships.

From the analysis, nine clusters can be obtained. The topics are summarised for each cluster, and the keywords in each cluster are listed.

Systemic risk, financial networks, network analysis, risk assessment, networks and interconnectedness are among the most highly co-occurring keywords with occurrence weights of 347, 67, 41, 33, 26, and 30, respectively. The 1768 keywords were able to form 9 clusters: cluster 1 (blue), cluster 2 (green), cluster 3 (red), cluster 4 (yellow), cluster 5 (purple), cluster 6 (orange), cluster 7 (brown) and cluster 8 (light blue).

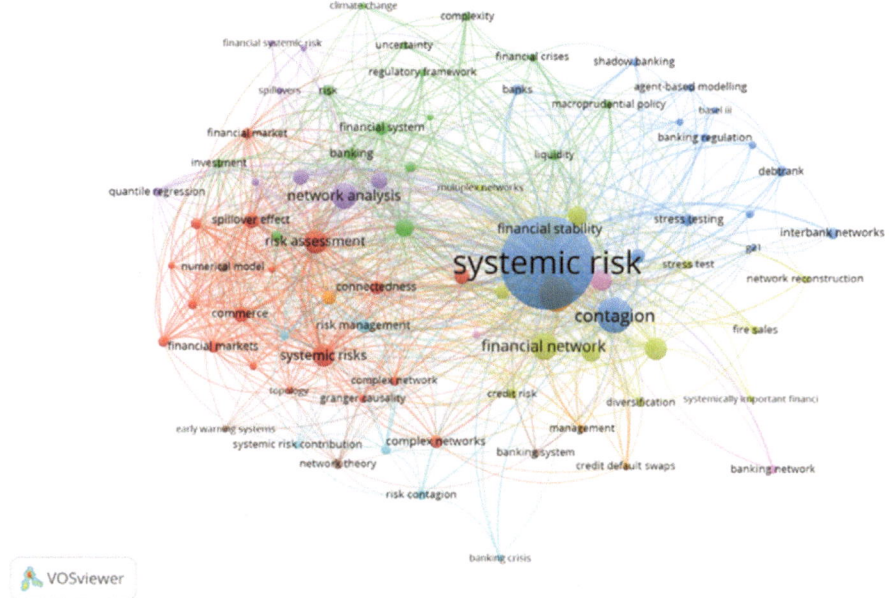

Fig. 2.4 Co-word analysis (VOSviewer)

In Fig. 2.5, the frequency of keywords is less than before 2018. The early studies between 2016 and 2018 focus on the banking crisis, interbank market and financial stability. In 2018 researchers introduced the systemic risk theme, risk assessment and interconnectedness. More recent studies after 2019 focus on the relationship between systemic risk network science. Indeed, the concept of systemic risk in network science is relatively new and appeared consistently between 2019 and 2020. The literature on systemic risk and network science is developing, and, specifically, it is concentrated on the importance of the financial networks that can alert to the prominent role institutions play in the system.

This emerging topic explains how a series of failures can propagate through the network in a cascading process (Sinha et al. 2013).

Using co-word analysis to draw clusters of keywords, we can also consider the themes, whose density and centrality help classify themes and map them in a two-dimensional diagram.

A thematic map (Fig. 2.6) is a very intuitive analysis that allows identifying themes according to the quadrant in which they are placed: upper-right quadrant: motor themes; lower-right quadrant: basic themes; lower-left quadrant: emerging or disappearing themes; upper left quadrant: very specialised/niche themes (Cobo et al. 2011).

The thematic map exploits the Keywords Plus field, associated with Thomson Reuters editorial experts and is supported by a semi-automated algorithm. Keywords Plus terms can capture an article's content with depth and variety (Della Corte et al.

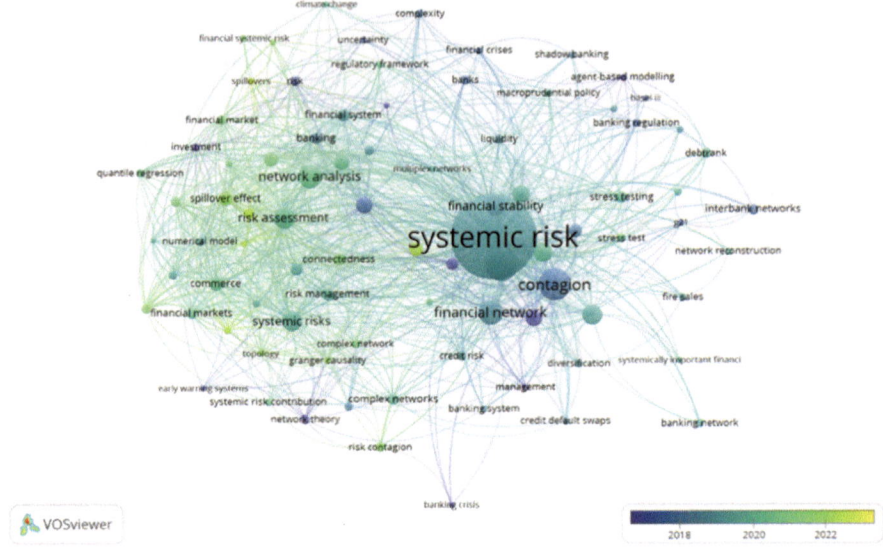

Fig. 2.5 Co-word analysis (VOSviewer)

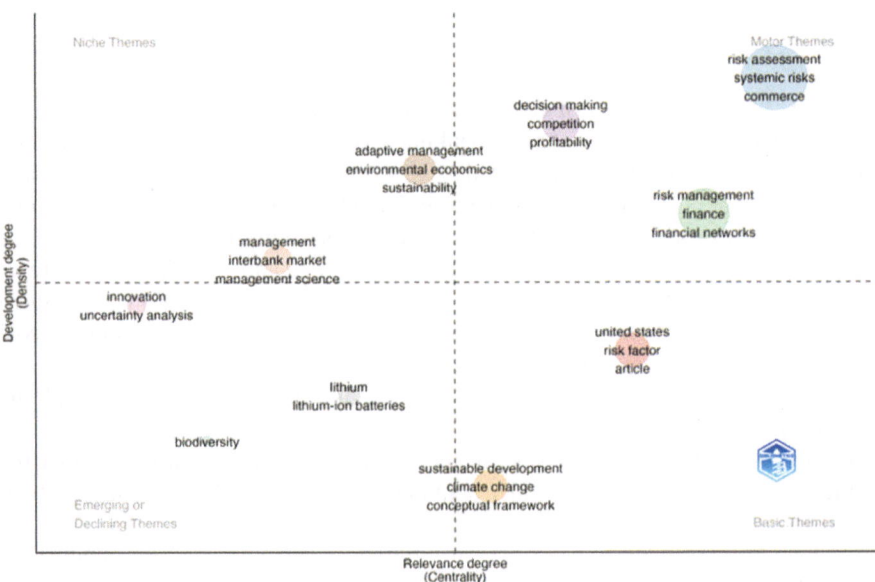

Fig. 2.6 Thematic map (bibliometrix)

2019). The upper-right quadrant shows the motor themes, which are more developed in the literature and are characterised by both high centrality and density. Among the "motor themes" the main concerns are risk management, finance, and financial networks. The upper-left quadrant shows high-density themes but with a low centrality. In this quadrant, it is possible to find the themes of management, interbank market and management science. In the lower-left quadrant are the emerging or declining themes. In this research, the themes of biodiversity, lithium and uncertainty analysis are emerging. Finally, the lower-right quadrant shows the themes that are basic and transversal. These themes concern general topics that are transversal to the different research areas of the field. In this area, we find sustainable development, climate change and conceptual framework that are recurring words associated with the Co-word analysis cluster.

Discussion of Results

As summarised above, there has been a significant increase in scientific production over time since 2007–08, confirming that the topics only received more attention after the Great Financial Crisis. Since 2012, there has been an increase in the number of papers, which is undoubtedly related to the intensification of the financial crisis and the 2020 pandemic crisis. These results highlight the interest of the international scientific community in the topic of systemic risk and network science. Focusing on the time horizon analysed, three phases have been identified. The first phase (2008–2013) is characterised by around forty publications, focusing on topics such as financial network theory, risk governance, the mechanisms that contributed to the Great Financial Crisis and systemic risk in the interbank market (Battiston et al. 2012; Billio et al. 2012; Georg 2013; Markose et al. 2012; Palma 2009; Van Asselt and Renn 2011). In the second phase (2014–2018), more than one hundred and seventy publications can be traced that highlight concepts related to systemic risk and financial networks, including: financial stability, contagion and interconnectedness. In particular, several studies focus on network structure as a tool for understanding and managing a wide range of phenomena, including systemic risk assessment (Acemoglu et al. 2015; Roukny et al. 2018; In't Veld and van Lelyveld 2014). The last period (2019–2023) is characterised by an exponential growth of scientific papers (more than three hundred papers), in which the concept of financial networks seems to predominate. In recent years, specific topics on climate change, the great challenge of this century (Cerqueti et al. 2021), big data and machine learning, cryptocurrencies, systemic digital risk and complex networks appear increasingly (Kou et al. 2019; Li et al. 2020a).

From the second phase (2014–2018), the literature review provides a first overview of climate risk in the financial system from a network science perspective. Climate change mitigation and adaptation pose governance challenges of unprecedented magnitude due to their long-term horizon, global nature and the enormous uncertainties they entail. The interconnectedness of systems dramatically increases the complexity of studying the impacts of climate events and requires a new design of models capable of analysing climate and weather damages (Helbing 2013). The role that financial and banking systems could play in inducing green investments has received increasing attention (Campiglio 2016; Mazzucato 2015). In particular, the results of the main research on this topic show that green policies can improve economic performance without putting pressure on the financial system (Monasterolo and Raberto 2017). In this context, the relationship between rapid decarbonisation policies and financial stability is at the top of the climate policy agenda (Lazarus and Tempest 2014; Weber and Mark Fulton 2015). Indeed, while the financial system can facilitate the transition to a green development path, it is also increasingly exposed to climate risks. In this context, the application of network science to the structure of relationships between financial institutions could be crucial for the stability of the whole system (Balint et al. 2017). Analysing this issue from a network perspective, Battiston et al. (2016) study the exposure of different classes of actors in the financial system using a well-known macro-network stress-test model (Bardoscia et al. 2015; Battiston et al. 2012). The results suggest that direct and indirect exposure to climate-related sectors accounts for a large proportion of investors' equity portfolios, especially for mutual and pension funds. Recently, however, it has been investigated whether the universe of investment funds shows systemic vulnerability to climate-related risks. Using supervisory data on Mexican banks and investment funds, Roncoroni et al. (2021) show how investment funds absorb losses and transmit the impact of climate shocks to other financial market participants. Gourdel and Sydow (2022) apply climate shocks to an integrated system of investment funds but focus on a thirty-day horizon rather than the long-term vulnerabilities of the sector. This shortcoming is addressed by Amzallag (2022), whose results suggest that despite the increasing focus on sustainable investments, systemic vulnerabilities remain and many funds' portfolio diversification approaches still do not adequately reflect climate risk.

Given the high degree of interconnectedness between actors in the financial system, the role of these relationships in relation to climate policy is one of the main challenges for the construction of agent network models.

Therefore, globalisation, by making our world even more exposed and vulnerable to risks, has highlighted the limits and costs of a highly interconnected world (Gambacorta et al. 2020; Romanosky 2016).

In this context, the interest of many researchers in analysing systemic digital risk has grown in recent years, in order to understand how technological developments and digital services are linked to complex and uncertain impacts and consequences of their development, causing multiple effects in society. Potential threats that can be defined as systemic risks include bankruptcy risks, cyber security, data misuse and privacy (Lupton 2015).

The digital economy, which includes companies that use information, data and internet technologies in their business models, is increasingly presenting itself as a hegemonic business model (Lynn et al. 2022). In terms of the analysing of interconnectedness in the networked digital economy, the digital economy is one of the few sectors considered to be even more interconnected than global finance. Moreover, this growth has been extremely rapid, not only in terms of the number of digitally connected devices increasing, but also in terms of new types of instruments.

The high level of complexity in the financial system played a key role in the outbreak of the 2008 financial crisis. Perrow (1984) provides a theoretical basis for distinguishing the risk properties of different types of complexity by differentiating between linear complexity and interactive complexity. Linear complexity implies a system composed of many parts that interact in a linear, visible and generally predictable way (Haldane 2009). In interactive complex systems, on the other hand, each part of the system is more likely to depend on any other part of the system in an unpredictable and irregular way. It is widely recognised that the period leading up to the Great Financial Crisis saw a significant increase in the complexity of the financial system. In particular, complex forms of securitisation led to an increase in the opacity of the interactions between different financial transactions. Similarly, several researchers argue that, in addition to high levels of interconnectedness, high levels of complexity are a key feature of the digital economy. In this respect, systemic digital risk manifests not only a high level of complexity, but above all a high level of interactive complexity, where extensive and unexpected connections between different parts of the system are possible. However, despite the persistence of cybersecurity failures and system fragility, the current business model follows the evolution of the digital economy, which is constantly pushing for further growth in network size and complexity.

The topic of digitisation has therefore gained particular scientific interest, also due to the rise of the cryptocurrency universe which has further increased the likelihood of hacking in the financial sector (Boissay et al. 2023).

Cryptocurrencies are digital or virtual currencies that are encrypted using cryptography. Although initially conceived to disrupt the centralised model of electronic payments in online commerce (Nakamoto 2008), cryptocurrencies have attracted strong interest since their popularisation, particularly as investment assets (Baur et al. 2018; Dyhrberg 2016; Umar and Gubareva 2020; Wu and Pandey 2014; Yermack 2015). However, given their volatile nature, the growth of cryptocurrencies as a new investment asset may create new risk spillovers to existing financial assets, potentially increasing the level of systemic risk in the financial system (Li et al. 2020b). In this regard, several studies have focused on spillovers between cryptocurrencies and other financial markets (Bouri et al. 2019; Katsiampa et al. 2019a, b; Koutmos 2018; Trabelsi 2018). Among them, Umar et al. (2021) contribute to the cryptocurrency literature by analysing the linkages between cryptocurrencies and the technology sector. Specifically, they analyse the spillovers and interdependences between the main markets of the IT sector in advanced and emerging economies. The findings suggest that there is a significant degree of interconnectedness between technology sectors globally. However, it is clear that the contribution to the cryptocurrency market remains insignificant. In fact, the crypto market appears to be less integrated with the technology sector and less exposed to systemic risk. These findings suggest that crypto may offer investors diversification benefits against the risk of the technology sector. On the other hand, other studies that have considered the contribution of cryptocurrencies to the financial system in terms of systemic risk have focused on the calculation and estimation of risk. In particular, Akhtaruzzaman et al. (2022) use the Conditional Value-at-Risk (CoVar) to measure the systemic risk of cryptocurrencies. The results suggest that the systemic risk peaked on 12 March 2020, but fell to a minimum the following day, demonstrating how the dynamics of systemic risk sharing among cryptocurrencies evolve rapidly.

In light of the discussion conducted, it can be argued that financial stability and recent innovations in climate finance and the digital revolution, when analysed with the help of complex network theory, can play a crucial role in the transformation of our society towards a more sustainable and secure model (Balint et al. 2017).

In this respect, it seems clear that the thread running through this discussion is the application of complex networks. In this regard, the field of financial networks represents one of the new applications of statistical physics that has attracted particular interest among researchers in recent years (Bardoscia et al. 2017).

In particular, the interest lies in their dynamics and evolution, which have led to theoretical advances such as the definition of the DebtRank to estimate the impact and propagation of shocks in the system (Battiston et al. 2012; Caldarelli 2022).

In financial markets research, the literature is growing rapidly, and with the application of the empirical approach, it provides a good basis for the development of new mathematical models, econometric analyses and a better understanding of large-scale phenomena such as contagion channels in the financial system, linkages between institutions and financial markets, and the analysis of systemic risk and financial stability (Battiston et al. 2016; D'Arcangelis and Rotundo 2016).

Among these, Bakkar and Nyola (2021) refer to the plethora of research that has highlighted how organisational complexity and geographical dispersion can provide diversification benefits and reduce systemic risks. The study analyses the effects of the Great Financial Crisis (2008–2009) and the European sovereign debt crisis (2010–2011) on the impact of complexity on stability for a sample of European banks. The main findings show that bank complexity, both organisationally and geographically, mitigates systemic risk, and that while bank complexity is associated with lower systemic risk in normal times, its effect is reversed when the banking system experiences global shocks.

Carreño and Cifuentes (2017), on the other hand, propose a framework for identifying the structure of a financial network and its evolution over time, applying it to an interbank market in order to identify the sources and channels of transmission of systemic risk in a network of banks. The results make it possible to identify the different groups of banks that could be vulnerable to shocks, as well as those that are more or less likely to generate shocks to the system. In addition, such information can be used to map systemic risks and related exposures and thus to prioritise surveillance.

Finally, Yang et al. (2023) apply the complex network model to the cryptocurrency market. The research analyses the risk-spreading capacity of cryptos and their contribution to systemic risk over the period 2018–2022. The results show that cryptocurrencies with higher risk-spreading capacity, as well as others that experience large price drops, contribute more to systemic risk.

In light of the discussion, it is possible to identify new challenges for scientific research related to systemic risk and network science. The first challenge concerns the need to study the relationship between climate risk and the application of complex networks. In this regard, it would be interesting to use network science and artificial intelligence models to investigate the nature of climate risk, to identify a link between financial systemic risks and climate change, and to see how the latter can influence global economic and financial crises. Another relevant challenge relates to understanding digital systemic risk. Growing attention to new technological developments has raised concerns about new systemic risks in the technology sector, including cybersecurity. Finally, the challenges posed by the application of complex network models to the innovative cryptocurrency and blockchain market should not be overlooked. In this respect, a review of the work has brought to light an interesting contribution by Chen et al. (2022), who propose a prototype risk map, the Johns Hopkins Blockchain Risk Map. In a context where blockchain has ushered in a new phase in the evolution of the internet from Web 2.0 to 3.0, the complex nature of this innovative technology has initiated new research to properly measure and communicate blockchain risks (Masera 2023). In order to promote better risk governance and develop appropriate regulation, the researchers in this paper propose a prototype risk map capable of identifying multidimensional risk exposures between different stakeholders. While the study is by no means exhaustive, the analysis encourages new research ideas to promote a better understanding of blockchain risks, including operational, decentralisation, security and systemic risks.

Conclusions

The bibliometric and systematic literature review in this chapter aims to provide a basic worldwide overview of research publications on systemic risk and network science, complementing the quantitative and qualitative analysis of previous narrative reviews on the topic. The literature on systemic risk has grown steadily over the years. It will continue to grow in order to better understand and manage risk in the financial system by modelling it in terms of financial networks.

The study of systemic risk in the context of network science represents a new research challenge for the coming years.

In particular, we have seen a growing interest in climate-related financial risks and the critical role of financial network models applied to climate stress testing and the transition to a low-carbon economy, which may have implications for financial stability. In addition, research in this area is increasingly focusing on systemic risks related to digitalisation and the cryptocurrency market. Finally, in the area of complex financial networks, the application of statistical physics to economic and social systems has been very successful in terms of results and impact. Indeed, it is now widely recognised that modelling the financial system as a network is a prerequisite for understanding a wide range of phenomena, including systemic risk.

This chapter also analyses the evolution of publications on systemic risk and networks, including influential aspects of the literature.

The results suggest that systemic risk and financial networks have experienced rapid growth over the last decade, which may contribute to a future new research agenda on the topic, providing different approaches to the study and a more comprehensive view of the phenomenon.

References

Acemoglu, D., Ozdaglar, A., Tahbaz-Salehi, A.: Systemic risk and stability in financial networks. Am. Econ. Rev. **105**, 564–608 (2015). https://doi.org/10.1257/aer.20130456

Akhtaruzzaman, M., Boubaker, S., Nguyen, D.K., Rahman, M.R.: Systemic risk-sharing framework of cryptocurrencies in the COVID–19 crisis. Financ. Res. Lett. **47**, 102787 (2022). https://doi.org/10.1016/j.frl.2022.102787

Albort-Morant, G., Ribeiro-Soriano, D.: A bibliometric analysis of international impact of business incubators. J. Bus. Res. **69**, 1775–1779 (2016). https://doi.org/10.1016/j.jbusres.2015.10.054

Amzallag, A.: Fund portfolio networks: a climate risk perspective. Int. Rev. Financ. Anal. **84**, 102259 (2022). https://doi.org/10.1016/j.irfa.2022.102259

Andersen, N.: Mapping the expatriate literature: a bibliometric review of the field from 1998 to 2017 and identification of current research fronts. Int. J. Hum. Resourc. Manage. **32**, 4687–4724 (2021). https://doi.org/10.1080/09585192.2019.1661267

Aria, M., Cuccurullo, C.: Bibliometrix: an R-tool for comprehensive science mapping analysis. J. Informet. **11**, 959–975 (2017). https://doi.org/10.1016/j.joi.2017.08.007

Backhaus, K., Lügger, K., Koch, M.: The structure and evolution of business-to-business marketing: a citation and co-citation analysis. Ind. Mark. Manage. **40**, 940–951 (2011). https://doi.org/10.1016/j.indmarman.2011.06.024

Baker, H.K., Kumar, S., Pattnaik, D.: Twenty-five years of *Review of Financial Economics*: a bibliometric overview. Rev. Financ. Econ. **38**, 3–23 (2020). https://doi.org/10.1002/rfe.1095

Baker, H.K., Kumar, S., Pandey, N.: Forty years of the *Journal of Futures Markets*: a bibliometric overview. J. Futur. Mark. **41**, 1027–1054 (2021). https://doi.org/10.1002/fut.22211

Bakkar, Y., Nyola, A.P.: Internationalization, foreign complexity and systemic risk: evidence from European banks. J. Financ. Stab. **55** (2021)

Balint, T., Lamperti, F., Mandel, A., Napoletano, M., Roventini, A., Sapio, A.: Complexity and the economics of climate change: a survey and a look forward. Ecol. Econ. **138**, 252–265 (2017). https://doi.org/10.1016/j.ecolecon.2017.03.032

Bardoscia, M., Battiston, S., Caccioli, F., Caldarelli, G.: DebtRank: a microscopic foundation for shock propagation. PLoS ONE **10**, e0130406 (2015). https://doi.org/10.1371/journal.pone.0130406

Bardoscia, M., Battiston, S., Caccioli, F., Caldarelli, G.: Pathways towards instability in financial networks. Nat. Commun. **8**, 14416 (2017). https://doi.org/10.1038/ncomms14416

Battiston, S., Puliga, M., Kaushik, R., Tasca, P., Caldarelli, G.: DebtRank: too central to fail? Financial networks, the FED and systemic risk. Sci. Rep. **2**, 541 (2012). https://doi.org/10.1038/srep00541

Battiston, S., D'Errico, M., Gurciullo, S.: Debtrank and the network of leverage. J. Altern. Investments **18**, 68–81 (2016). https://doi.org/10.3905/jai.2016.18.4.06

Baur, D.G., Hong, K., Lee, A.: Bitcoin: medium of exchange or speculative assets? J. Int. Finan. Markets. Inst. Money **54**, 177–189 (2018)

Benoit, S., Colliard, J.-E., Hurlin, C., Pérignon, C.: Where the Risks Lie: A Survey on Systemic Risk. Social Science Research Network, Rochester, NY (2015)

Billio, M., Getmansky, M., Lo, A.W., Pelizzon, L.: Econometric measures of connectedness and systemic risk in the finance and insurance sectors. J. Financ. Econ. **104**, 535–559 (2012). https://doi.org/10.1016/j.jfineco.2011.12.010

Boissay F., Borio C., Leonte C.S., Shim I.: Prudential policy and financial dominance: exploring the link. BIS Quart. Rev. (2023)

Bouri, E., Shahzad, S.J.H., Roubaud, D.: Co-explosivity in the cryptocurrency market. Financ. Res. Lett. **29**, 178–183 (2019). https://doi.org/10.1016/j.frl.2018.07.005

Calabretta, G., Durisin, B., Ogliengo, M.: Uncovering the intellectual structure of research in business ethics: a journey through the history, the classics, and the pillars of 'Journal of Business Ethics.' J. Bus. Ethics **104**, 499–524 (2011)

Caldarelli, G.: Senza uguali. Comprendere con le reti un mondo che non ha precedenti. Egea (2022)

Campiglio, E.: Beyond carbon pricing: the role of banking and monetary policy in financing the transition to a low-carbon economy. Ecol. Econ. **121**, 220–230 (2016)

Carreño, J.G., Cifuentes, R.: Identifying complex core–periphery structures in the interbank market. J. Netw. Theory Financ. (2017)

Cerqueti, R., Ciciretti, R., Dalò, A., Nicolosi, M.: ESG investing: a chance to reduce systemic risk. J. Financ. Stab. **54** (2021). https://doi.org/10.1016/j.jfs.2021.100887

Chen, J., Tong, Q.Q., Verma, H., Sharma, A., Dahbura, A., Liew, J.K.-S.: The complexity of blockchain risks simplified and displayed: introduction of the Johns Hopkins blockchain risk map. J. Financ. Data Sci. (2022). https://doi.org/10.3905/jfds.2022.1.117

Cobo, M.J., López-Herrera, A.G., Herrera-Viedma, E., Herrera, F.: An approach for detecting, quantifying, and visualizing the evolution of a research field: a practical application to the fuzzy sets theory field. J. Informet. **5**, 146–166 (2011). https://doi.org/10.1016/j.joi.2010.10.002

D'Arcangelis, A.M., Rotundo, G.: Complex networks in finance. In: Commendatore, P., Matilla-García, M., Varela, L.M., Cánovas, J.S. (eds.) Complex Networks and Dynamics: Social and Economic Interactions, pp. 209–235. Springer International Publishing, Cham (2016)

Della Corte, V., Del Gaudio, G., Sepe, F., Sciarelli, F.: Sustainable tourism in the open innovation realm: a bibliometric analysis. Sustainability. **11**, 6114 (2019). https://doi.org/10.3390/su11121 6114

Dičpinigaitienė, V., Novickytė, L.: Application of systemic risk measurement methods: a systematic review and meta-analysis using a network approach. Quant. Finance Econ. **2**(4), 798–820 (2018)

Donthu, N., Kumar, S., Pattnaik, D.: Forty-five years of *Journal of Business Research*: a bibliometric analysis. J. Bus. Res. **109**, 1–14 (2020). https://doi.org/10.1016/j.jbusres.2019.10.039

Donthu, N., Kumar, S., Mukherjee, D., Pandey, N., Lim, W.M.: How to conduct a bibliometric analysis: an overview and guidelines. J. Bus. Res. **133**, 285–296 (2021). https://doi.org/10.1016/j.jbusres.2021.04.070

Durisin, B., Puzone, F.: Maturation of corporate governance research, 1993–2007: an assessment. Corp. Govern. Int. Rev. **17**, 266–291 (2009). https://doi.org/10.1111/j.1467-8683.2009.00739.x

Dyhrberg, A.H.: Bitcoin, gold and the dollar—a GARCH volatility analysis. Financ. Res. Lett. **16**, 85–92 (2016). https://doi.org/10.1016/j.frl.2015.10.008

Ellegaard, O., Wallin, J.A.: The bibliometric analysis of scholarly production: how great is the impact? Scientometrics **105**, 1809–1831 (2015). https://doi.org/10.1007/s11192-015-1645-z

Gambacorta, L., Frost, J., Gambacorta, R.: The Matthew effect and modern finance: on the nexus between wealth inequality, financial development and financial technology. CEPR Discussion Papers (2020)

Georg, C.-P.: The effect of the interbank network structure on contagion and common shocks. J. Bank. Finance **37**, 2216–2228 (2013). https://doi.org/10.1016/j.jbankfin.2013.02.032

Gourdel, R., Sydow, M.: Non-banks contagion and the uneven mitigation of climate risk. https://papers.ssrn.com/abstract=4305521 (2022)

Haldane, A.G.: Rethinking the Financial Network. Presented at the Financial Student Association, Amsterdam (2009)

Helbing, D.: Globally networked risks and how to respond. Nature **497**, 51–59 (2013). https://doi.org/10.1038/nature12047

Hu, C., Song, M., Guo, F.: Intellectual structure of market orientation: a citation/co-citation analysis. Mark. Intell. Plan. **37**, 598–616 (2019). https://doi.org/10.1108/MIP-08-2018-0325

In't Veld, D., van Lelyveld, I.: Finding the core: Network structure in interbank markets. J. Banking Finance **49**, 27–40 (2014). https://doi.org/10.1016/j.jbankfin.2014.08.006

Jackson, M.O., Pernoud, A.: Systemic risk in financial networks: a survey. arXiv:2012.12702 [physics, q-fin] (2020)

Katsiampa, P., Corbet, S., Lucey, B.: High frequency volatility co-movements in cryptocurrency markets. J. Int. Financ. Markets, Inst. Money **62**, 35–52 (2019a). https://doi.org/10.1016/j.int fin.2019.05.003

Katsiampa, P., Corbet, S., Lucey, B.: Volatility spillover effects in leading cryptocurrencies: a BEKK-MGARCH analysis. Finance Res. Lett. **29**, (2019b). https://doi.org/10.1016/j.frl.2019.03.009

Khan, A., Hassan, M.K., Paltrinieri, A., Dreassi, A., Bahoo, S.: A bibliometric review of Takaful literature. Int. Rev. Econ. Financ. **69**, 389–405 (2020). https://doi.org/10.1016/j.iref.2020.05.013

Khan, A., Goodell, J.W., Hassan, M.K., Paltrinieri, A.: A bibliometric review of finance bibliometric papers. Finance Res. Lett. 102520 (2021). https://doi.org/10.1016/j.frl.2021.102520

Kim, J., Mcmillan, S.: Evaluation of internet advertising research: a bibliometric analysis of citations from key sources. J. Advert. **37**, 99–112 (2008). https://doi.org/10.2753/JOA0091-3367370108

Knobloch, K., Yoon, U., Vogt, P.M.: Preferred reporting items for systematic reviews and meta-analyses (PRISMA) statement and publication bias. J. Craniomaxillofac. Surg. **39**, 91–92 (2011). https://doi.org/10.1016/j.jcms.2010.11.001

Kou, G., Chao, X., Peng, Y., Alsaadi, F.E., Herrera-Viedma, E.: Machine learning methods for systemic risk analysis in financial sectors. Technol. Econ. Dev. Econ. **25**, 716–742 (2019). https://doi.org/10.3846/tede.2019.8740

Koutmos, D.: Return and volatility spillovers among cryptocurrencies. Econ. Lett. **173**, 122–127 (2018). https://doi.org/10.1016/j.econlet.2018.10.004

Krippendorff, K. Principles of design and a trajectory of artificiality. J. Prod. Innov. Manage. **28** (2004)

Krippendorff, K.: Content analysis: An Introduction to Its Methodology. Sage, Thousand Oaks (2011)

Kumar, S., Lim, W.M., Pandey, N., Christopher Westland, J.: 20 years of *Electronic Commerce Research*. Electron. Commer. Res. **21**, 1–40 (2021a). https://doi.org/10.1007/s10660-021-09464-1

Kumar, S., Sureka, R., Lim, W.M., Kumar Mangla, S., Goyal, N.: What do we know about business strategy and environmental research? Insights from *Business Strategy and the Environment*. Bus. Strategy Environ. **30**, 3454–3469 (2021b). https://doi.org/10.1002/bse.2813

Lazarus, M., Tempest, K.: Estimating international mitigation finance needs: A top-down perspective. Presented at the November 24 (2014)

Li, J., Li, J., Zhu, X., Yao, Y., Casu, B.: Risk spillovers between FinTech and traditional financial institutions: evidence from the U.S. Int. Rev. Financ. Anal. **71**, (2020a). https://doi.org/10.1016/j.irfa.2020.101544

Li, K.K., Huang, B., Tam, T., Hong, Y.: Does the COVID-19 pandemic affect people's social and economic preferences? Evidence China (2020b). https://doi.org/10.2139/ssrn.3690072

Linnenluecke, M.K., Chen, X., Ling, X., Smith, T., Zhu, Y.: Research in finance: a review of influential publications and a research agenda. Pac. Basin Financ. J. **43**, 188–199 (2017). https://doi.org/10.1016/j.pacfin.2017.04.005

Lupton, D.: Digital Bodies (2015). https://doi.org/10.2139/ssrn.2606467

Lynn, T., Rosati, P., Conway, E., Curran, D., Fox, G., O'Gorman, C.: The digital economy and digital business. In: Digital Towns: Accelerating and Measuring the Digital Transformation of Rural Societies and Economies, pp. 69–89. Springer International Publishing, Cham (2022)

Markellof, R., Warner, G., Wollin, E.: Modeling systemic risk to the financial system. MITRE Corporation Technical Paper, 12-1870 (2012)

Markose, S., Giansante, S., Shaghaghi, A.R.: 'Too interconnected to fail' financial network of US CDS market: topological fragility and systemic risk. J. Econ. Behav. Organ. **83**, 627–646 (2012). https://doi.org/10.1016/j.jebo.2012.05.016

Masera, R.: Web 1.0, 2.0, 3.0; InfoSphere; Metaverse: an overview. Monetary, Financial, Societal and Geopolitical Transformation Cusps (2023). https://doi.org/10.2139/ssrn.4337362

Mazzucato, M.: The Green Entrepreneurial State. https://papers.ssrn.com/abstract=2744602 (2015)

Mezei, J., Sarlin, P.: RiskRank: measuring interconnected risk. Econ. Model. **68**, 41–50 (2018). https://doi.org/10.1016/j.econmod.2017.04.016

Monasterolo, I., Raberto, M.: Is there a role for central banks in the low-carbon transition? A stock-flow consistent modelling approach. https://papers.ssrn.com/abstract=3075247 (2017)

Nakamoto, S.: Bitcoin: A Peer-to-Peer Electronic Cash System (2008)

Neveu, A.R.: A survey of network-based analysis and systemic risk measurement. J. Econ. Interac. Coord. **13**, 241–281 (2018). https://doi.org/10.1007/s11403-016-0182-z

Pacelli V.: Rischio sistemico e scienza delle reti. BANCARIA **12** (2021)

Palma, J.G.: The revenge of the market on the rentiers. Why neo-liberal reports of the end of history turned out to be premature. Camb. J. Econ. **33**, 829–869 (2009). https://doi.org/10.1093/cje/bep037

Pati, D., Lorusso, L.N.: How to write a systematic review of the literature. HERD **11**, 15–30 (2018). https://doi.org/10.1177/1937586717747384

Perrow, C.: Normal Accidents: Living with High-Risk Technologies. Basic Books, New York (1984)

Poledna, S., Martínez-Jaramillo, S., Caccioli, F., Thurner, S.: Quantification of systemic risk from overlapping portfolios in the financial system. J. Financ. Stab. **52**, 100808 (2021). https://doi.org/10.1016/j.jfs.2020.100808

Ramos-Rodrígue, A.-R., Ruíz-Navarro, J.: Changes in the intellectual structure of strategic management research: a bibliometric study of the *Strategic Management Journal*, 1980–2000. Strateg. Manag. J. **25**, 981–1004 (2004). https://doi.org/10.1002/smj.397

Romanosky, S.: Examining the costs and causes of cyber incidents. J. Cyber. Secur. tyw001 (2016). https://doi.org/10.1093/cybsec/tyw001

Roncoroni, A., Battiston, S., D'Errico, M., Hałaj, G., Kok, C.: Interconnected banks and systemically important exposures. J. Econ. Dyn. Control **133** (2021). https://doi.org/10.1016/j.jedc.2021.104266

Roukny, T., Battiston, S., Stiglitz, J.E.: Interconnectedness as a source of uncertainty in systemic risk. J. Financ. Stab. **35**, 93–106 (2018). https://doi.org/10.1016/j.jfs.2016.12.003

Samiee, S., Chabowski, B.R.: Knowledge structure in international marketing: a multi-method bibliometric analysis. J. Acad. Mark. Sci. **40**, 364–386 (2012). https://doi.org/10.1007/s11747-011-0296-8

Silva, W., Kimura, H., Sobreiro, V.A.: An analysis of the literature on systemic financial risk: a survey. J. Financ. Stab. **28**, 91–114 (2017). https://doi.org/10.1016/j.jfs.2016.12.004

Singh, V.K., Singh, P., Karmakar, M., Leta, J., Mayr, P.: The journal coverage of web of science, scopus and dimensions: a comparative analysis. Scientometrics **126**, 5113–5142 (2021). https://doi.org/10.1007/s11192-021-03948-5

Sinha, S., Thess, M., Markose, S.: How unstable are complex financial systems? Analyzing an inter-bank network of credit relations. New Econ. Windows **13**, 59–76 (2013). https://doi.org/10.1007/978-88-470-2553-0_5

Tamala, J.K., Maramag, E.I., Simeon, K.A., Ignacio, J.J.: A bibliometric analysis of sustainable oil and gas production research using VOSviewer. Cleaner Eng. Technol. **7**, 100437 (2022). https://doi.org/10.1016/j.clet.2022.100437

Trabelsi, N.: Are there any volatility spill-over effects among cryptocurrencies and widely traded asset classes? JRFM **11**, 1–17 (2018)

Umar, Z., Trabelsi, N., Alqahtani, F.: Connectedness between cryptocurrency and technology sectors: international evidence. Int. Rev. Econ. Financ. **71**, 910–922 (2021). https://doi.org/10.1016/j.iref.2020.10.021

Umar, Z., Gubareva, M.: A time–frequency analysis of the impact of the Covid-19 induced panic on the volatility of currency and cryptocurrency markets. J. Behav. Exp. Finance. **28** (2020)

Van Asselt, M.B.A., Renn, O.: Risk governance. J. Risk Res. **14**, 431–449 (2011). https://doi.org/10.1080/13669877.2011.553730

Weber, C., Mark Fulton, W.: Consultant: carbon asset risk. Discussion Framework (2015)

Wu, C., Pandey, V.: The value of Bitcoin in enhancing the efficiency of an investor's portfolio. J. Financ. Plan. **27**, 44–52 (2014)

Yang, M.-Y., Wang, C., Wu, Z.-G., Wu, X., Zheng, C.: Influential risk spreaders and their contribution to the systemic risk in the cryptocurrency network. Finance Res. Lett. **57** (2023)

Yermack, D.: Chapter 2—is bitcoin a real currency? An economic appraisal. In: Lee Kuo Chuen, D. (ed.) Handbook of Digital Currency, pp. 31–43. Academic Press, San Diego (2015)

Zupic, I., Čater, T.: Bibliometric methods in management and organization. Organ. Res. Methods **18**, 429–472 (2015). https://doi.org/10.1177/1094428114562629

Chapter 3
A Holistic Journey into Systemic Risk: Theoretical Background, Transmission Channels and Policy Implications

Vincenzo Pacelli, Lucianna Cananà, Anirban Chakraborti, Caterina Di Tommaso, and Matteo Foglia

Abstract Systemic risk represents a critical challenge in modern financial systems characterized by complex interconnections. This chapter comprehensively analyses systemic risk, exploring its measurement, models, determinants, interconnections, and the key variables influencing its dynamics. One of the central focuses of this chapter is to explore the transmission channels through which systemic risk propagates. By analyzing various channels, including contagion risk, credit risk, liquidity risk, market risk, operational risk, and macroeconomic risk, the chapter unveils the mechanisms through which disruptions can spread across financial institutions, markets, and economies. The interconnected nature of these channels is also emphasized to showcase the amplification of systemic risk. The interconnections between financial institutions and markets are crucial factors of systemic risk. We discuss the significance of network analysis and emphasize the importance of considering both visible and hidden (shadow) interconnections when assessing systemic risk. By identifying the vulnerabilities and interdependencies within the financial system, policymakers could then develop targeted measures to mitigate systemic risks. The chapter highlights the need for proactive monitoring, enhanced risk management practices, and coordinated regulatory efforts across jurisdictions. These policy implications

V. Pacelli (✉) · L. Cananà · C. Di Tommaso · M. Foglia
University of Bari Aldo Moro, Bari, Italy
e-mail: vincenzo.pacelli@uniba.it

L. Cananà
e-mail: lucianna.canana@uniba.it

C. Di Tommaso
e-mail: caterina.ditommaso@uniba.it

M. Foglia
e-mail: matteo.foglia@uniba.it

A. Chakraborti
Jawaharlal Nehru University, New Delhi, India
e-mail: anirban@jnu.ac.in

© The Author(s) 2025   43
V. Pacelli (ed.), *Systemic Risk and Complex Networks in Modern Financial Systems*,
New Economic Windows, https://doi.org/10.1007/978-3-031-64916-5_3

could then strengthen the financial system's resilience and reduce the likelihood of systemic crises.

Keywords Systemic risk · Financial networks · Shadow links · Systemic risk channel

What is Systemic Risk?

"Systemic risks are for financial market participants what Nessie, the monster of Loch Ness, is for the Scots (and not only for them): Everyone knows and is aware of the danger. Everyone can accurately describe the threat. Nessie, like systemic risk, is omnipresent, but nobody knows when and where it might strike. There is no proof that anyone has really encountered it, but there is no doubt that it exists."—Sheldon and Maurer (1998)

The complete and exhaustive definition of systemic risk has been a research challenge for several years. Understanding its "genome" is a prerequisite to anticipating and responding optimally to banking and financial crises and avoiding contagion to the real economy. The willingness to define systemic risk implies the primary purpose of looking for a quantitative method of measuring risk to predict and control its evolution. The basic idea is that managing what cannot be measured over time is impossible,[1] which is of great importance in the context of financial stability.

Hellwig and Admati (2014) and Ellis et al. (2022) show that the ambiguous definition of systemic risk can suggest a different approach to mitigate it, and these can imply different aims and consequences on the real economy. As we have analyzed in the first chapter, systemic risk can be defined as the risk of uncontrolled propagation of a crisis of a single player or area of an economic system to a broader system. So, systemic risk is about how the financial system's collapse can get disrupted. Therefore, the probability of a collapse of the financial system is prompted by the downside co-movements (unidirectional and multidirectional) of assets and/or by a widespread drought of liquidity (Benoit et al. 2017; Acharya et al. 2017; Pacelli et al. 2022). The attention is on the potential collapse of the entire financial system, or a part of it, rather than on the failure of individual components that independently do not threaten to bring the rest of the system down. The systemic risk can be referred to as an exogenous shock that affects many financial institutions simultaneously. On the other hand, sometimes the notions refer to the "chain reaction" driven by the debtor's default, which in turn implies the bankruptcy of its creditors. This cascade effect does not depend on the size of the original trigger but on the endogenous

---

[1] In the words of Sir William Thomson: "I often say that when you can measure something that you are speaking about, express it in numbers, you know something about it; but when you cannot measure it when you cannot express it in numbers, your knowledge is of the meagre and unsatisfactory kind: it may be the beginning of knowledge, but you have scarcely, in your thoughts advanced to the stage of science, whatever the matter might be." (From lecture to the Institution of Civil Engineers, London (3 May 1883), "Electrical Units of Measurement", Popular Lectures and Addresses (1889), Vol. 1, 80–81).

self-fulfilling process of diffusion which makes the crisis implode. So, the risk of contagion that causes a systemic crisis can, therefore, depend on the size of the original shock, its breadth, or even the interconnections of the affected actors. However, over the past years, a generally accepted definition of systemic risk has not been accepted. For example, according to Kaufman (1995), systemic risk is: "The probability that cumulative losses will accrue from an event that sets in motion a series of successive losses along a chain of institutions or markets comprising a system […] That is, systemic risk is the risk of a chain reaction of falling interconnected dominos". For the Bank for International Settlements (BIS 1994): "the risk that the failure of a participant to meet its contractual obligations may in turn cause other participants to default with a chain reaction leading to broader financial difficulties". The Board of Governors of the Federal Reserve System (BGFR 2001): "In the payments system, the systemic risk may occur if an institution participating on a private large dollar payments network were unable or unwilling to settle its net debt position. If such a settlement failure occurred, the institution's creditors on the network might also be unable to settle their commitments. Serious repercussions could, as a result, spread to other participants in the private network, to other depository institutions not participating in the network, and to the nonfinancial economy general".

During the financial crisis, the definition has become more specific. ECB (2009) defined systemic risk as the risk "that financial instability becomes so widespread that it impairs the functioning of a financial system to the point where economic growth and welfare suffer materially". According to this definition, systemic risk is the risk of multiple institutions failing and collapsing multiple markets; systemic risk does not arise if the initial shock spreads and only generates losses in several institutions or the failure of one institution. In this sense, the emphasis is placed on systemic risk arising from the possibility of contagion, which is possible when institutions are strongly interconnected. The G20 Report (Board 2009) defines systemic risk as "a risk of disruption to financial services that is caused by an impairment of all or parts of the financial system and has the potential to have serious negative consequences for the real economy". Two assumptions underlying the definition: (i) the shock affects a substantial part of the financial system, and (ii) systemic financial events have to be high. The emphasis on real effects reflects the vision that the main concern for economic policy is the production of real goods and services. The impact on the real economy can be broadcast mainly via three channels. First and foremost, there are problems with payment systems: a loss of widespread trust can induce people to withdraw a large number of deposits, causing the bankruptcy of banks that lack liquidity but would instead be solvents. Second, credit disruptions can create severe reductions in the provision of funds to finance profitable investment opportunities in the non-financial sector. Thirdly, the collapse in asset prices, driven, for example, by a general decline in asset prices, can induce bankruptcies of households and financial and commercial enterprises and reduce economic activity through a fall in wealth and increasing uncertainty.

Since the 2007 crisis, the academic literature has provided a multitude of expressions related to systemic risk. Adrian and Brunnermeier (2016) define systemic risk as: "The risk that institutional distress spreads widely and distorts the supply of credit

and capital to the real economy". Acharya (2009): "[The risk] of widespread failures of financial institutions or freezing up of capital markets that can substantially
reduce the supply of such intermediated capital to the real economy". In the words of
Abdymomunov (2013), "In general, systemic risk is perceived as the risk of a negative shock, severely affecting the entire financial system and the real economy. This
shock can have different causes and triggers, such as a macroeconomic shock, a shock
caused by the failure of an individual market participant that affects the entire system
due to tight interconnections, or a shock caused by information disruption in financial
markets". The author argues that financial stress occurs when the market behavior
changes due to growth of uncertainty, modifying the expectations and estimates for
potential losses and asset value. In addition, Patro et al. (2013) define systemic risk
as the probability of a severe decline in the financial system caused by a strong
and far-reaching event, such as the failure of a financial institution, that negatively
affects not only financial markets but the economy as a whole. Then, for the authors,
systemic risk is a condition in which the entire financial system is under stress at the
same time, resulting in a credit and liquidity crisis. According to Danielsson et al.
(2012), systemic risk can be defined as the sum of market volatility risk arising from
fundamental changes and endogenous feedback from market participants. In fact,
for the authors, systemic risk is endogenous. It depends on the behavior of market
participants, and this perception depends on the perceived risk. If market participants anticipate a higher risk in the future, they will act on this assumption. This
definition recalls Robert K. Merton's theory of self-fulfilling prophecies (1948). A
self-fulfilling prophecy is a prediction that comes true simply because it has been
expressed. Prediction and event are in a circular relationship, according to which the
prediction generates the event and the event verifies the prediction. The self-fulfilling
prophecy is therefore a set of mental mechanisms that we implement to confirm our
assumptions or theories. If we believe that something is real (or could become real
in the near future), we will act as if it were (or if it already was) and, in doing so, as a
consequence of our actions and attitudes, that situation becomes real. Merton's theory
of self-fulfilling prophecies undoubtedly draws inspiration from the formulation that
another famous American sociologist, William Thomas, had given of what has gone
down in history as the Thomas Theorem which states: "If men define certain situations as real, they are real in their consequences." From this perspective, markets
are therefore able to self-determine the onset of a crisis or make it systemic, through
their forecasts and the behaviors resulting from them.

Recently, Poledna et al. (2020) conceptualized systemic risks as the latent potential for a threat or hazard to disseminate disruptions or losses across various
interconnected components within complex systems.

Another acceptable definition of systemic risk that we use in this book is
Schwarcz's definition (Schwarcz 2008): "The risk that an economic shock such
as market or institutional failure triggers (through a panic or otherwise) either the
failure of a chain of markets or institutions or a chain of significant losses to financial
institutions, resulting in increases in the cost of capital or decreases in its availability,
often evidenced by substantial financial market price volatility".

These different notions of systemic risk differ and have other implications. For example, during the U.S. crisis (1980), financial institutions found themselves in difficulties because they had followed similar strategies and were exposed similarly to the risks of a rise in interest rates and a fall in real estate markets. The spread within the financial sector did not play a role, just as the crisis did not significantly affect the real economy. While the Swedish crisis of 1992, very similar in characteristics to the American one of the '80s, induced a serious credit squeeze, making the Swedish recession of the early '90s, the most acute since the Great Depression of the '30s. Finally, the financial crisis of 2007–2009, the sovereign debt crisis (2011–14), the COVID-19 pandemic, and the Silicon Valley Bank defaults driven by contagion mechanisms, i.e., the propagation of shocks, significantly impacted the real economy.

Based on these comparisons of systemic risk diagnosis, all definitions seem to agree on three components of systemic risk: the trigger event, the propagation of shock, and the impact on the real economy and financial system through direct and indirect channels. Following Taylor (2010) the trigger event can be derived from the following:

- Public sector → monetary policy surprise—a monetary policy that kept interest rates too low for too long and an ad hoc bailout policy that led to fear and panic ("Whatever it takes").
- External shock → natural disaster or terrorist event—9/11 terrorist attack; COVID-19 pandemic.
- Financial markets → default of a large financial institution—Lehman Brothers fails; SVB collapse.

According to Hellwig and Admati (2014), there are three determinants of propagation (contagion) of shock:

- via Physical exposures—banks are interconnected via claims and liabilities;
- via Information-based—systemic run;
- via Market and price—the spiral of co-sell assets when one distressed bank.

We can conclude by assuming that: (i) systemic risk is a risk that concerns a large part of the financial system or a significant number of financial institutions; (ii) the transmission (propagation-amplification mechanism) is a key element; (iii) the systemic risk hurts the real economy. And this is to clarify the conceptual framework that we follow in this chapter.[2]

---

[2] Refer to the first chapter (essay) of this book by V. Pacelli for a complete and detailed analysis of the transmission channels, causes and propagation mechanisms of systemic risk.

Channels of Transmission

As discussed in the previous section, we can define systemic risk as the risk of widespread disruption or failure of an entire financial system or market, often triggered by an event or shock that can rapidly spread across various interconnected institutions, sectors, or countries. Therefore, we can identify several channels through which systemic risk can be transmitted into the financial system and real economy.

Contagion occurs when distress or failures in one institution or market spread to other interconnected institutions or markets. It can be triggered by direct linkages, such as interbank lending or counterparty relationships, where the failure of one institution affects others (Hasman 2013; Choudhury and Daly 2021; Lin and Zhang 2023). Contagion can also occur through indirect linkages, such as investor panic or loss of confidence, which can lead to a broader market sell-off (Kiss et al. 2018; Dosumu et al. 2023).

When financial institutions, particularly banks, establish intricate networks of financial instruments like claims, liabilities, and derivatives, they become tightly interconnected (Bardoscia et al. 2021). This interconnectedness can create a vulnerability where a shock to one institution has the potential to rapidly propagate to others (Farmer et al. 2020). In this case, a bank facing substantial losses due to its exposure to a troubled asset can trigger a series of adverse repercussions. Other banks with direct or indirect dealings with the distressed bank may suffer from losses as well, either because they hold similar risky assets or have extended credit to the troubled institution. This ripple effect can spread throughout the financial system, leading to a chain reaction of negative consequences.

The interconnectedness phenomenon amplifies the initial shock's impact, potentially transforming a localized problem into a systemic crisis (Roncoroni et al. 2021). The recent pandemic crisis has demonstrated that different types of events can trigger systemic crises by amplifying the effect of external phenomena on the financial markets (Baber 2020). Therefore, during times of financial distress, confidence in the stability of financial institutions can wane, prompting investors and depositors to withdraw their funds in fear of contagion (Giansante et al. 2023). As a result, this may cause liquidity shortages for affected banks, further exacerbating the crisis and posing systemic risks to the broader economy.

To mitigate the risks of propagation, regulatory authorities, and policymakers often emphasize the importance of monitoring and managing systemic risks (Foglia et al. 2023). Implementing measures to enhance transparency, risk assessment, and stress testing can help identify vulnerabilities and bolster the resilience of financial institutions to withstand shocks (Rizwan et al. 2020). Additionally, promoting better risk management practices within banks and encouraging diversification of assets can reduce the potential for contagion, ultimately contributing to a more robust and stable financial system and avoiding the correlated information-based contagion. This contagion refers to the propagation of financial distress triggered by disseminating negative information, rumors, or general perception of instability within the financial system. In these situations, fear and uncertainty can spread rapidly among investors, depositors, and market participants, leading to a phenomenon known as a "run" on financial institutions. This mass withdrawal of deposits can quickly deplete a bank's liquidity, making it challenging for the institution to meet its financial obligations. As the news of the initial institution's troubles spreads, it can cause a domino effect, where other depositors start losing confidence in other financial institutions as well. The fear of losing their savings prompts them to withdraw their funds from those institutions too, creating a self-fulfilling prophecy of instability.

The consequences of information-based contagion can be severe and far-reaching. The liquidity crisis faced by the affected institutions can hinder their ability to lend money to businesses and individuals, leading to a credit crunch and a slowdown in economic activity. Moreover, the loss of confidence in the broader financial system can trigger a vicious cycle of sell-offs in the markets, causing asset prices to plummet and exacerbating the overall crisis.

During the EU sovereign crisis, the propagation or contagion of shocks was closely linked to the health of the banks within the financial system. This crisis highlighted the significant issues faced by banks, particularly concerning Non-performing loans (NPLs) or distressed assets. The contagion effect in this case was triggered by the declining value of specific assets, causing a widespread decline in market prices and setting off a chain reaction of negative consequences throughout the financial system. As the crisis unfolded, the declining value of distressed assets created a crisis of confidence among investors and financial institutions. The uncertainty surrounding the banks' exposures to NPLs and distressed assets led to heightened risk aversion, with financial institutions becoming cautious about lending and investing (Accornero et al. 2017; Thornton and Di Tommaso 2021). The reluctance to extend credit further exacerbated the economic downturn, impacting businesses and consumers alike. The declining value of their assets eroded their capital positions and financial health. To mitigate their losses and reduce their risk exposure, these institutions were compelled to sell investments at depressed prices, through for example securitizations. However, in some cases, the massive selling of distressed assets in the market only worsened the situation (Kiesel et al. 2020; Dell'Atti et al. 2023), as it further drove down asset prices and intensified the crisis. This negative feedback loop of declining asset prices and financial institutions' behavior contributed to a vicious cycle of the market downturn, exacerbating the crisis's systemic impact. The weakened balance sheets of financial institutions due to their exposure to distressed assets created concerns about their solvency and overall stability, amplifying market uncertainties (Aiyar et al. 2015). Unluckily, the high levels of NPL are a common factor in many banking crises (Ari et al. 2021). As underlined by the Financial Stability Review in November 2022 (ECB 2022), the macroeconomic conditions are important in propagating the crisis. In general, macroeconomic factors, such as recessions, interest rate fluctuations, currency crises, or natural disasters are key factors in propagating the crises. Adverse macroeconomic conditions can impact the financial health of institutions, increase default rates, or lead to sharp market declines (ECB 2022). Economic interdependencies and global linkages can magnify the transmission of macroeconomic risk across countries and regions (Di Tommaso et al. 2023).

It is important to note that these transmission channels are often interconnected, and a single event or shock can trigger multiple channels simultaneously, amplifying the systemic risk. Financial regulators and policymakers closely monitor these channels and work to implement measures to mitigate systemic risk and enhance the financial system's resilience.

Determinant of Systemic Risk: Theoretical and Empirical Backgrounds

The literature on the determinants of systemic risk is vast and multifaceted, focusing on several key factors, from a micro and macro lens. This summary highlights key findings related to these determinants.

Banks fundamental characteristics

- Total Assets: The size of an institution's balance sheet is a significant determinant of systemic risk. Larger institutions can strongly impact the financial system if they fail (Benoit et al. 2017; Pacelli 2021).

    - Size Matters: Larger financial institutions often have a more extensive network of customers, counterparties, and interconnectedness with other financial institutions. Therefore, when a large institution encounters financial difficulties or fails, its size magnifies the potential for systemic repercussions. Empirical studies have shown that the size of financial institutions is positively correlated with systemic risk. Larger institutions are often considered "too big to fail" because their failure could have catastrophic consequences (Foglia and Angelini 2021). During the 2008 financial crisis, the failure of several large institutions, such as Lehman Brothers and AIG, exemplified how their size exacerbated systemic risk.

- Leverage: High leverage magnifies gains and losses, making it a critical determinant of systemic risk. Highly leveraged institutions are more likely to experience distress and contribute to systemic risk (e.g., Acharya and Thakor 2016; Poledna et al. 2014)
- Asset quality. (NPLs): Elevated levels of Non-Performing Loans (NPLs) signal poor asset quality and credit risk, increasing systemic risk. If NPLs accumulate across financial institutions, it can signal broader economic problems and financial stress, increasing systemic risk. Empirical studies have shown that elevated levels of NPLs in the banking sector are associated with increased systemic risk (Accornero et al. 2017; Thornton and Di Tommaso 2021).
- Profitability: ROA and ROE are crucial for assessing bank profitability and risk-taking behavior. Declining profitability indicators can be warning signs of financial stress. Empirical research has shown that bank profitability indicators, especially ROA, are closely monitored by regulators and analysts for signs of financial

health and risk-taking behavior (Smith et al. 2003; Weiß et al. 2014). For example, during the 2008 financial crisis and Eurozone debt crisis (2010–2013), some banks reported significant declines in profitability indicators, particularly ROA, due to exposure to toxic assets and deteriorating credit quality. These deteriorations in profitability were a precursor to systemic problems in the banking sector.

Financial structure characteristics

- Interconnectedness: The degree of interconnectedness among financial institutions is a central determinant of systemic risk. High interconnectivity increases the risk that distress in one institution will spread throughout the system. The financial network structures and interbank lending play a crucial role in understanding systemic risk. Several empirical studies have confirmed the importance of interconnectedness as a determinant of systemic risk (e.g., Hasman 2013; Cai et al. 2018; Lin and Zhang 2023). Researchers often measure interconnectedness using network analysis, identifying central financial network nodes crucial for transmitting shocks. For example, studies have used measures like "centrality" to determine which financial institutions play key roles in transmitting systemic risk (see Billio et al. 2012; Foglia and Angelini 2021).

  - Direct Financial Linkages: Institutions often have direct financial relationships with one another, such as loans, investments, or trading partnerships. When these institutions are tightly interconnected, the failure of one can lead to losses or liquidity problems for others. For example, the collapse of Lehman Brothers (2007), SVB, and Credit Suisse (2023) highlighted the extent of direct financial linkages in the global financial system and the subsequent contagion.
  - Counterparty Exposure: Interconnectedness also exists through counterparty risk in financial contracts, such as derivatives and swaps. If a significant counterparty defaults, it can trigger a chain reaction of payment obligations, potentially straining other market participants and the broader financial system.

- Financial Innovation and New Financial Entities: Advances in financial instruments, innovations, and new financial intermediaries (FinTech) can introduce complexity and opacity to financial markets (Pacelli et al. 2022). While innovation can enhance market efficiency, it can also create systemic vulnerabilities if market participants do not fully understand the risks associated with new products.

Macroeconomic environment

- Boom and Bust Cycles: Economic booms, characterized by robust GDP growth and low unemployment, can foster excessive risk-taking behavior in financial markets (Festić et al. 2011). Financial institutions may become complacent during these periods and extend credit to borrowers with higher default risk. This leads to an increase in vulnerabilities in the financial system. When the economic cycle reverses, and a bust occurs, systemic risk intensifies as these vulnerabilities are exposed. Focusing on the financial market, asset bubbles, driven by excessive speculation and irrational exuberance, can pose substantial systemic risk. These bubbles can form in various asset classes, including real estate and equities, and their eventual burst can lead to severe market disruptions and financial crises.

- Monetary Policy and Interest Rates: Central banks' monetary policy decisions (such as changes in interest rates) directly impact economic cycles. Expansionary monetary policy, with low interest rates, can stimulate borrowing and investment but may also incentivize riskier behavior. On the other hand, tightening monetary policy can lead to a slowdown in economic activity, potentially triggering financial distress (Colletaz et al. 2018; Pacelli et al. 2022).
- Government Debt and bank risk. This type of risk is related to significant bank exposure to sovereign bonds. This exposure creates a dangerous link between sovereign debt and the banking system, known as the "diabolical loop", which acts as a channel for transmitting and amplifying financial shocks (Brunnermeier et al. 2011; Shambaugh et al. 2012; Foglia et al. 2023). The diabolical loop is a process in which concerns about the solvency of sovereign governments fuel concerns about the solvency of banks and vice versa, creating a dangerous cycle that can accelerate financial crises and lead to liquidity and solvency crises. The loop can be summarized as follows steps. (1) Bank Exposure to Sovereign Debt: Banks hold a significant amount of bonds issued by their national governments. These government bonds are considered safe assets and constitute a substantial portion of banks' portfolios. However, this high exposure makes them extremely sensitive to fluctuations in the prices of government bonds. (2) Concerns about Sovereign Solvency: When doubts arise about the solvency of a sovereign government, such as due to high public debt or economic troubles, the prices of its government bonds may begin to decline. This price decline raises concerns about potential losses for banks holding such bonds in their portfolios. (3) Impact on Banks: Growing concerns about the solvency of sovereign governments can negatively affect confidence in banks, leading to deposit outflows and increased funding costs for the banks themselves. This financial pressure can prompt banks to limit lending and exercise caution with counterparties, further weakening economic activity. (4) Reinforcing Sovereign Concerns: The weakening of the banking system, in turn, heightens concerns about sovereign governments as the likelihood of intervention to save struggling banks increases. This intervention may require significant public resources, increasing public debt and further exacerbating the financial crisis. The effect of the diabolical loop was evident during the European sovereign debt crisis between 2010 and 2012. The rapid rise in yields of government bonds in countries like Greece, Ireland, Portugal, Spain, and Italy was fueled by this cycle of mutual distrust between sovereign governments and banks. The resulting financial instability significantly impacted the European economy, including credit contraction and deteriorating economic conditions.

External shocks

- Globalization and Interconnectedness: The increasing globalization of financial markets and economies has amplified the transmission of external shocks. A crisis in one part of the world can quickly spread to other regions through interconnected financial institutions and markets. This interdependence can magnify systemic risk, making it essential for regulators to monitor cross-border exposures.

- Trade and Supply Chain Disruptions: Trade tensions and disruptions in global supply chains can have macroeconomic implications. These disruptions can negatively impact economic growth, corporate profitability, and financial stability. As seen during the COVID-19 pandemic, supply chain disruptions can trigger systemic risk by affecting multiple sectors simultaneously (Rizwan et al. 2020).
- Geopolitical Risks: Geopolitical events introduce uncertainty and volatility into financial markets. These events can erode investor confidence, leading to market sell-offs and financial instability. Geopolitical risks have the potential to amplify systemic risk in an interconnected world (Ding and Zhang 2021; Wang et al. 2022)
- Climate change risk. Climate risk includes physical risks related to the material impacts of climate change (such as floods, water stress, fires, and cyclones) and transition risks associated with adaptation to a low-carbon and more sustainable economy. These risks can affect the soundness of financial institutions (Di Tommaso et al. 2023). From a financial-economic perspective, transition risk is relevant because a disorderly transition to a low-carbon economy could rapidly devalue carbon-intensive assets. This could generate severe financial risks, i.e., cause the emergence of non-performing loans (NPLs). For example, Semieniuk et al. (2021) highlight four channels through which climate risks may impact the financial system: the banking channel (e.g., an increase in NPLs could lead to credit rationing and thus a reduction in investment), the investment channel, the consumption channel, and the government debt channel.

Measuring Systemic Risk and Financial Stability

According to Angelini (2015), systemic risk can arise and spread within a given class of financial institutions (the bank run), among markets (stock, currency, securities), and companies in different sectors and geographic areas. As we showed, systemic risk is not easy to define, and its measurement is difficult given the many definitions of financial stability (Bisias et al. 2012). Hence, there are many methods to measure systemic risk. Following the financial crisis, indicators and tools for measuring systemic risk have increased exponentially. However, there is still no consensus on the best set of indicators to be used to prevent the systemic crisis. Billio et al. (2012) argue that a single systemic risk measure is undesirable, as such a strategy cannot reach all mechanisms that could generate a crisis. A framework for monitoring and managing financial stability will need to include both a diversity of perspectives and a continuous process to reassess the evolution of the structure of the financial system, adapting its systemic risk measures to these changes. Appropriately aggregated, combining different systemic risk measures has greater predictive power in explaining macroeconomic shocks than a single systemic risk measure (Giglio et al. 2016).

Hence, *how can systemic risk be measured*? According to Abdymomunov (2013), there is a diverse point of view on systemic shocks. Therefore, different methods and definitions of systemic risk are necessary. According to Huang et al. (2009, 2012), banks' balance sheet information, such as earnings, liquidity, level of non-performing loans, and capital quality, are the essential variables that the regulators have to control. On the other hand, Brunetti et al. (2019) show how stock market variables can predict systemic events.

The classification provided by Bongini and Nieri (2014), Silva et al. (2017) and Ellis et al. (2022) helps investigate the different approaches to estimating and measuring systemic risk. According to Bisias et al. (2012), the measurement methods and the optimum choice of indices change and depend on what is essential to monitor. In particular, they indicate four different classifications of the taxonomy of systemic risk measures: (i) by data requirements, (ii) by supervisory scope, (iii) by event/ decision time horizon, and (iv) by research method.

Overall, the systemic risk measurement can be subdivided into five broad categories: (i) theoretical models of bank runs (e.g., Flannery 1996; Rochet and Tirole 1996; Allen and Gale 2000; Battiston et al. 2012; Gertler and Kiyotaki 2015); (ii) empirical network models (e.g., Billio et al. 2012; Diebold and Yılmaz 2014; Härdle et al. 2016; Musmeci et al. 2017); (iii) indicator-based systemic risk approach proposed by the Basel Committee on Banking Supervision; (iv) market-based models (e.g., Acharya et al. 2012; Adiran and Brunnermeier, 2016; Brownlees and Engle 2017) and (v) Composite Indexes, such as Financial Stress Index (Duca and Peltonen 2013; Oet et al. 2015), CISS (Hollo et al. 2012), and Financial stability index (Creel et al. 2015).

Indicator-Based Measure: Regulatory Approach

The Basel Committee on Banking Supervision (BCBS 2011) proposes a method to measure systemic risk. It is an indicator based on 5 dimensions: size, interconnectedness, substitutability, cross-jurisdictional activity, and complexity. This indicator aims to identify the global systemically important banks (G-SIBs). The individual financial institutions' systemic risk is measured based on the individual score, divided by the sum of the scores for all banks in the sample. Each item has the same importance (see Table 3.1).

According to this methodology, the systemic importance of a bank is positively correlated to the degree of interconnectedness (lending between banks). Therefore, the problem of one bank also affects the others, namely the domino effect. Furthermore, the systemic importance is positively correlated to its complexity, e.g., corporate structure, business models, and operations of the bank itself. Instead, systemic importance is negatively correlated with the degree of substitutability. If the bank specializes in one type of service, its failure would affect the entire financial system. Size plays the dominant role in the level of systemic importance (Brownlees and Engle 2017; Pacelli et al. 2022). High levels of assets imply a high impact on the

Table 3.1 Indicator-based measure

Category and weighting	Individual indicator
Cross-jurisdictional activity (20%)	• Cross-jurisdictional claims • Cross-jurisdictional liabilities
Size (20%)	• Total exposures as defined for use in the Basel III leverage ratio
Interconnectedness (20%)	• Intra-financial system assets • Intra-financial system liabilities • Wholesale funding ratio
Substitutability (20%)	• Assets under custody • Payments cleared and settled through payment systems • Values of underwritten transactions in debt and equity markets
Complexity (20%)	• OTC derivatives notional value • Level 3 assets • Held for trading and available for sale value

Source Basel Committee on Banking Supervision, Global Systemically Important Banks: Assessment Methodology and the Additional Loss Absorbency Requirement (BCBS 2011)

entire financial system in the event of a malfunction. To measure size, reference is made to the leverage ratio (Basel III). International operations also have a positive effect. Many international transactions (cross-border assets and liabilities) imply that the default of the bank could have a greater impact on the financial system.

However, the methodology proposed by the Basel Committee has raised a lot of criticism (Masciantonio 2015; Benoit et al. 2017; Trapanese 2022). One of the main challenges is the oversimplification of systemic risk through the use of a static number of indicators. These approaches often employ fixed thresholds or cutoff points to determine the systemic importance of institutions or the overall level of systemic risk. However, systemic risk is dynamic and can evolve rapidly. Fixed thresholds may fail to capture changes in risk over time or during periods of financial instability. A more adaptive and flexible approach is needed to account for the evolving nature of systemic risk. Data availability and quality pose another significant challenge. Indicator-based measures rely on outdated methods to assess systemic risk. However, obtaining reliable data across various institutions can be challenging, particularly during financial stress. Incomplete data can lead to biased assessments and distort systemic risk measurement. Moreover, risk-shifting behavior is another concern associated with indicator-based measures. Banks may strategically adjust their activities (risk exposures) to stay below regulatory thresholds. Indicator-based measures may not fully capture these risk-shifting behaviors, which can undermine the effectiveness of regulatory interventions in mitigating systemic risk.

Systemic Risk and Financial Stability: Two Sides of the Same Coin

In recent years, central banks, regulatory, and international institutions have relied on a set of aggregate indicators to assess vulnerability in the financial system. Financial stability is hard to measure due to many latent interactions between the financial system, derivatives markets, the real economy, and so on. Given that several episodes of financial crises have been triggered by severe cyclical downturns, it is natural for financial stability authorities to monitor macroeconomic vulnerability. The purpose of the Composite Index is to measure financial instability by combining individual stress indicators into a single number (Hollo et al. 2012) due to the multidimensional dimension of the phenomenon. Furthermore, these indicators can be interpreted as a measure of systemic risk, trying to quantify and summarize the stress in the financial system in a single statistic (Cambón and Estévez 2016). Typically, they are built using a simple mean approach or weighted sum of indicators or by principal component analysis. Pioneering work is the study of Illing and Liu (2003), which built an index of financial stress for Canada to provide a quantitative score for the macroeconomic financial stress, including measures for bank risks, exchange rates, and capital market. However, the crisis raised awareness about interactions between financial and business cycles, therefore careful monitoring of developments in individual financial institutions should be an integral part of macroeconomic supervision and policy design (Claessens et al. 2011). To this end, most indexes have been developed, such as the Financial Condition Index (Hatzius et al. 2010), the Financial Stress Index (Oet et al. 2015), CISS (Hollo et al. 2012), and the Financial Stability Index (Creel et al. 2015). These indices are used as indicators for crisis prediction, attending to the study of the evolution of financial risk conditions (or financial stress index) at the macro level, including variables such as GDP, inflation, bond, and stock volatility. Generally, the indexes are constructed in general as a weighted sum of several financial variables.

One of the most important indicators is the "systemic stress" indicator (CISS) developed by the ECB (Hollo et al. 2012). The index aggregates five sub-indices: (i) the banking and non-banking sector, (ii) the money market, (iii) the stock market, (iv) the securities, and finally, (v) the currencies markets. The CISS (Fig. 3.1) is built taking into account and assigns greater weight to scenarios where financial stress simultaneously affects multiple market segments.

Another group of indicators refers to the banking sector, like the Z-score (e.g., Goodhart et al. 2006; Čihák and Schaeck 2010; Tonzer 2015). In this approach, accounting data are proxies for fundamental bank attributes that measure a bank's financial vulnerability. Goodhart et al. (2006) affirm that financial crisis monitoring can be done with an indicator of banking sector profitability and the probability of default. Čihák and Schaeck (2010) found that the bank's return on equity and corporate leverage are good indicators of systemic banking problems. The percentage of Non-Performing Loans (NLPs) and the capital adequacy ratio help identify banking turmoil. Also, a recent strand of literature attempts to estimate financial

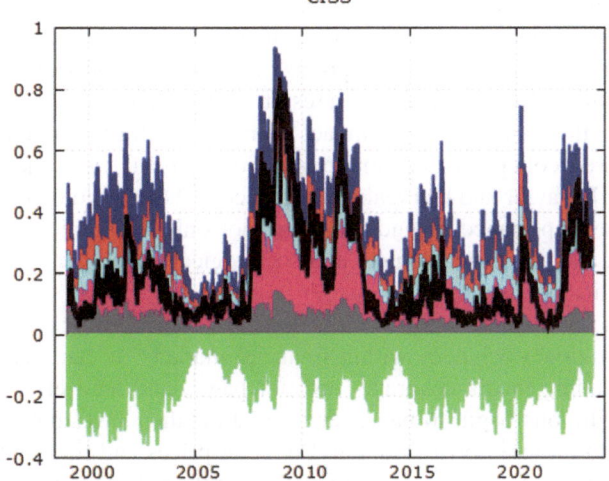
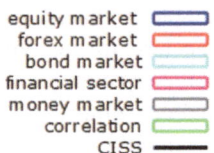

Fig. 3.1 Dynamic evolution of CISS index

stability indicators by CAMELS methodology (e.g., Arena, 2008; Shaddady and More 2019). This study suggests how CAMELS indicators are successful in anticipating distress phenomena. For this purpose, different indicators that described the financial system's vulnerability were used. A simple aggregate indicator of banking sector stability can be constructed as a weighted average of partial indicators of the financial soundness of banks. Usually, an aggregate micro-prudential indicator includes different groups of variables that reflect the health of banks. The indicators are typical: (i) capital adequacy, (ii) assets quality, (iii) earnings and profitability, (iv) liquidity, and (v) sensitivity to market risk.[3]

However, the prediction of financial crises has to consider the important aspects of human behavior and beliefs. As underlined by *Merton's concept of self-fulfilling prophecy* (Merton 1948) and *Thomas's theorem* (Thomas 1938) on perception and interpretation, the role of human behavior, beliefs, and expectations can play an important role in shaping financial crises. Both sociologists highlight the powerful role that beliefs and expectations can play in shaping individual and group behaviors. Are the financial crises the logical consequences of the worsening of economic and financial conditions, or are they caused by the behavior of individuals? In this line, the *Thomas's theorem* (Thomas 1938) highlights the role of perception and interpretation in shaping social reality: "*If men define situations as real, they are real in their consequences*". In his paper, Thomas argued that the way people perceive and interpret situations, regardless of whether those perceptions are accurate or not,

---

[3] These measures, so-called financial soundness indicators (FSI), are collected by the International Monetary Fund (IMF) and measure the health of a country's financial sector. FSIs are divided into two sets: core set and encouraged set. The first set includes banking sector indicators (bank health statistics), while the encouraged set includes additional banking statistics (data on other sectors and markets).

can have very real and tangible effects on their behavior and actions. It suggests that people act based on their subjective understanding of a situation, and these actions can lead to outcomes that align with their perceptions. Therefore, if individuals perceive economic conditions as worsening and, as a result, change their investment and consumption behavior, this can negatively impact economic performance and potentially lead to the feared collapse of economic situations. In this case, their perception influenced their behavior and had real consequences. These sociological perspectives emphasize the importance of understanding not only economic and financial fundamentals but also the psychological and sociological factors that can contribute to market volatility and crises.

All these elements influence financial stability indicators. The self-fulfilling crisis is a complex and often irrational behavior within financial markets. It can stem from various sources, including media coverage, investor panic, and herd effect. When individuals and institutions begin to panic, and withdraw their investments based on fear, it can lead to a rapid downward spiral in asset prices and financial instability, even if the economy's underlying fundamentals are relatively sound (like financial indicators). Essentially, the belief that a crisis is imminent can trigger the crisis the indicator is attempting to predict. Hence, financial stability indicators may struggle to incorporate or predict the self-fulfilling crisis because it is driven by human psychology and behavior, factors that are difficult to quantify accurately. Therefore, these financial composite indexes may provide a false sense of security by not accounting for the potential for sentiment-driven crises.

Market-Based Measures

The use of aggregate indicators is a natural starting point for assessing systemic risk, but there are two main issues. First, many indicators are built on monthly data and are subject to reporting delays. As the current crisis has shown, tensions in the financial sector can develop very quickly. Since these indicators are for old construction, capturing the state of the financial system with "delay" is certainly of limited use in assessing risks rapidly. Second, indicators focus on financial system aggregates and provide little information on the status of individual financial institutions, particularly their interconnectedness. Third, do not capture potential mood swings in financial markets. To this end, another recent strand of literature focuses on the possible spillover effect across financial institutions. These methodologies combine market and balance sheet data using public data. Acharya et al. (2012) developed the Marginal Expected Shortfall (MES) that indicates the marginal contribution of a financial firm to systemic risk, i.e., its tendency to be undercapitalized when all financial systems are undercapitalized. This contribution is measured by the extreme expected loss of the financial system; the "Expected Shortfall" (Acharya et al. 2017). The CoVaR (Conditional Value at Risk) model, proposed by Adiran and Brunnermeier (2016), measures the system loss conditional on each institution's

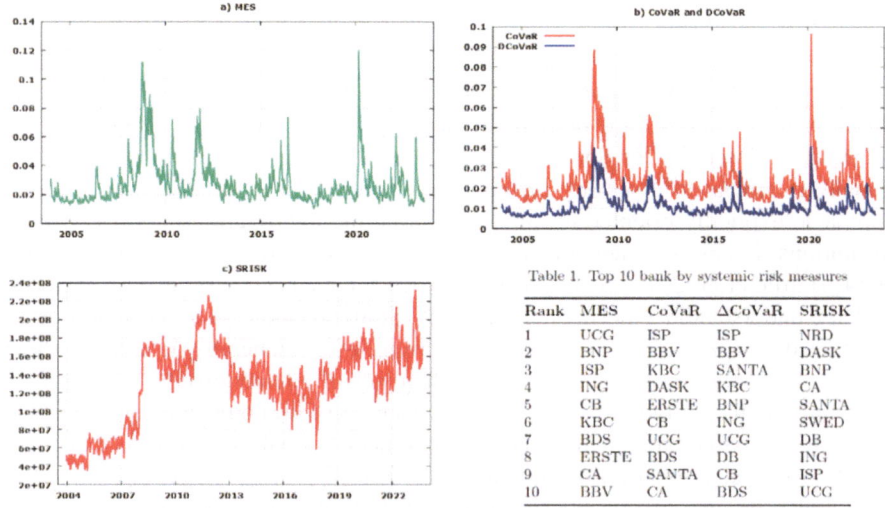

Table 1. Top 10 bank by systemic risk measures

Rank	MES	CoVaR	ΔCoVaR	SRISK
1	UCG	ISP	ISP	NRD
2	BNP	BBV	BBV	DASK
3	ISP	KBC	SANTA	BNP
4	ING	DASK	KBC	CA
5	CB	ERSTE	BNP	SANTA
6	KBC	CB	ING	SWED
7	BDS	UCG	UCG	DB
8	ERSTE	BDS	DB	ING
9	CA	SANTA	CB	ISP
10	BBV	CA	BDS	UCG

Fig. 3.2 Market-based systemic risk measures based on a G-SIB European sample from December 2004 to September 2023. **a** MES index. **b** CoVaR and DCoVaR index. **c** SRISK measure. Table 3.1 shows the top 10 banks for each systemic risk measure

distress. Unlike VaR, CoVaR tries to capture the spillover effects between financial firms. Brownlees and Engle (2017) propose the Conditional Capital Shortfall Index for Systemic Risk Measurement (SRISK) to measure each financial institution's systemic risk contribution. More in-depth, the SRISK is the expected capital shortage that a financial company would face in the event of a systemic event. Financial firms with the highest SRISK contribute more to the under-capitalization of the financial sector during a crisis and are, therefore, at higher systemic risk. Figure 3.2 shows the market-based measures (MES; CoVaR, SRISK) for a G-SIB European.

As we can see, the measures capture tense moments (high peaks) well. However, as the Table shows, each ranks the banks differently. The main limitation of market measures of systemic risk is that they only capture one aspect of the risk (Giudici and Parisi 2018; Brogi et al. 2021). In addition, the structural diversity of methodologies leads to different results in terms of bank classification. For example, Kleinow et al. (2017) and Foglia and Angelini (2021) found that different market measures produce different rankings of banks. This is because these measures are sensitive to the extreme distribution of stock returns, which can make the bank ranking heterogeneous. The table shows the top 10 banks ranked by five systemic risk measures at each period. The different rankings produced by the risk measures suggest that these measures cannot accurately identify the most systemically important financial institutions over a certain period. This confirms the criticism of the literature (Danielsson et al. 2012; Benoit et al. 2017; Brogi et al. 2021; Foglia and Angelini 2021), who argue that these individual measures of systemic risk are not suitable for supervisory purposes. Therefore, increasingly in recent years, network models have been proposed to overcome these limitations.

Network Analysis

A purely quantitative analysis often does not reflect the real complexity of the systemic risk, leading to a partial estimate of the probability of default. Understanding the interconnectedness of financial institutions domestically and internationally is crucial to assessing systemic risk (Bardoscia et al. 2021). A different approach, explicitly oriented towards estimating the interrelationships between all institutions, is based on network models (Billio et al. 2012; Diebold and Yılmaz 2014; Härdle et al. 2016). A network approach for financial systems is a powerful tool for understanding the financial market. Studying these links is important to measure the capital losses of a contagion event. More generally, it is known that market prices are formed by complex mechanisms of interactions that often reflect speculative behavior rather than by the fundamentals of the companies to which they refer. Models based only on market data may reflect "partial" components that could lead to a partial estimation of systemic risk. This weakness suggests that models should also be enriched by considering the structure of the financial system as a whole. In particular, Billio et al. (2012) use the quarterly returns of hedge funds, banks, and insurance companies to develop several interconnection measures based on the Granger causality test. The results show the predominant role of banks in transmitting shocks compared to other financial institutions. Similarly, Diebold and Yılmaz (2014) estimate returns by applying an auto-regressive vector model (VAR), showing the connecting links between the major financial institutions in the United States.

Figure 3.3 shows the two types of "famous" financial networks. As we can see, such a representation of spillover gives us insight into which banks input and receive the most risk, i.e., the one that plays the hub role. These measures are able to capture the connections, thus the peculiarity of the too-interconnected-to-fail banking systems. However, focusing only on one system (layer) may affect the estimation of systemic risk, as empirically explained by Poledna et al. (2015, 2021). Therefore, new network models involve multilayer network models, where each layer represents a peculiar node feature.

Multilayer Design

The network representation described in the introduction captures only one type of interaction. However, economic actors (banks, firms, and countries) are involved simultaneously and can interact in a variety of ways (Poledna et al. 2015; Musmeci et al. 2017; Aldasoro and Alves 2018; Wang et al. 2021). For example, a shock that affects the banking system can affect the real economy through the credit channel, and a shock from the real economy (firms and households) can affect financial stability. This suggests that the financial system can be modeled as a multilayer network, where each layer represents a different type of interaction. For example, one layer

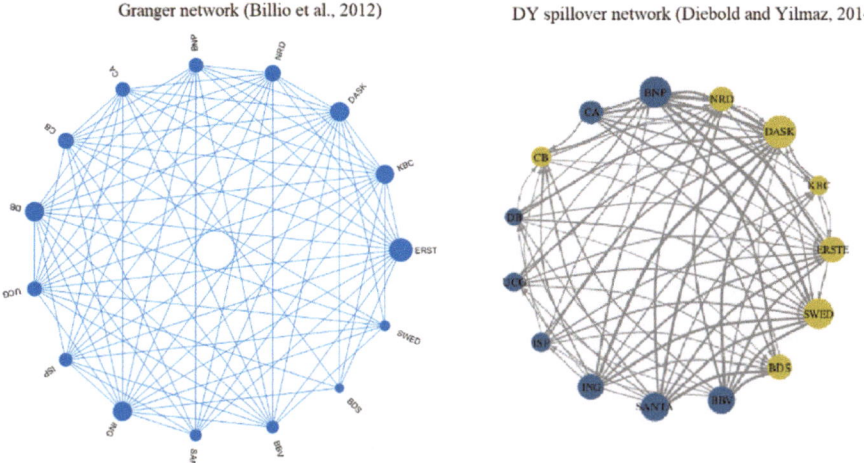

Fig. 3.3 Bank network based on a G-SIB European sample from December 2004 to September 2023. **a** Granger causality network based on Billio et al. (2012); **b** return spillover network based on Diebold and Yilmaz (2012, 2014) framework

could represent banks' lending relationships, while another could represent firms' trading relationships (Fig. 3.4).

In recent years, the adoption of multilayer networks has gained significant traction in financial research for modeling the intricate connections, associations, and exchanges among financial participants and markets. This methodology proves instrumental in analyzing diverse dimensions of financial systems, encompassing interbank affiliations, the propagation of stock return volatility, and the interplay between financial news and interactions within the stock market (Poledna et al. 2021; Bardoscia et al. 2021). The multilayer network approach has several advantages over traditional single-layer network models. First, it allows us to capture the different ways in which shocks can propagate through the financial system. For example, a shock to the banking system could lead to a decline in lending, which could lead to a decline in investment and economic growth. This could, in turn, lead to a decline in the value of assets, which could further weaken the banking system. Second, the multilayer network approach allows us to take into account the different types of relationships between economic actors. For example, we can distinguish between lending relationships, trading relationships, and ownership relationships. This can help us to understand the dynamics of systemic risk better. By considering the different types of interactions between economic actors, we can develop better models of how shocks propagate through the financial system. This can help us to develop better policies to mitigate systemic risk.

However, it is also important to note that the relationships between economic agents (nodes) in a network can be different and tend to change over time (Pacelli 2021). This means that the ideal network model should be able to capture the dynamics of risks. This straightforward theoretical model proposed in the next section

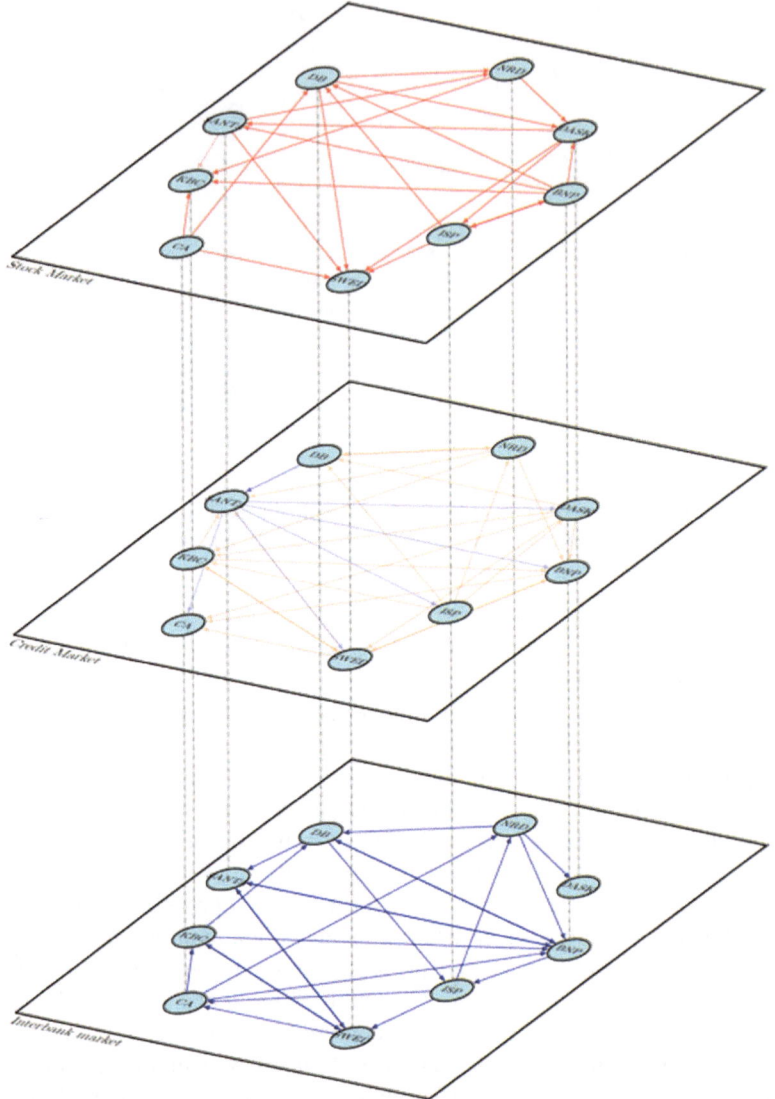

Fig. 3.4 Simulated multilayer network based on a G-SIB European sample from December 2004 to September 2023. The network is composed of three layers: stock market risk layer, credit risk layer, and interbank layer, respectively

is a time-varying multilayer network model because it can capture the dynamics of how shocks propagate through the financial system. The model allows us to consider how the network can change after an exogenous shock, such as a bank failure or a decline in asset prices. It is based on the idea that the financial system can be decomposed into interconnected layers. Each layer represents a different type of interaction between economic actors. The model also assumes that the network can change over time. This is because the relationships between economic actors can change in response to shocks. For example, a bank failure could lead to a decline in lending, which could then lead to a decline in the number of links between banks. Our framework is a simplification of the real world, but it can be used to study the dynamics of systemic risk. By considering how the network can change after an exogenous shock, the model can help us understand how shocks can propagate through the financial system and lead to systemic instability.

Theoretical Dynamic Network Modeling Approaches

Systemic risk, identified as the potential for financial institutions to trigger a harmful contagion mechanism from the financial economy to the broader real economy, is characterized by the risk of disruption to financial services on a large scale. A commonly adopted definition involves the impairment of the entire or parts of the financial system, with serious negative consequences for the real economy. This risk extends beyond individual institutions and manifests through various leakage mechanisms, as observed in the 2008 subprime crisis (e.g., liquidity shrinkage, asset fire sales, and derivatives market value decline). Economic networks, viewed traditionally as robust to *localized* external perturbations driven by *aggregate* shocks like monetary and productivity fluctuations, faced challenges to this perspective after the 2007–09 financial downturn. (Gabaix 2011) demonstrated theoretically and empirically that firm-level shocks could drive *aggregate fluctuations*, challenging the notion that economic cycles are solely propelled by broad shocks. Additionally, Acemoglu et al. (2012) proposed that *shock propagation* through sectoral linkages contributes to aggregate volatility. These seminal works emphasize the role of *granular* economic entities in explaining *systemic instability*, providing an alternative view to the traditional understanding of economic crises.

In the recent past, active researchers in the field of "Econophysics," which included physicists, economists, mathematicians, and financial engineers, had showcased (Abergel et al. 2013) the models and analyses of systemic risk, network dynamics, and related topics. The central theme of the book was systemic risk, a scenario where financial institutions could trigger a dangerous contagion mechanism affecting both the financial and real economies. This risk, once confined to the monetary market, had expanded significantly, notably during the 2008 subprime crisis, and hence the book emphasized the importance of understanding and controlling systemic risk as a crucial societal and economic challenge.

In one of the chapters, the author (Demange 2013) introduced a straightforward and unified model for understanding the spread of defaults among financial institutions. It also puts forth measures to assess the risk a bank poses to the overall system. The paper contended that existing standard contagion processes may not encompass crucial features of financial contagion, prompting the need for a more comprehensive model. In another chapter, the authors (Lautier and Raynaud 2013) employed graph theory to analyze integration and systemic risk in derivative markets. The research involved constructing graphs to investigate various classes of underlying assets (such as energy products, metals, financial assets, and agricultural products) over a twelve-year period. Minimum Spanning Trees (MSTs) from graph theory, was utilized for understanding the dynamic behavior of the price system in a high-dimensional analysis. MSTs were particularly valuable for studying systemic risk as they represented the shortest and most probable paths for the propagation of a "price shock". The chapter explored the topology of MSTs, examined their time-dependent correlation-based graphs, and analyzed the evolution and stability of these graphs over time. In a more recent work, the authors (Kumar et al. 2021) studied distress propagation in connected networks, particularly focusing on economic production networks. They showed using computational models and empirical data analyses, that the rate and extent of distress propagation depend on the network's topology. The research examined economic networks at different levels, such as individual agents, firms/sectors, and countries, each revealing emergent dynamical properties. While microscopic levels showed immunity, macroscopic levels were more prone to failure. The proposed dynamical interaction model characterized distress propagation across different network modules, initiated at various "shock epicenters". "Vulnerable" modules contributed to significant "destabilization". Using empirical data from an Indian administrative state, the study revealed a network with *hub-and-spoke* structures and moderate *disassortativity*. The novelty of their research was in analyzing the production network at different levels, demonstrating that 'too-big-to-fail' modules surpass 'too-central-to-fail' modules in distress propagation.

Thus, using different theoretical models, this essay reviewed the analyses and simulations to gain insights into the relative importance and effects of different determinants. It explored the sensitivity of systemic risk to changes in specific variables and their interactions. The transmission channels of systemic risk are interconnected, and a single event or shock could activate multiple channels simultaneously, thereby amplifying the overall systemic risk. The findings highlight the significance of certain factors in driving systemic risk, providing policymakers with actionable information to develop targeted interventions. Financial regulators and policymakers need to actively monitor these channels and implement measures to mitigate systemic risk, in order to enhance the resilience of the financial system.

A Theoretical Model on Financial Structure Characteristics: A Liquidity Risk Case Study

In this chapter, we have approached systemic risk from a multidimensional perspective. We have highlighted how this risk is quite challenging to estimate and model, as it depends on microeconomics, macroeconomics, social and market sentiment, and sometimes hidden link (node) factors. Based on the different components of systemic risk highlighted above, in this subsection, we model systemic risk based on liquidity risk using a mathematical modeling approach. Indeed, in recent times, the world banks have been worried about the liquidity risk problem. This phenomenon represents an alarming financial risk to banks that can create disastrous consequences in case of negligence. The susceptible infected-removed (SIR) is the simplest epidemiologic model composed of compartments that describe reality very well. At this end, we propose this epidemiological model as a first approach to better described the situation. The approach we propose aims to describe the interconnections when assessing systemic risk. A set of applications—traditional and new ones—is described to show the importance of this model. The approach we propose aspires to analyze the transmission channels through which systemic risk propagates. Specifically, we study a model that describes various channels, such as contagion risk, credit risk, based on the susceptible-exposed-infected-removable (SEIR) epidemic model simulation with a delay of credit risk. We consider an epidemiological approach (Kang et al. 2018) to construct the SEIR contagion model with a delay of credit risk. Initially, as said before, these mathematical models were studied to describe the spread of infectious diseases. We analyze a SEIR model with delay on scale-free networks for the measurement of the systemic risk. As described in previous sections, there are very few studies in the literature about the SEIR model with delay on scale-free networks. These models divide the population into compartments with certain characteristics. We consider an extension of the SIR model, that is, a modified (Kanno 2015) SEIR model in which the exposed compartment is added. They are infected not yet infectious. The model is inspired by research by Kanno (2015), but, as mentioned before, the compartment exposed E is added. Once a node arrives at the infected state, it has either recovered and is no longer susceptible or has died. On a network graph, the nodes are banks that are in one of the states, meanwhile, the edges are the interbank contracts between banks. We describe the interbank network as a graph and $k = 1;$ $2; ...; n$ denotes the degree of a bank in the interbank network. It's important to keep in mind that the major banks, including mega-bank groups, have a central role as counterparties in the interbank market.

According to Zhang et al. (2021), we classify the status of banks at time t into four categories: susceptible banks $S_k(t)$, exposed banks $E_k(t)$, infected banks $I_k(t)$, and immune banks $R_k(t)$. We suppose that susceptible banks $S_k(t)$, exposed banks $E_k(t)$ infected banks $I_k(t)$, and immune banks $R_k(t)$ are the density of banks in the four states at the time t moreover $S_k(t) + E_k(t) + I_k(t) + R_k(t) = 1$ without keeping count of new banks entering. The model consists of four ordinary differential equations illustrating the interaction between Susceptible (S), exposed (E), Infected (I), and Recovered or Resistant (R) affected by systemic risk.

Θ $I_k(t)$ is the probability that a bank at a node is in default and depends on $I(t) = I_1(t), I_2(t), ...$ This template is used to account for the incubation period during which the virus cannot be transmitted. The interconnected nature of these channels is also emphasized to show the amplification of systemic risk. The model can be written as the following SEIR model with delay on scale-free networks for $t > \tau$:

$$\begin{cases} \frac{dS_k(t)}{dt} = b - \beta k \Theta(I_k(t))S_k(t) - bS_k(t) \\ \frac{dE_k(t)}{dt} = \beta k \Theta(I_k(t))S_k(t) - \beta k e^{-b\tau}\Theta(I_k(t-\tau))S_k(t-\tau) - bE_k(t) \\ \frac{dI_k(t)}{dt} = \beta k e^{-b\tau}\Theta(I_k(t-\tau))S_k(t-\tau) - (b+\gamma)I_k(t) \\ \frac{dR_k(t)}{dt} = -bR_k(t) + \gamma I_k(t) \end{cases} \quad (3.1)$$

It is well known that banks have a set of risk indicators that are most important for their strategy.

At this end, we define, according to Mourad et al. (2022), the general liquidity ratio which is the ratio of the number of assets in short-term (less than one year) and the number of liabilities in the same period, this ratio must always be greater than 1;

- $S_k(t)$: the set of banks at instant t, which have the general ratio > 1 that is they have low credit risk and have not been infected by default risk at time t.
- $E_k(t)$: the set of banks at instant t, which have the general ratio > 1 but has suffered the impact and so that are infected but not yet able to pass the contagion.
- $I_k(t)$: the set of banks at instant t, which have the general ratio < 1 that is they have been infected by the risk and have credit risk at time t.
- $R_k(t)$: all the banks at the instant t, which have gone bankrupt or banks resistant (immune banks).

where $\beta > 0$ is the spread rate that is it indicates the speed at which contagion spreads. In other words, $\beta > 0$ measures the contagiousness of an infected bank. $\gamma > 0$ is the bankruptcy rate. This parameter represents the rate at which an infected bank will go bankrupt.

On the other hand, we suppose that the new banks that enter the market are all susceptible and that banks leaving the market are compensated by those entering, so we call b, $0 < b < 1$, the percentage of the banks that enter and leave the market.

So, in the first equation of the model in Eq. (3.1), the first and the third term are the number of new entries banks and banks leaving the market of susceptible nodes with degree k at time t, respectively, while the second term describes the probability that a susceptible bank with degree k is infected by contact with infected banks at time t.

$\theta(I_k(t))$ is the probability that a randomly chosen bank at a target node has defaulted at time t and depends on $I(t) = I_1(t), I_2(t), \ldots$ Furthermore, we indicate $k\Theta(I_k(t))$ the expected number of defaulted banks associated with the banks linked to a defaulted bank with the degree k.

We note that $\tau > 0$ is the latent time that the exposed banks have financial immunity. We suppose it is a fixed constant.

The second equality of Eq. 3.1 accounts for the densities of the exposed banks with degree k who enter the exposed state and are still in the exposed state at time t. We indicate $ke^{-b\tau}$ the probability of exposed banks at time t who were infected at time $\tau > 0$ but they did not be out of the market and so return to the susceptible banks according to the rate $ke^{-b\tau}$.

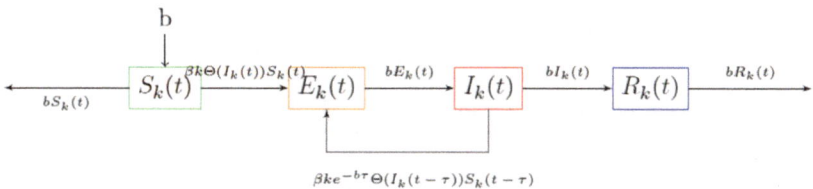

Finally, we can find the basic reproduction number R_0 because it is an important characteristic threshold in the study of whatever epidemic transmission. Furthermore, we note that the delay plays an important role in the basic reproduction number of model 1, in fact we can prove that R_0 monotonously decreases as t increases and this means that the longer the latent period τ of the risk credit is, the smaller the risk of default of bank is. It should be noted that the model considered could be extended by considering exogenous interventions such as those of the policymakers to reduce and/or contain the risk of default.

References

Abdymomunov, A.: Regime-switching measure of systemic financial stress. Ann. Finance **9**(3), 455–470 (2013)

Abergel, F., Chakrabarti, B.K., Chakraborti, A., Ghosh, A. (eds.): Econophysics of Systemic Risk and Network Dynamics. Springer Milano (2013)

Accornero, M., Alessandri, P., Carpinelli, L., Sorrentino, A.M.: Non-Performing Loans and the Supply of Bank Credit: Evidence from Italy. Bank of Italy Occasional Paper, vol. 374 (2017)

Acemoglu, D., Carvalho, V.M., Ozdaglar, A., Tahbaz-Saleh, A.: The network origins of aggregate fluctuations. Econometrica **80**, 1977–2016 (2012)

Acharya, V.V., Thakor, A.V.: The dark side of liquidity creation: leverage and systemic risk. J. Financ. Intermed. **28**, 4–21 (2016)

Acharya, V., Engle, R., Richardson, M.: Capital shortfall: a new approach to ranking and regulating systemic risks. Am. Econ. Rev. **102**(3), 59–64 (2012)

Acharya, V.V., Pedersen, L.H., Philippon, T., Richardson, M.: Measuring systemic risk. Rev. Financ. Stud. **30**(1), 2–47 (2017)

Acharya, V.V.: A theory of systemic risk and design of prudential bank regulation (2009)

Adiran, T., Brunnermeier, M.K.: CoVaR. Am. Econ. Rev. **106**(7), 1705 (2016)

Aiyar, M.S., Bergthaler, M.W., Garrido, J.M., Ilyina, M.A., Jobst, A., Kang, M.K.H., Moretti, M.M.: A Strategy for Resolving Europe's Problem Loans. International Monetary Fund (2015)

Aldasoro, I., Alves, I.: Multiplex interbank networks and systemic importance: an application to European data. J. Financ. Stab. **35**, 17–37 (2018)

Allen, F., Gale, D.: Financial contagion. J. Polit. Econ. **108**(1), 1–33 (2000)

Angelini, P.: Macroprudential policies: a discussion of the main issues (No. 271). Bank of Italy, Economic Research and International Relations Area (2015)

Arena, M.: Bank failures and bank fundamentals: a comparative analysis of Latin America and East Asia during the nineties using bank-level data. J. Bank. Finance **32**(2), 299–310 (2008)

Ari, A., Chen, S., Ratnovski, L.: The dynamics of non-performing loans during banking crises: a new database with post-COVID-19 implications. J. Bank. Finance **133**, 106140 (2021)

Baber, H.: Spillover effect of COVID19 on the global economy. Transl. Mark. J. (TMJ) **8**(2), 177–196 (2020)

Bardoscia, M., Barucca, P., Battiston, S., Caccioli, F., Cimini, G., Garlaschelli, D., Caldarelli, G.: The physics of financial networks. Nat. Rev. Phys. **3**(7), 490–507 (2021)

Battiston, S., Puliga, M., Kaushik, R., Tasca, P., Caldarelli, G.: Debtrank: too central to fail? financial networks, the fed and systemic risk. Sci. Rep. **2**(1), 541 (2012)

BBBS: Global Systemically Important Banks: Assessment Methodology and the Additional Loss Absorbency Requirement. Basel Committee on Banking Supervision, Basel (2011)

Benoit, S., Colliard, J.E., Hurlin, C., Pérignon, C.: Where the risks lie: a survey on systemic risk. Rev. Finance **21**(1), 109–152 (2017)

BGFR: Policy statement on payments system risk. Technical report, Board of Governors of the Federal Reserve System, Washington, D.C. (2001)

Billio, M., Getmansky, M., Lo, A.W., Pelizzon, L.: Econometric measures of connectedness and systemic risk in the finance and insurance sectors. J. Financ. Econ. **104**(3), 535–559 (2012)

BIS (Bank for International Settlements): 64th Annual Report. Technical report (1994b)

Bisias, D., Flood, M., Lo, A.W., Valavanis, S.: A survey of systemic risk analytics. Annu. Rev. Financ. Econ. **4**(1), 255–296 (2012)

Board, F.S.: Guidance to assess the systemic importance of financial institutions, markets and instruments: initial considerations. Report to G20 Finance Ministers and Governors (2009)

Bongini, P., Nieri, L.: Identifying and regulating systemically important financial institutions. Econ. Notes: Rev. Bank. Finance Monetary Econ. **43**(1), 39–62 (2014)

Brogi, M., Lagasio, V., Riccetti, L.: Systemic risk measurement: bucketing global systemically important banks. Ann. Finance **17**(3), 319–351 (2021)

Brownlees, C., Engle, R.F.: SRISK: a conditional capital shortfall measure of systemic risk. Rev. Financ. Stud. **30**(1), 48–79 (2017)

Brunetti, C., Harris, J.H., Mankad, S., Michailidis, G.: Interconnectedness in the interbank market. J. Financ. Econ. **133**(2), 520–538 (2019)

Brunnermeier, M., Garicano, L., Lane, P.R., Pagano, M., Reis, R., Santos, T., Vayanos, D.: European safe bonds (ESBies). Euro-nomics.com **26** (2011)

Cai, J., Eidam, F., Saunders, A., Steffen, S.: Syndication, interconnectedness, and systemic risk. J. Financ. Stab. **34**, 105–120 (2018)

Cambón, M.I., Estévez, L.: A Spanish financial market stress index (FMSI). Span. Rev. Financ. Econ. **14**(1), 23–41 (2016)

Choudhury, T., Daly, K.: Systemic risk contagion within US states. Stud. Econ. Financ. **38**(4), 836–860 (2021)

Čihák, M., Schaeck, K.: How well do aggregate prudential ratios identify banking system problems? J. Financ. Stab. **6**(3), 130–144 (2010)

Claessens, S., Kose, M.A., Terrones, M.E.: Financial cycles: what? How? When?. In: NBER International Seminar on Macroeconomics, vol. 7, No. 1, pp. 303–344. University of Chicago Press, Chicago, IL (2011)

Colletaz, G., Levieuge, G., Popescu, A.: Monetary policy and long-run systemic risk-taking. J. Econ. Dyn. Control **86**, 165–184 (2018)

Creel, J., Hubert, P., Labondance, F.: Financial stability and economic performance. Econ. Model. **48**, 25–40 (2015)

Danielsson, J., Shin, H.S., Zigrand, J.P.: Endogenous and systemic risk. In: Quantifying Systemic Risk, pp. 73–94. University of Chicago Press (2012)

Dell'Atti, S., Di Tommaso, C., Pacelli, V.: NPLs securitizations, CDS spreads and spillover effect: evidence from the European banking system. Glob. Bus. Rev. 09721509231159738 (2023)

Demange, G.: Diffusion of defaults among financial institutions. In: Abergel, F., Chakrabarti, B.K., Chakraborti, A., Ghosh, A. (eds.) Econophysics of Systemic Risk and Network Dynamics, pp. 3–17. Springer Milano (2013)

Diebold, F.X., Yilmaz, K.: Better to give than to receive: predictive directional measurement of volatility spillovers. Int. J. Forecast. **28**(1), 57–66 (2012)

Diebold, F.X., Yılmaz, K.: On the network topology of variance decompositions: measuring the connectedness of financial firms. J. Econometrics **182**(1), 119–134 (2014)

Ding, Z., Zhang, X.: The impact of geopolitical risk on systemic risk spillover in commodity market: an EMD-based network topology approach. Complexity **2021**, 1–17 (2021)

Dosumu, O.E., Sakariyahu, R., Oyekola, O., Lawal, R.: Panic bank runs, global market contagion and the financial consequences of social media. Econ. Lett. **228**, 111170 (2023)

Duca, M.L., Peltonen, T.A.: Assessing systemic risks and predicting systemic events. J. Bank. Finance **37**(7), 2183–2195 (2013)

ECB: Financial stability review. Technical report, European Central Bank (2009)

ECB: Financial stability review. Technical report, European Central Bank (2022)

Ellis, S., Sharma, S., Brzeszczyński, J.: Systemic risk measures and regulatory challenges. J. Financ. Stab. **61**, 100960 (2022)

Farmer, J. D., Kleinnijenhuis, A.M., Nahai-Williamson, P., Wetzer, T.: Foundations of system-wide financial stress testing with heterogeneous institutions (2020)

Festić, M., Kavkler, A., Repina, S.: The macroeconomic sources of systemic risk in the banking sectors of five new EU member states. J. Bank. Finance **35**(2), 310–322 (2011)

Flannery, M.J.: Financial crises, payment system problems, and discount window lending. J. Money, Credit, Bank. **28**(4), 804–824 (1996)

Foglia, M., Angelini, E.: The triple (T3) dimension of systemic risk: identifying systemically important banks. Int. J. Financ. Econ. **26**(1), 7–26 (2021)

Foglia, M., Pacelli, V., Wang, G.J.: Systemic risk propagation in the Eurozone: a multilayer network approach. Int. Rev. Econ. Financ. **88**, 332–346 (2023)

Gabaix, X.: The granular origins of aggregate fluctuations. Econometrica **79**, 733–772 (2011)

Gertler, M., Kiyotaki, N.: Banking, liquidity, and bank runs in an infinite horizon economy. Am. Econ. Rev. **105**(7), 2011–2043 (2015)

Giansante, S., Manfredi, S., Markose, S.: Fair immunization and network topology of complex financial ecosystems. Physica A **612**, 128456 (2023)

Giglio, S., Kelly, B., Pruitt, S.: Systemic risk and the macroeconomy: an empirical evaluation. J. Financ. Econ. **119**(3), 457–471 (2016)

Giudici, P., Parisi, L.: Corisk: credit risk contagion with correlation network models. Risks **6**(3), 95 (2018)

Goodhart, C.A., Sunirand, P., Tsomocos, D.P.: A model to analyse financial fragility. Econ. Theor. **27**, 107–142 (2006)

Härdle, W.K., Wang, W., Yu, L.: Tenet: tail-event driven network risk. J. Econometrics **192**(2), 499–513 (2016)

Hasman, A.: A critical review of contagion risk in banking. J. Econ. Surv. **27**(5), 978–995 (2013)

Hatzius, J., Hooper, P., Mishkin, F.S., Schoenholtz, K.L., Watson, M.W.: Financial conditions indexes: a fresh look after the financial crisis (No. w16150). National Bureau of Economic Research (2010)

Hellwig, M., Admati, A.: The Bankers' New Clothes: What's Wrong with Banking and What to Do about It-Updated Edition. Princeton University Press (2014)

Hollo, D., Kremer, M., Lo Duca, M.: CISS-a composite indicator of systemic stress in the financial system. Working Paper Series No 1426/March 2012, European Central Bank (2012)

Huang, X., Zhou, H., Zhu, H.: A framework for assessing the systemic risk of major financial institutions. J. Bank. Finance **33**(11), 2036–2049 (2009)

Huang, X., Zhou, H., Zhu, H.: Assessing the systemic risk of a heterogeneous portfolio of banks during the recent financial crisis. J. Financ. Stab. **8**(3), 193–205 (2012)

Illing, M., Liu, Y.: An index of financial stress for Canada (No. 2003–14). Bank of Canada. J. Financ. Stab. **5**(3), 224–255 (2003)

Kang, H., Sun, M., Yu, Y., Fu, X., Bao, B.: Spreading dynamics of an SEIR model with delay on scale-free networks. IEEE Trans. Netw. Sci. Eng. **7**(1), 489–496 (2018)

Kanno, M.: The network structure and systemic risk in the Japanese interbank market. Jpn. World Econ. **36**, 102–112 (2015)

Kaufman, G.G.: Comment on systemic risk. Res. Financ. Serv. Bank. Financ. Mark. Systemic Risk **7**, 47–52 (1995)

Kiesel, F., Manz, F., Schiereck, D.: The conditional stock market response to banks' distressed asset sales on CDS availability. Appl. Econ. **52**(56), 6123–6135 (2020)

Kiss, H.J., Rodriguez-Lara, I., Rosa-Garcia, A.: Panic bank runs. Econ. Lett. **162**, 146–149 (2018)

Kleinow, J., Moreira, F., Strobl, S., Vähämaa, S.: Measuring systemic risk: a comparison of alternative market-based approaches. Financ. Res. Lett. **21**, 40–46 (2017)

Kumar, A., Chakrabarti, A.S., Chakraborti, A., Nandi, T.: Distress propagation on production networks: coarse-graining and modularity of linkages. Physica A **568**, 125714 (2021)

Lautier, D., Raynaud, F.: Systemic risk and complex systems: a graph-theory analysis. In: Abergel, F., Chakrabarti, B.K., Chakraborti, A., Ghosh, A. (eds.) Econophysics of Systemic Risk and Network Dynamics, pp. 19–37. Springer Milano (2013)

Lin, S., Zhang, H.: Interbank contagion risk in China under an ABM approach for network formation. Eur. Financ. Manag. **29**(2), 458–486 (2023)

Masciantonio, S.: Identifying and tracking global, eu, and eurozone systemically important banks with public data. Appl. Econ. Q. **61**(1), 25–64 (2015)

Merton, R.K.: The self-fulfilling prophecy. Antioch Rev. **8**(2), 193–210 (1948)

Mourad, H., Fahim, S., Burlea-Schiopoiu, A., Lahby, M., Attioui, A.: Modeling and mathematical analysis of liquidity risk contagion in the banking system. J. Appl. Math. (2022)

Musmeci, N., Nicosia, V., Aste, T., Di Matteo, T., Latora, V.: The multiplex dependency structure of financial markets. Complexity (2017)

Oet, M.V., Dooley, J.M., Ong, S.J.: The financial stress index: identification of systemic risk conditions. Risks **3**(3), 420–444 (2015)

Pacelli, V., Miglietta, F., Foglia, M.: The extreme risk connectedness of the new financial system: European evidence. Int. Rev. Financ. Anal. **84**, 102408 (2022)

Pacelli, V.: Rischio sistemico e scienza delle reti. Bancaria **12** (2021)

Patro, D.K., Qi, M., Sun, X.: A simple indicator of systemic risk. J. Financ. Stab. **9**(1), 105–116 (2013)

Poledna, S., Thurner, S., Farmer, J.D., Geanakoplos, J.: Leverage-induced systemic risk under Basle II and other credit risk policies. J. Bank. Finance **42**, 199–212 (2014)

Poledna, S., Molina-Borboa, J.L., Martínez-Jaramillo, S., Van Der Leij, M., Thurner, S.: The multi-layer network nature of systemic risk and its implications for the costs of financial crises. J. Financ. Stab. **20**, 70–81 (2015)

Poledna, S., Martínez-Jaramillo, S., Caccioli, F., Thurner, S.: Quantification of systemic risk from overlapping portfolios in the financial system. J. Financ. Stab. **52**, 100808 (2021)

Poledna, S., Rovenskaya, E., Dieckmann, U., Hochrainer-Stigler, S., Linkov, I.: Systemic risk emerging from interconnections: the case of financial systems. In: Hynes, W., Lees, M., Müller, J. (eds.) Systemic Thinking for Policy Making: The Potential of Systems Analysis for Addressing Global Policy Challenges in the 21st Century. OECD Publishing, Paris, France (2020)

Rizwan, M.S., Ahmad, G., Ashraf, D.: Systemic risk: the impact of COVID-19. Financ. Res. Lett. **36**, 101682 (2020)

Rochet, J.C., Tirole, J.: Interbank lending and systemic risk. J. Money, Credit, Bank. **28**(4), 733–762 (1996)

Roncoroni, A., Battiston, S., D'Errico, M., Hałaj, G., Kok, C.: Interconnected banks and systemically important exposures. J. Econ. Dyn. Control **133**, 104266 (2021)

Schwarcz, S.L.: Systemic risk. Geo. Lj **97**, 193 (2008)

Semieniuk, G., Campiglio, E., Mercure, J.F., Volz, U., Edwards, N.R.: Low-carbon transition risks for finance. Wiley Interdiscip. Rev. Clim. Change **12**(1), e678 (2021)

Shaddady, A., Moore, T.: Investigation of the effects of financial regulation and supervision on bank stability: the application of CAMELS-DEA to quantile regressions. J. Int. Finan. Markets. Inst. Money **58**, 96–116 (2019)

Shambaugh, J.C., Reis, R., Rey, H.: The euro's three crises [with comments and discussion]. Brookings Papers on Economic Activity, pp. 157–231 (2012)

Sheldon, G., Maurer, M.: Interbank lending and systemic risk: an empirical analysis for Switzerland. Revue Suisse d Economie Politique Et De Statistique **134**, 685–704 (1998)

Silva, W., Kimura, H., Sobreiro, V.A.: An analysis of the literature on systemic financial risk: a survey. J. Financ. Stab. **28**, 91–114 (2017)

Smith, M.B.D., De Nicolo, M.G., Boyd, J.H.: Crisis in Competitive Versus Monopolistic Banking Systems. International Monetary Fund (2003)

Taylor, J.B.: Defining systemic risk operationally. Ending government bailouts as we know them, pp. 33–57 (2010)

Thomas, W.I.: The child in America. Рипол Классик (1938)

Thornton, J., Di Tommaso, C.: The effect of non-performing loans on credit expansion: do capital and profitability matter? Evidence from European banks. Int. J. Financ. Econ. **26**(3), 4822–4839 (2021)

Di Tommaso, C., Foglia, M., Pacelli, V.: The impact and the contagion effect of natural disasters on sovereign credit risk. An empirical investigation. Int. Rev. Financ. Anal. **87**, 102578 (2023)

Tonzer, L.: Cross-border interbank networks, banking risk and contagion. J. Financ. Stab. **18**, 19–32 (2015)

Trapanese, M.: Regulatory Complexity, Uncertainty, and Systemic Risk: are Regulators Hehogs or Foxes?. Bank of Italy Occasional Paper, p. 698 (2022)

Wang, G.J., Chen, Y.Y., Si, H.B., Xie, C., Chevallier, J.: Multilayer information spillover networks analysis of China's financial institutions based on variance decompositions. Int. Rev. Econ. Financ. **73**, 325–347 (2021)

Wang, Y., Bouri, E., Fareed, Z., Dai, Y.: Geopolitical risk and the systemic risk in the commodity markets under the war in Ukraine. Financ. Res. Lett. **49**, 103066 (2022)

Weiß, G.N., Bostandzic, D., Neumann, S.: What factors drive systemic risk during international financial crises? J. Bank. Finance **41**, 78–96 (2014)

Zhang, Y.C., Li, Z., Zhou, G.B., Xu, N.R., Liu, J.B.: The evolution model of public risk perception based on pandemic spreading theory under perspective of COVID-19. Complexity **2021**, 1–10 (2021)

Chapter 4
Macro-Prudential Policies to Mitigate Systemic Risk: An International Overview

Vincenzo Pacelli and Maria Melania Povia

Abstract In the wake of the global financial crisis, a clear awareness has emerged that systemic risk can only be addressed and managed through macro-prudential policies. Many countries have embarked on new reforms of the financial regulatory system with macro-prudential mechanisms, technical standards and instruments aimed at mitigating systemic risk. The main purpose of this essay is to summarise the main reforms in Europe, the United States, China, Islamic countries and Japan and to explain how these reforms fit together. In addition, through a comparative analysis of regulatory structures, we assess the effectiveness of macro-prudential tools in preventing and mitigating potential systemic risks and safeguarding financial stability. The results show a growing and widespread focus on prudential policies and architectures to mitigate systemic risk, although the different economies analysed start from very different positions.

Keywords Systemic risk · Macroprudential policies · Financial stability

Introduction

As previously noted in other chapters in this volume, the rapid succession of different global crises has highlighted the increasing vulnerability of modern financial systems to systemic risk. This has led to a renewed focus on policies and supervisory architectures aimed at mitigating systemic risk. As mentioned in other chapters in this volume, systemic risk manifests itself through contagion mechanisms that arise from a shock, whether endogenous or exogenous to the financial system, and spread throughout the financial system and the real economy. This chapter will not deal with the definition and nature of systemic risk, as this has been analysed in other essays in this

V. Pacelli (✉) · M. M. Povia
University of Bari Aldo Moro, Bari, Italy
e-mail: vincenzo.pacelli@uniba.it

M. M. Povia
e-mail: maria.povia@uniba.it

V. Pacelli (ed.), *Systemic Risk and Complex Networks in Modern Financial Systems*,
New Economic Windows, https://doi.org/10.1007/978-3-031-64916-5_4

volume. Instead, it will focus on international regulatory and supervisory policies and infrastructures to mitigate systemic risk and prevent the spread of crises and contagion.

In recent decades, the growing and diversifying interconnectedness of different actors in the system and the resulting increased systemic complexity have led to new and more complex challenges for policymakers and regulators. Since the 2007 financial crisis, it has become clear that a micro-prudential approach to the supervision of the financial system, which focuses exclusively on individual actors without paying sufficient attention to the multiple interconnections between them, is extremely superficial and potentially detrimental to financial stability (Pacelli 2021). To understand how a crisis can propagate and prevent contagion, it is essential to analyse the interconnections and links between the actors in the system. A purely macro-prudential supervisory approach at the international level is necessary to support this view from a regulatory perspective.

Since the onset of the financial crisis in 2007, there has been much research on the different paths of regulatory reform. Initially, scholars focused on identifying the causes of the crisis and the factors that determined the propagation of shocks at the international level.[1] Subsequently, the debate focused on a critical assessment of the new regulatory framework provided by the Basel III Accord. The call for greater macro-prudential openness represents a discontinuity with the two previous Accords, which failed to prevent the onset and subsequent propagation of the crisis. Many studies have blamed the Basel II Accord for the 2007 financial crisis. This is due to the lack of perfect harmonisation at the international level, inadequate measures to prevent and manage liquidity risk, and insufficient cooperation between micro and macro-prudential supervision (Crouhy et al. 2008; Drumond 2009; Moosa 2010; Onado 2009; Resti and Sironi 2011).

As regulatory reforms and supervisory policies are constantly evolving at the international level, it is particularly relevant to conduct a comparative analysis of the current supervisory architectures in major economies. The aim of this essay is to analyse the evolution of supervisory policies and systems in Europe, the United States, China, Islamic countries and Japan aimed at mitigating systemic risk and contagion mechanisms. The chapter examines the effectiveness of different macro-prudential tools in preventing and mitigating potential systemic risks and safeguarding financial stability. Section "The Necessary Shift from Purely Micro-Prudential Regulatory and Supervisory Schemes to Macro-Prudential Approaches" summarises the international evolution of regulatory schemes, providing a basis for a timely comparison of different supervisory systems. Section "A Comparative Analysis of International Financial Regulatory Structures to Mitigate Systemic Risk" presents and describes the main supervisory frameworks designed to mitigate systemic risk in Europe, the United States, China, Islamic countries and Japan. Section "Conclusions" presents the results of the comparative analysis and suggests concluding remarks.

---

[1] See the second chapter of this volume for a thorough review of the literature.

The Necessary Shift from Purely Micro-Prudential Regulatory and Supervisory Schemes to Macro-Prudential Approaches

As mentioned above, the macro-prudential supervisory and regulatory approach aims to prevent the spread of financial instability and thereby safeguard the stability of the financial system. Macro-prudential supervisory policy complements micro-prudential policy. It combines the supervision of individual players in the system with the analysis of interconnections and links between them. The aim is to prevent a crisis from spreading by identifying and managing common exposures, economic-financial and ownership links, risk concentrations and interdependencies that determine contagion risks. Essentially, macro-prudential policies aim to maintain financial stability by employing a range of tools and policies to limit contagion mechanisms and mitigate systemic risk (Tomuleasa 2015).

The term "macro-prudential" originated in 1979 during a meeting of the Cooke Committee, which preceded the Basel Committee on Banking Supervision (BCBS). It was created to describe how problems related to a specific institution could have negative effects on the financial system as a whole (Baker 2013; Blundell-Wignall and Roulet 2014; Clement 2010). It was not until 1986 that the Bank for International Settlements (BIS) defined macro-prudential policy as a set of policies aimed at promoting the soundness of the financial system as a whole and the payments system (BRI 1986). Although there is no commonly agreed definition of "macro-prudential policy", it is generally understood to be aimed at preserving financial stability by managing risks that affect the financial system as a whole, a view shared by many scholars (Agénor and Pereira da Silva 2012; Cerutti et al. 2017; Engel 2016; Galati and Moessner 2013). According to the Bank for International Settlements (2010), macro-prudential policies are regulatory measures aimed at maintaining the stability of the financial system, reducing the potential for systemic risk in the face of internal and external shocks, and ensuring the effectiveness and proper functioning of the financial system. From a macro-prudential perspective, safeguarding individual financial institutions is necessary but not sufficient to ensure the stability of the financial system as a whole (Ghosh 2016; Zdzienicka et al. 2015). As mentioned above, micro-prudential policies aim to reduce the risk of failure of individual financial institutions. Macro-prudential policies, on the other hand, aim to reduce contagion and mitigate the risks and costs of systemic crises. This is achieved by analysing the interconnectedness between financial institutions and the procyclicality of the financial system (Claessens et al. 2013). In other words, the two policies should therefore be seen in a logic of complementarity and cooperation rather than opposition: an effective micro-prudential policy is essential to provide regulators with information on the risk assessment of banks, allowing macro-prudential policy to intervene and preemptively mitigate the potential contagion effects of a shock within a system (Gualandri and Noera 2014; Vinals 2013). To this end, the institutional framework of macro-prudential policy is based on the definition of the architecture of the competent authorities and the mandate given to them, with governance that ensures independence, accountability and credibility. The definition of institutional supervisory architectures may depend on several factors: the structure of the financial system, historical reasons and political constraints. However, the common perspective on the role of macro-prudential policy is based on the ultimate goal of limiting the costs of financial crises and aiming at the stability of the financial system as a whole by preventing the emergence of systemic risks.

A Comparative Analysis of International Financial Regulatory Structures to Mitigate Systemic Risk

Macro-prudential policy, which emerged as a response to the 2007 financial crisis, has taken on an increasingly operational profile in recent years.

As has been widely emphasised, the increasing interconnectedness of the various actors in the system and the consequent increase in systemic complexity have led to a constantly evolving framework in which multiple sources of systemic risk can combine dynamically, even in new forms. For this reason, the adoption of macroprudential policies requires the ability to assess and measure systemic risks in advance, taking into account their multidimensional and complex nature. Most international institutional solutions are still being defined. However, the main and current supervisory structures in Europe, the United States, China, Islamic countries and Japan aimed at mitigating systemic risk are described below.

Europe

The financial crisis of 2007–2008, which began in the United States with the subprime mortgage crisis and then spread to Europe with the sovereign debt crisis, was an important test of the validity and effectiveness of the rules and institutional architecture of banking supervision and, in particular, of banking crisis management at the European level. The numerous banking crises that occurred in the years following the 2008 crisis revealed serious shortcomings in the regulation and supervision of the financial sector and significant contradictions in the institutional arrangements for dealing with unstable situations.

The need to overcome the shortcomings and inefficiencies revealed during the financial crisis led the EU institutions to intervene in the search for more effective institutional solutions and rules. In 2011, the European Union provided itself with two prudential pillars: one of a micro-prudential nature, based on the European System of Financial Supervisors and centred on the three European Supervisory Authorities (ESA)[2]; the other purely macro-prudential, entrusted to the European Systemic Risk

---

[2] European Banking Authority (EBA), European Securities and Markets Authority (ESMA), European Insurance and Occupational Pensions Authority (EIOPA), operational since 2011.

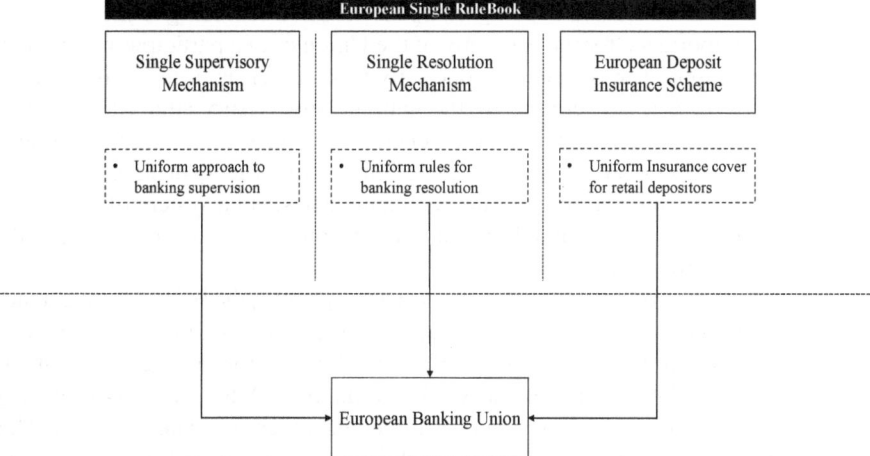

Fig. 4.1 The three pillars of the European Banking Union

Board (Gualandri and Noera 2014). From a micro-prudential point of view, the far-reaching project of the European Banking Union is characterised by the centralisation of decision-making in the hands of European authorities, in cooperation with national resolution authorities, and by the application of the Single Rulebook, a single set of European rules. The European Banking Union is thus a far-reaching regulatory and institutional project, which constitutes the European response to the international financial crisis, and which aims at constituting the process of economic and financial unification of the euro area, restoring confidence in the banking sector and supporting economic growth, while safeguarding financial stability. The Banking Union project rests on three pillars: the Single Supervisory Mechanism, entrusted to the European Central Bank and National Supervisory Authorities; the Single Resolution Mechanism, a centralised crisis management system; and the European Deposit Insurance Scheme (Boccuzzi 2015; Onado 2017; Porretta 2021) (Fig. 4.1).

The Single Supervisory Mechanism (SSM) is the result of a long reform process, introduced into European law by Regulation 1024/2013 of 15 October 2013. As defined in the Guide to Banking Supervision, the objectives of the SSM are the safety and stability of the European banking system, the integration and stability of the financial system, and increased responsibility for banking supervision in the euro area countries. The MUV represents the institutional implementation of the shift from the traditional principles of cooperation and coordination between national authorities to the entrustment of banking supervision to the ECB. In particular, the MUV has no legal personality and supervisory powers and decisions are entrusted to the ECB and the Competent National Supervisory Authorities (NCAs) of the participating countries. The operational scheme outlined in Article 6 of Regulation 1024/2013 provides for the exercise of direct and decentralised supervision of the banking sector. In the first case, direct supervision is exercised by the ECB, in cooperation with the competent national authorities, over significant banks and banking groups.

The significance of banks and banking groups is assessed according to the following criteria: size, importance for the economy of the Union or of a participating Member State and the importance of cross-border activities. As far as decentralised supervision is concerned, it is exercised by the national supervisory authorities for less significant banks in accordance with the general guidelines and instructions issued by the ECB. As mentioned above, the Single Supervisory Mechanism is not a mere transfer of powers from national to European authorities, but a system of joint exercise of supervisory powers aimed at Community integration and maintaining the stability of the Eurosystem.

In addition to the normal prudential supervision of banks, the ECB's remit extends to intervention when a bank or banking group is in difficulty. In particular, the exercise of supervision and prudential oversight of the banking sector represents a strategic moment of reconciliation between the MUV and the BRRD, the legal basis for early intervention in problem situations and for the management of banking crises. This includes the Single Resolution Mechanism (SRM), a harmonised European resolution mechanism based on the Single Resolution Fund (SRF) and a Single Resolution Authority (Single Resolution Board), and a set of rules for crisis management of credit institutions and investment firms, the Recovery and Resolution Framework (Carmassi et al. 2010; Hadjiemmanuil 2015). The European crisis management framework is contained in the Banking Recovery and Resolution Directive (BRRD 2014/59/EU), which entered into force in January 2015. The Directive is the regulatory implementation of the Key Attributes for Effective Resolution Regime for Financial Institutions, formulated by the Financial Stability Board in 2011 in response to the impact of the failure of systemically important financial institutions during the Great Financial Crisis of 2007. In particular, the Recovery and Resolution Framework is a far-reaching discipline with a strategic scope and a strong preventive orientation, providing resolution authorities with harmonised powers and technical tools for crisis management to be applied at the first signs of fragility of the credit institution. The operational scheme of the resolution framework is divided into three operational phases, distinguished by the intensity of the intervention: prevention, early intervention, resolution. The first phase, preparation and prevention, is represented by the set of instruments and measures for a bank or banking group aimed at avoiding or reducing the likelihood of the occurrence of stress and crisis situations. The preparatory measures include the preparation of the recovery plan (BRRD, Article 5–9) and the resolution plan (BRRD, Article 10–14), which allow intermediaries to plan, in the normal course of business, the activities to be carried out in the event of an emergency situation and thus to define the recovery or resolution modalities. The second phase of early intervention, on the other hand, provides for appropriate measures to be taken by supervisors to promptly remedy deteriorating situations in the capital, profitability, financial and asset quality of banks, in order to facilitate the restoration of normal business and to avoid further deterioration that may require resolution intervention. The last phase, in order of intensity of intervention, is Resolution, which provides a set of specific tools to enable an orderly winding-up of the institution in the event of insolvency. The resolution process is the culmination of a highly innovative regulatory path that has significantly changed the regulatory and institutional framework of European countries. The innovative character lies in the definition of "resolution", which refers to the set of measures and instruments adopted to resolve a pathological, insolvency or near-insolvency situation. In this respect, the introduction of the Single Resolution Mechanism has made it possible to reduce the need for public financial support, to avoid negative effects on financial stability and to resolve intermediaries that are "too big to fail" or "too interconnected to fail" and cannot be allowed to fail because of their size, operational complexity and interconnectedness in the financial system.

The third pillar of the Banking Union concerns the harmonisation of deposit guarantee schemes in different Member States (Directive 2014/49/EU). Under the current approach, Deposit Guarantee Schemes (DGS) intervene during the resolution process to help cover losses that would be borne by depositors. According to Article 109 of the BRRD, Deposit Guarantee Schemes play an important role in the renewed banking crisis management system, with the possibility of multiple interventions based on the principle of least cost. In particular, there is support for a bank in crisis to avoid its failure, so-called preventive interventions, and support for sales in the course of a compulsory liquidation, so-called alternative interventions in the strict sense. However, Article 11 of the DGSD provides that Member States may authorise Deposit Guarantee Schemes to take alternative measures to those mentioned above in order to prevent the failure of a bank. Although the regulation of DGSs implements the principle of harmonisation, it is not yet uniform and leaves unresolved uncertainties with regard to national specificities. These include: the level of coverage; the timing and modalities of reimbursements; the financial resources; the modalities of contributions; and the role of the funds in crisis procedures. Thus, the main aspects of the Community discipline that have so far been left to national legislation suggest a more or less identical, if not similar, intervention to that carried out in the past. However, the European Commission's recent proposal to revise the BRRD, published on 18 April 2023, confirms the active role of deposit guarantee schemes in the management of banking crises and calls for their strengthening in order to safeguard depositors' confidence and the stability of the financial system. Moreover, the Commission's proposal preserves the discretion of individual Member States to implement preventive and alternative measures in national legislation, in contrast to the reimbursement of depositors and the financing of resolution, which remain mandatory interventions (Pallini et al. 2023).

Finally, the second macro-prudential pillar with which the European Union has equipped itself in the wake of the crisis is entrusted to the European Systemic Risk Board (ESRB). The establishment of the ESRB is the culmination of a long reform process aimed at creating an institutional framework for macro-prudential policy at the European level, which operates at two levels: under the ESRB and under the competent national authorities (Ehrmann and Schure 2020). The ESRB has the legal responsibility for the prevention and mitigation of systemic risk in the European Union; its task is to issue systemic risk warnings and to trigger the application of macro-prudential instruments in the various Member States through recommendations and guidelines. The operational implementation of macro-prudential policies is the responsibility of national authorities, but it is up to the ESRB to define the governance framework and the operational aspects with which Member States must comply. The primary responsibility for macro-prudential policy in the EU lies with national authorities for two main reasons. The first lies in the nature of systemic risk, which can arise from aspects that differ from one country's economic and financial configuration to another. A second motivation can be traced back to the time when the ESRB was established, when supervisory policy and bank resolution arrangements were still in the hands of national authorities. The completion and uniform harmonisation of the European Banking Union should help to significantly reduce the current regulatory gaps. The operational scheme of the ESRB under the macro-prudential pillar can be divided into two main areas: the organisational aspects and the governance of macro-prudential policy in the Member States, and the intermediate objectives and reference instruments. With regard to the first area, the fields of intervention of the Authority concern the identification and monitoring of potential risks of financial instability and the planning of interventions. The second area concerns the definition of the policy strategy, i.e. the complex operational framework of macro-prudential policy, which includes: the shock transmission mechanisms, the intermediate and final objectives and instruments, as well as the indicators for monitoring the emergence of systemic risks (Díez-Esteban et al. 2022). An examination of the European institutional architecture reveals a gradual shift towards greater convergence between the legal systems of European countries in the regulation and implementation of banking crisis management. Indeed, the European response to the subprime mortgage crisis was clearly inadequate and highlighted the need to rethink European banking crisis discipline. Thus, after a long reform process, the new regulatory framework for banking supervision and crisis management has taken shape as part of the broader European project aimed at strengthening the integration process, establishing a new European governance in the financial sector and monitoring the emergence of systemic risks.

United States

In recent years, the US financial regulatory structure has evolved in response to financial crises and the need to adapt to new market developments and financial product innovations. The financial crisis that began in the summer of 2007 and developed into the worst economic recession since the Great Depression exposed significant weaknesses in the US financial regulatory system. Until the onset of the crisis in 2007, financial regulation and supervision focused on the supervision of individual intermediaries, neglecting the linkages between them (Desai and Downs 2022; Orhan et al. 2020; Pacelli 2021; Sykes 2018). The inherent characteristics of the regulatory system have thus exposed its weaknesses and inadequacies in crisis prevention, management and resolution. Excessive fragmentation has led to regulatory differentiation and inconsistencies, allowing arbitrage and ineffectiveness of prudential standards. Against this background, there was a need to reform the architecture of financial supervision in order to make it more appropriate and efficient to act in an integrated manner in the various sectors of the financial system, through greater coordination between institutions. However, the legislator's intervention did not affect the specific characteristics of the multi-agency supervisory model but rather made some corrections to the system, such as strengthening some of its centralising elements (Pierini 2017). In particular, on 21 July 2010, the Dodd-Frank Wall Street Reform and Consumer Protection Act (Act No. 111–203) was passed, which included important macro-prudential approaches for US financial stability. In particular, the Act provides for the creation of a Council composed of supervisors from various US regulatory agencies, including the Federal Reserve Board and the Bureau of Consumer Financial Protection, with significant powers and responsibilities (Webel 2013). The main functions of these bodies are: to designate banks, financial activities and non-bank financial firms as systemically important; to recommend higher prudential standards for any activity or firm that the Council determines contributes to systemic risk; to collect information on institutions and market practices that could contribute to systemic risk; to monitor on an ongoing basis the economic effects of regulatory constraints to reduce systemic risk, including the costs and benefits of imposing limits on the size, organisational complexity and interconnectedness of large financial institutions; to monitor domestic and international regulatory developments and analyse their impact on systemic risk in the US financial market.

The operational scheme of the new US financial regulatory framework is thus based on four pillars that are required to operate in an integrated manner, but with distinct and interdependent responsibilities: (i) macro-prudential supervision; (ii) micro-prudential supervision; (iii) the crisis management and resolution framework; and (iv) consumer protection (Masera 2011). As part of the first pillar of macro-prudential supervision, the Dodd-Frank Act gave the Financial Stability Oversight Council (FSOC) a broad range of powers and responsibilities to enhance cooperation among financial regulators and address potential risks of instability in the US financial system (Murphy and Bernier 2012). The FSOC is charged with identifying risks to US financial stability, promoting market discipline and transparent disclosure, and monitoring situations of systemic instability. The FSOC is chaired by the Secretary of the Treasury and consists of nine members representing major financial institutions and one member with expertise in the insurance industry. The scope of the Financial Stability Oversight Council extends to micro-prudential supervision and the crisis management and resolution framework. Thus, the exercise of macro-prudential supervision of the financial system represents a strategic point of reconnection between the first and third pillars of the supervisory framework. In addition, the Council has the authority to identify systemically important financial institutions and bring them under the Fed's micro-prudential supervision. The expansion of the FSOC's supervisory mandates represents a clear and obvious departure from the European approach.

From a micro-prudential perspective, the Federal Reserve System (Fed) is the central bank of the United States. It performs important functions to promote the efficient functioning of the US economy and, more generally, the public interest. The Fed exercises monetary authority to promote maximum employment and stable prices, promotes the soundness and stability of individual financial institutions, and monitors their impact on the financial system. The Fed also oversees the safety and efficiency of payment systems and consumer protection. In general, it is responsible for supervising and regulating all systemically important banks, bank holding companies and non-bank financial institutions. The Federal Deposit Insurance Corporation (FDIC) is an independent agency established by Congress to ensure stability and confidence in the US financial system (Labonte 2012). To accomplish its mission, the FDIC supervises financial institutions and oversees complex resolution processes to ensure consumer protection. The Office of the Comptroller of the Currency (OCC) regulates and supervises all national banks and federal savings associations, as well as branches and agencies of foreign banks. The OCC is an independent agency of the US Department of the Treasury. It is the primary prudential regulator and conducts proactive, risk-based supervision based on the principles of integrity, competence, cooperation, and independence.

The National Credit Union Administration (NCUA), established by the US Congress, is an independent federal agency that protects the deposits of federally insured credit unions and oversees the cooperative financial system through regulation and supervision. The Securities Investor Protection Corporation (SIPC), established by the Securities Investor Protection Act, is one of the primary investor protection agencies in the United States. Unlike other federal and state agencies that deal with investment fraud, SIPC's goal is to recover securities and customer assets held by brokerage firms that are insolvent or in financial distress.

Finally, the Federal Insurance Office (FIO) was established to oversee the insurance industry and to work with the FSOC and state insurance regulators in the supervision of systemically important institutions (Mclain 2014; Williams 2019).

As noted above, the Dodd-Frank Act also corrected key aspects of the regulatory framework that led to the 2007 crisis, including the introduction of the Collins Amendment. The latter provided for the application of capital requirements to all systemically important[3] non-bank financial institutions and companies, and stipulated that capital requirements for depository institutions, which are different from banks and use advanced risk measurement approaches, should be proportional to their size and activities. In 2018, the Dodd-Frank Act was reviewed and amended to calibrate it towards a more specific rejection of the proportionality principle. In particular, for community banks, a full exemption from certain prudential standards and stress tests was provided for banks with less than USD 10 billion in assets. The fourth pillar of the new US institutional framework, in the area of crisis management and resolution, introduces new obligations to prepare recovery and resolution plans, technically called living wills. Specifically, Section 165 of the Dodd-Frank Act requires all systemically important institutions, both banking and non-banking, to prepare and regularly update living wills. The preparation of the plans represents the prevention phase, during which institutions engage in self-regulatory activities to identify measures and tools to deal with disruptions. The plans are reviewed by the Federal Reserve in conjunction with the Federal Deposit Insurance Corporation and, if they do not contain effective provisions for dealing with potential instability,

---

[3] With more than USD 50 billion in assets.

the institutions are asked to amend them. To the extent that the individual financial institution has been unable to make the required changes to the plan, the authorities intervene by imposing stricter capital, leverage, liquidity levels, and restrictions on the institution's activities. In addition, under the provisions of the Resolution Mechanism for Significant Institutions, the Board may place a financial institution into receivership if there is substantial evidence that the firm is in a situation of default or is likely to default. Although the Bankruptcy Code is the default legal framework for resolving a failing bank holding company, the Dodd-Frank Act creates an alternative resolution framework through the Orderly Liquidation Authority (OLA). The OLA is a specialised resolution regime for large and complex financial firms, designed to ensure that these institutions are subject to the resolution process in an orderly manner and without threatening financial stability (Merler 2018).

Finally, the last pillar of the Dodd-Frank Act, introduced in response to the financial crisis, is the Bureau of Consumer Financial Protection (BCFP), which is responsible for the regulation and supervision of financial products and services. In general, the BCFP aims to protect consumers from unlawful practices in the trade of financial services by improving the transparency and integrity of the financial system (Carpenter 2012).

China

The Great Financial Crisis of 2007 also called into question the regulatory adequacy of China's financial system. Although the response to the global crisis was timely compared to Western economies, China was not spared from the economic and financial consequences of the crisis (Wang and Sun 2013). In fact, with the rapid development of financial markets after the international financial crisis, China gradually moved towards a sector-based regulatory model, with separate regulators for banks, insurance companies and financial markets (Barth et al. 2006; Pizzirusso 2013; Song and Xiong 2018). A new and lengthy process of financial system reform began only since the 14th annual session in 2023 of the National People's Congress (NPC). The main pillars of the new framework were the setting of new macroeconomic targets, the definition of a plan to encourage foreign investment and stimulate economic growth, and the approval of a plan to reorganise certain government institutions. In other words, the new framework provides not only for greater economic cooperation, but also for the consolidation of certain functions among the various financial regulators. Under the second pillar, the NPC plenary approved a new reform framework with the establishment of the National Administration of Financial Regulation (NAFR). The NAFR replaces the previous supervisory body, the China Banking and Insurance Regulatory Commission (CBIRC) and will also assume responsibility for consumer and financial investor protection (Fig. 4.2). In particular, the NAFR's functions are not limited to direct supervision of banks and non-bank financial institutions but extend to both market conduct and prudential regulation. In this area, the main supervisory powers over the financial system have been transferred from the People's Bank of China to the NAFR. Moreover, the new regulatory framework has not changed the role of the China Securities Regulatory Commission (CSRC), which has continued to exercise its supervisory powers over the securities market since 1992. Although the main functions of the CSRC were not revised by the new reform, the role of regulating and supervising the corporate bond market, previously exercised by the National Development and Reform Commission, was included.

The revision of the legal framework for the supervision of China's financial system also affected the operational framework for macro-prudential policy. In particular, prior to the reform, the Financial Stability and Development Committee (FSDC) was a financial regulatory body that, under the leadership of the State Council, ensured the strengthening of the PBC's macro-prudential policies and systemic risk mitigation role, and promoted the development of China's financial system (Bin et al. 2022; Huang 2010). Recent changes to the macro-prudential policy framework include the abolition of the FSDC and the establishment of the Central Financial Commission (CFC) and the Central Financial Working Committee (CFWC). The former is the supervisory body responsible for the design, development and supervision of the financial sector. The latter is responsible for overseeing the Chinese government's powers over the financial system. In sum, the CFC and CFWC represent the culmination of a broad reorganisation of China's supervisory bodies aimed at giving the government direct control and supervision over the financial system (Foo et al. 2023; He and Wei 2023).

Against this backdrop, the 14th Annual Session of the National People's Congress highlighted the Chinese government's determination to intensify economic cooperation and open up the Chinese market to foreign investors. In addition, financial regulators drew up a new plan to restructure various financial institutions operating in China. Given the breadth of the reform plan, it may take some time and further changes to complete the consolidation process. Therefore, it will be necessary to

Fig. 4.2 Financial regulatory system reform

examine the new measures critically and analytically in the coming years to understand how these changes in the institutional set-up may have directly affected the business operations of international investors in China and the stability of the financial system.

Islamic Countries

The term Islamic finance refers to the complex of organisational structures, transactions and financial contracts that conform to the dictates of Islamic law (Adamo 2008; Hamaui and Mauri 2009). It is based on a set of principles that distinguish it from conventional finance and represents an alternative financial system characterised by its strong dependence on sacred texts and the considerable influence of religion on law. Indeed, Islamic banking is based on *Shariah* law, which prohibits interest on loans, investment in sectors that do not meet Islamic requirements, speculation and gambling, and promotes profit and loss sharing (PLS), the payment of a religious tax, the supervision of Islamic banking by a religious council and, finally, the ethical nature of assets and investments (Arsyianti 2019; Hassan and Mollah 2018; Miglietta 2009).

The rapid expansion of Islamic banking and capital markets, measured in terms of the number of institutions and the complexity and sophistication of products and services, has sparked growing interest among scholars and regulators in the need to provide a legal framework for the financial system through the creation of international regulatory and supervisory organisations. The operation of financial systems in Islamic countries is based on the adoption of one of the following forms: fully Islamic, dual systems and neutral or partial inclusion. In the first case, countries allow the operation of financial and banking institutions only if they comply with *Shariah* principles, thus prohibiting the operation of conventional financial institutions. In the case of the dual system, countries allow Islamic finance and conventional finance to coexist, while encouraging an increase in the market share of Islamic finance. Finally, the neutral system maintains a fair and impartial nature (Addi and Bouoiyour 2023; Kepli and Yazid 2013; Zulkhibri 2019). As highlighted earlier, Islamic financial institutions operate according to *Shariah* principles and the relevant regulatory framework regulates their operations and contributes to achieving the objectives of efficiency, stability, risk mitigation and economic development. However, the integration of Islamic finance into the international financial environment has required the introduction of a legal framework with international standards that can ensure the sound and prudent management of financial institutions compared to international regulatory frameworks.

The long process of harmonising *Shariah* principles has encouraged the establishment of regulatory organisations to develop international standards and guidelines. These include the Accounting and Auditing Organisation for Islamic Financial Institutions (AAOIFI), the Islamic Financial Services Board (IFSB), the International Islamic Liquidity Management Corp (IILM), the Islamic International Rating

Agency (IIRA) and the International Islamic Financial Market (IIFM) (Elasrag 2014). At the forefront of the regulatory institutional set-up is the Shariah Board, a committee of Islamic researchers and scholars whose role is to verify the compliance of Islamic financial institutions' products with *Shariah* principles. The Accounting and Auditing Organisation for Islamic Financial Institutions (AAOIFI) is an independent, non-profit international organisation dedicated to the promotion and development of Islamic legal standards. Its main objective is to develop accounting, auditing, governance and ethical models for the activities of Islamic financial institutions, taking into account not only international standards and practices but also the need to comply with *Shariah* principles. It also reviews and adapts conventional standards to the Islamic financial system and provides guidance to central banks in Muslim countries to promote compliance with its regulatory guidelines. For the development of the Islamic financial sector, the Islamic Finance Service Board (IFSB), together with the AAOIF, aims to promote the development of a transparent Islamic financial services industry and to provide guidance on the supervision and regulation of institutions offering Islamic financial products. In recent years, the main focus at the international level has been the ratification of standards on capital adequacy, risk management and corporate governance. Standard-setting has enabled supervisors to pursue the principles of soundness, stability and integrity in the Islamic financial sector and to develop a wide range of technical standards and guidance in close cooperation with the Basel Committee and the International Monetary Fund. Specifically, the IFSB's operational scheme includes: (i) the development of standards in accordance with Islamic law to ensure a sound banking and financial system; (ii) the development of internal risk management structures; (iii) the identification and monitoring of risk exposures of Islamic financial instruments; (iv) and the definition of standards for the supervision, control and regulation of the Islamic financial system. The IIFM is an international standard-setting body for the Islamic financial services industry and promotes the harmonisation of *Shariah*-compliant financial contracts. In the area of cross-border liquidity management, the International Islamic Liquidity Management Corporation (IILM) was established by central banks, financial authorities and multilateral organisations to develop and issue *Shariah*-compliant financial instruments and facilitate cross-border Islamic liquidity management. Finally, the IIRA promotes the development of the capital market and financial system by providing tools and services to support Islamic finance.

The operational scheme of Islamic supervision is organised according to an approach that is different from conventional supervision, while pursuing the same objectives of operational independence of supervisory bodies, development of a sound legal framework and an accountable governance structure. On the other hand, supervisors in many jurisdictions do not seem to have harmonised procedures for the supervision of Islamic banks, which would require a cross-sectoral approach that analyses the interrelations and linkages between different Islamic financial sectors with a view to preventing the propagation of systemic risk. On the specific front of systemic risk mitigation, therefore, the regulatory and policy infrastructures of Islamic countries still appear to lag behind those in the US and Europe and would therefore require further modernisation towards a greater focus on the determinants of risk of contagion between countries, sectors and actors in the financial system.

Japan

In contrast to the international experience, the resilience of the Japanese regulatory infrastructure was tested a few years before the outbreak of the financial crisis in 2007 (Tamaki 2008).

In particular, the non-performing loans (NPL) crisis of the late 1990s was a clear test signal for Japanese regulators to overhaul the regulatory framework and reorient it towards the goal of systemic stability (Hoshi and Patrick 2000). The reform, which was completed before the outbreak of the Great Crisis, allowed for greater resilience and robustness of the banking sector, allowing for greater neutralisation and mitigation of systemic risk. In other words, the stability and greater risk aversion of the Japanese banking system are two of the main factors that have contributed to greater systemic resilience and moderate macro-prudential stringency (Lopez and Bruni 2019).

Japan's operational banking supervision system is based on the Financial Services Agency (FSA). The FSA is the authority responsible for ensuring the stability of the financial system, the protection of depositors and investors, and the development of the financial services sector. It achieves this through policy-making and the inspection and supervision of financial institutions. The operational structure of the FSA consists of three main bodies: the Strategy Development and Management Bureau, the Policy and Markets Bureau and the Supervision Bureau. In general, the FSA is entrusted with the tasks of general supervision of the financial system, including the establishment of an efficient regulatory system and the early detection of unstable situations. The FSA operates under the guidance of the Government, which has general supervisory, inspection and sanctioning powers. In this regard, it is envisaged that the authority, in order to ensure sound and prudent management, may require banks to prepare a plan for the improvement of their operations through a list of specific measures to be taken within a specified period. In addition, the Government may decide directly to suspend all or part of the intermediation. Other institutions, including the Bank of Japan (BOJ) and the General Directors of the Ministry of Finance's local financial offices, are also involved in the implementation of the overall supervisory objectives. In light of the recent 2020 reform, a task force has been established to improve coordination between FSA's inspections and BOJ's supervision of banking operations. The reform is based on the joint supervision of financial institutions by the FSA and the BOJ, thereby reducing their respective burdens by implementing measures to promote data integration and the sharing of supervisory results (Venter 2022).

In addition to representing the Japanese banking system, the Bank of Japan contributes to maintaining the stability of the financial system through money issuance and exchange control. As noted above, the exercise of supervision over financial institutions is considered a government function. In fact, the BOJ is not a regulator and has no authority to exercise prudential supervision over banks. However, as a lender of last resort, the BOJ can provide liquidity to institutions in the event of insolvency, and thus has the power to enter into agreements under which it may be empowered to supervise financial institutions.

Finally, the Deposit Insurance Corporation of Japan (DIC) is an organisation established under the Deposit Insurance Act (DIA) to administer the deposit insurance system. In the event of a bank failure, the main role of the DIC is to protect depositors and the financial system as a whole. The DIA regulatory framework divides resolution procedures into three categories:

- ordinary procedures: relating to ordinary resolution schemes;
- financial crisis management procedures: related to bank deposit insurance rescue schemes;
- ordinary resolution procedures: based on the Key Attributes of Effective Resolution Regimes for Financial Institutions adopted by the Financial Stability Board (FSB) and involving a regime that is broadly similar in effect to the bail-in regime.

The distinguishing feature of the operational scheme of the Japanese supervisory system is the evidence that the financial supervisory agencies are under the political power, and therefore there is no independent control entrusted to the central bank, except within the limits of the supervisory powers over the stability of the financial system.

Conclusions

The onset and rapid spread of the 2007 financial crisis highlighted the limitations of the regulatory and supervisory architecture in the financial system of the world's major economies. The need to overcome regulatory inefficiencies has led to a broadening of the analysis from a purely micro-prudential perspective to a macro-prudential one that adequately takes into account the complex relationships that characterise financial systems. The comparative analysis in this chapter has revealed significant differences in the regulatory architecture of different economies and the different responses of regulators to the consequences of the 2007 financial crisis. As already highlighted, the NPL crisis in Japan in the late 1990s was a critical event that led to radical changes in the financial architecture and a complete overhaul of supervision in Japan. The experience of the crisis and the new regulatory framework that emerged from it fostered a stronger focus on macro-prudential policy on financial stability and risk aversion. The *ante litteram* application of the new regulatory guidelines to the Japanese financial system contributed to more effective mitigation and neutralisation of contagion risks during the 2007 financial crisis. In contrast to the Japanese experience, the US and European regulatory systems were more surprised by the evolution of the 2007 crisis and, to varying degrees, proved inadequate to deal with it effectively. The rapid spread of the crisis across sectors, countries and individual players highlighted the need to rethink the macro-structure of regulation and supervision in the years immediately following the crisis. In response to the subprime mortgage crisis and the subsequent sovereign debt crisis, the regulatory architecture in the US and Europe underwent a structural overhaul. As has been widely discussed, the US regulatory architecture is characterised by a multi-agency structure in which banks are supervised by multiple regulators. The three-pillar structure of the European Banking Union is a regulatory framework designed to manage a single supervisory system. Although the US and European regulatory architectures are structurally different, there is a deep convergence in the regulatory and supervisory objectives of the two economies. In particular, both regulatory frameworks are designed to complement the micro-prudential and macro-prudential approaches to supervision, both of which are designed to promote financial stability. EU-US coordination on financial regulation and supervision has been a benchmark for building a sound global financial system and for reviewing financial regulation in other economic systems. Among them, in the early years of overhauling the regulatory framework of its financial system, China adopted a regulatory system heavily influenced by the US experience and based on the multi-agency regulatory model. However, China's regulatory reform is not complete, and the recent crises show that important steps remain to be taken. Significant delays in modernising the regulatory and supervisory structures to mitigate systemic risk are also evident in Islamic countries, where the rapid expansion of the financial industry has not been matched by the establishment of an adequately efficient, transparent and internationally recognised regulatory framework. An important first step towards international financial integration has been the establishment of various institutions that have played a key role in pursuing the harmonisation of rules based on *Shariah* principles, but even in these countries several important steps remain to be taken. Indeed, significant systemic risks lurk in the regulatory and supervisory inconsistencies at the international level. The comparative analysis of the main international financial regulatory infrastructures reveals a growing and widespread focus on supervisory policies and architectures aimed at mitigating systemic risk, although the different economies analysed start from very different positions. While the supervisory frameworks in different countries reflect the structure of the financial system, the historical nature and the political constraints, there is a solid and convincing convergence towards the objectives of stability and systemic risk mitigation. Indeed, the rapid succession of crises at the global level has raised awareness everywhere of the growing vulnerability of modern economic and financial systems to systemic risks and of the relative need to address and manage these risks through modern and appropriate regulatory and supervisory frameworks.

References

Adamo, R.: Il boom della finanza islamica. Da alcune operazioni di cartolarizzazione spunti di riflessione. Bancaria (2008)

Addi, A., Bouoiyour, J.: Interconnectedness and extreme risk: evidence from dual banking systems. Econ. Model. **120**, 106150 (2023). https://doi.org/10.1016/j.econmod.2022.106150

Agénor, P.R., Pereira da Silva, L.A.: Macroeconomic stability, financial stability, and monetary policy rules. Int. Finance **15**, 205–224 (2012). https://doi.org/10.1111/j.1468-2362.2012.013 02.x

Arsyianti, L.D.: The role of Shariah supervisory board in Islamic financial industry. Al-Infaq: Jurnal Ekonomi Islam **1** (2019). https://doi.org/10.32507/ajei.v1i1.394

Baker, A.: The gradual transformation? The incremental dynamics of macroprudential regulation. Regul. Gov. **7**, 417–434 (2013). https://doi.org/10.1111/rego.12022

Barth, J.R., Zhou, Z., Arner, D., Hsu, B.F.C., Wang, W.: Financial Restructuring and Reform In Post-WTO China. Wolters Kluwer (2006)

Bin, H., Liansheng, Z., Juncheng, L.: China's systemic financial risk: basic dimensions, key areas, and evolving trends. Soc. Sci. China **43**, 102–124 (2022). https://doi.org/10.1080/02529203. 2022.2093069

Blundell-Wignall, A., Roulet, C.: Macro-prudential policy, bank systemic risk and capital controls. OECD J. Financ. Market Trends **2013**, 7–28 (2014)

Boccuzzi, G.: L'Unione bancaria europea: nuove istituzioni e regole di vigilanza e di gestione delle crisi bancarie. Bancaria Editrice (2015)

BRI: BRI (1986)

BRRD, 2014/59/EU: BRRD, 2014/59/EU

Carmassi, J., Lamanda, C., Luchetti, E., Maino, R., Masera, R., Mazzoni, G., Micossi, S., Forti, N.: Preventing and managing future crises. Bancaria Editrice (2010)

Carpenter, D.H.: The consumer financial protection bureau (CFPB): a legal analysis. In: The Consumer Financial Protection Bureau: Overview and Analyses, pp. 1–38 (2012)

Cerutti, E., Claessens, S., Laeven, L.: The use and effectiveness of macroprudential policies: new evidence. J. Financ. Stab. **28**, 203–224 (2017)

Claessens, S., Ghosh, S.R., Mihet, R.: Macro-prudential policies to mitigate financial system vulner-abilities. J. Int. Money Financ. **39**, 153–185 (2013). https://doi.org/10.1016/j.jimonfin.2013. 06.023

Clement, P.: The Term 'Macroprudential': Origins and Evolution. https://papers.ssrn.com/abstract= 1561624 (2010)

Crouhy, M., Turnbull, S.M., Jarrow, R.A.: The subprime credit crisis of 07. (2008). https://doi.org/ 10.2139/ssrn.1112467

Desai, C.A., Downs, D.H.: Banking regulatory constraints and personal bankruptcy filings in the US. J. Financ. Regul. **8**, 75–103 (2022). https://doi.org/10.1093/jfr/fjab011

Díez-Esteban, J.M., Farinha, J.B., García-Gómez, C.D., Mateus, C.: Does board composition and ownership structure affect banks' systemic risk? European evidence. J. Bank. Regul. **23**, 155–172 (2022)

Drumond, I.: Bank capital requirements, business cycle fluctuations and the Basel accords: a synthesis. J. Econ. Surv. **23**, 798–830 (2009). https://doi.org/10.1111/j.1467-6419.2009.00605.x

Ehrmann, M., Schure, P.: The European systemic risk board—governance and early experience. J. Econ. Policy Reform. **23**, 290–308 (2020). https://doi.org/10.1080/17487870.2019.1683011

Elasrag, H.: Corporate governance in Islamic finance: basic concepts and issue. https://papers.ssrn. com/abstract=2442014 (2014)

Engel, C.: Macroprudential policy under high capital mobility: policy implications from an academic perspective. J. Japan. Int. Econ. **42**, 162–172 (2016)

Foo, T., Davies, E., Gregory, K., Yang, T., Liao, Y., Yang, Y., Zhao, P., Chen, J.: China establishes new financial regulator—among major priorities in post-COVID era. Clifford Chance (2023)

Galati, G., Moessner, R.: Macroprudential policy—a literature review. J. Econ. Surv. **27**, 846–878 (2013). https://doi.org/10.1111/j.1467-6419.2012.00729.x

Ghosh, S.: Macroprudential policies, crisis and risk-taking: evidence from dual banking systems in GCC countries. J. Islamic Acc. Bus. Res. **7**, 6–27 (2016). https://doi.org/10.1108/JIABR-03-2014-0011

Gualandri, E., Noera, M.: Rischi sistemici e regolamentazione macroprudenziale. In: Lo stato della finanza. Scritti in onore di Marco Onado, pp. 15–62. Il Mulino (2014)

Hadjiemmanuil, C.: Bank resolution financing in the banking union. LSE Law, Society and Economy Working Papers, vol. 6 (2015)

Hamaui, R., Mauri, M.: Economia e Finanza Islamica. Il Mulino (2009)

Hassan, A., Mollah, S.: Islamic Finance: Ethical Underpinnings, Products, and Institutions. Palgrave Macmillan (2018)

He, Z., Wei, W.: China's financial system and economy: a review. Ann. Rev. Econ. **15**, 451–483 (2023). https://doi.org/10.1146/annurev-economics-072622-095926

Hoshi, T., Patrick, H.: The Japanese financial system: an introductory overview. In: Hoshi, T., Patrick, H. (eds.) Crisis and change in the Japanese financial system, pp. 1–33. Springer, US, Boston, MA (2000)

Huang, H.: Institutional structure of financial regulation in China: lessons from the global financial crisis. J. Corp. Law Stud. **10**, 219–254 (2010). https://doi.org/10.1080/14735970.2010.11419825

Kepli, Z., Yazid, M.: The legal and regulatory framework (2013)

Labonte, M.: Financial regulatory reform: systemic risk and the federal reserve. In: Systemic Risk: Oversight and Reform Considerations, pp. 11–47 (2012)

Lopez, C., Bruni, F.: The macroprudential policy framework needs to be global. International Financial Architecture for Stability and Development/Crypto-assets and Fintech (2019)

Masera, R.: Reforming financial systems after the crisis: a comparison of EU and USA. https://papers.ssrn.com/abstract=1815138 (2011)

Mclain, L.T.: Asset management firms, financial stability and the FSOC: elements and considerations (2014)

Merler, S.: Varieties of Banking Union: Resolution Regimes and Backstops in Europe and the US. Istituto Affari Internazionali (IAI) (2018)

Miglietta, F.: Le fonti dell'Islam e le scuole coraniche. In: Fondi sovrani arabi e finanza islamica, pp. 76–82. Egea (2009)

Moosa, I.A.: Basel II as a casualty of the global financial crisis. Basel II as a casualty of the global financial crisis. J. Banking Regul. 95–114 (2010)

Murphy, E.V., Bernier, M.B.: Financial stability oversight council: a framework to mitigate systemic risk. In: Financial Stability Oversight Council: Overview and Select Studies, pp. 1–48 (2012)

Onado, M.: La crisi finanziaria internazionale: le lezioni per i regolatori. BS. (2009). https://doi.org/10.1435/29376

Onado, M.: Alla ricerca della banca perduta. Il Mulino (2017)

Orhan, A., Ferhan Benli, V., Castanho, R.A.: Assessing the systemic risk between American and European financial systems. Prague Econ. Papers **29**, 649–671 (2020). https://doi.org/10.18267/j.pep.756

Pacelli, V.: Rischio sistemico e scienza delle reti. BANCARIA **12** (2021)

Pallini, A., De Cesare, M., Calrco, F., Grasso, G., De Lisa, R.: La proposta di nuove regole europee sulle crisi bancarie: cosa cambia per i sistemi di garanzia dei depositi? **9** (2023)

Pierini, A.: L'Unione Europea verso l'unione bancaria: indicazioni e dilemmi derivanti dal confronto con il modello di vigilanza sui mercati creditizi degli Stati Uniti d'America. Diritto pubblico comparato ed europeo 173–230 (2017). https://doi.org/10.17394/85827

Pizzirusso, G.: Il sistema bancario cinese. Giappichelli (2013)

Porretta, P.: Integrated Risk Management. Regole, rischi, capitale, liquidità e nuove opportunità strategiche. Egea (2021)

Resti, A., Sironi, A.: La crisi finanziaria e Basilea 3: origini, finalità e struttura del nuovo quadro regolamentare. Carefin Working Papers. Centre for Applied Research in Finance (2011)

Song, Z.M., Xiong, W.: Risks in China's Financial System. https://www.nber.org/papers/w24230 (2018)

Sykes, J.B.: Regulatory reform 10 years after the financial crisis: systemic risk regulation of non-bank financial institutions. In: Financial Crises and Programs: Developments, Analyses and Research, pp. 111–188 (2018)

Tamaki, N.: Bank Regulation in Japan. IFO DICE Rep. **6**, 9–13 (2008)

Tomuleasa, I.I.: Macroprudential policy and systemic risk: an overview. Procedia Econ. Finance **20**, 645–653 (2015). https://doi.org/10.1016/S2212-5671(15)00119-7

Venter, Z.: Macroprudential policy under uncertainty. Port. Econ. J. **21**, 161–209 (2022). https://doi.org/10.1007/s10258-021-00194-8

Vinals, J.: Making Macroprudential Policy Work. IMF (2013)

Wang, B., Sun, T.: How effective are macroprudential policies in China? (2013)

Webel, B.: The dodd-frank wall street reform and consumer protection act: issues and summary. In: The Dodd-Frank Act: Rulemaking Coordination and Impact of Rules, pp. 117–147 (2013)

Williams, G.M.: FSOC and the systemic risk of nonbank companies. Banking Law J. **136**, 415–419 (2019)

Zdzienicka, A., Chen, S., Kalan, F.D., Laseen, S., Svirydzenka, K.: Effects of monetary and macroprudential policies on financial conditions: evidence from the United States. IMF Working Papers 2015 (2015). https://doi.org/10.5089/9781513519159.001.A001

Zulkhibri, M.: Macroprudential policy and tools in a dual banking system: insights from the literature. Borsa Istanbul Rev. **19**, 65–76 (2019). https://doi.org/10.1016/j.bir.2018.04.001

Chapter 5
Systemic Risks and Multilayer Financial Networks: From Contagion to Mitigation

Maria Cristina Quirici and Roberto Moro-Visconti

Abstract The global financial system's interconnectedness has increased due to globalization, technological advancements and the integration of financial markets. Financial institutions and markets across different countries are more closely linked than ever before; while this interconnectedness facilitates global trade and investment, it also means that financial turmoil can quickly spread from one country to another. Systemic risk is the possibility that an event at the company level could trigger severe instability or collapse an entire industry or economy. The fall of Lehman Brothers in 2008 showed that the failure of a single entity could have far-reaching effects on the global financial system. This chapter innovatively interprets the financial system as a complex network formed by the relationships among various "nodes": banks, financial institutions, markets, and consumers. These networks are intricate and opaque, making it challenging to understand and predict how risks and failures in one part of the system can affect the rest with a domino impact. In managing systemic risk, regulators and policymakers play a vital role, implementing stricter regulatory frameworks, overseeing financial institutions more closely, and developing mechanisms to identify and mitigate risks early. This chapter shows that effective strategies to mitigate systemic risk involve better risk assessment models, more robust regulatory frameworks, and international cooperation among regulatory bodies. Stress testing, capital adequacy requirements, and monitoring of "too big to fail" institutions, as well as of "too interconnected to fail" ones, are part of these strategies, that may usefully consider network theory to link economic agents to their edging patterns.

Keywords Financial system · Stability · Financial shock · Network theory · Regulation · Policymakers

M. C. Quirici (✉)
University of Pisa, Pisa, Italy
e-mail: maria.cristina.quirici@unipi.it

R. Moro-Visconti
Università Cattolica del Sacro Cuore, Milan, Italy
e-mail: roberto.moro@unicatt.it

© The Author(s) 2025
V. Pacelli (ed.), *Systemic Risk and Complex Networks in Modern Financial Systems*,
New Economic Windows, https://doi.org/10.1007/978-3-031-64916-5_5

Introduction

The global financial system, involving both stock and credit markets (Foglia et al. 2024) has become highly interconnected over the past few decades. This elevated interconnectedness has resulted in an increased frequency of financial crises, characterized by a faster transmission of turmoil between countries. As demonstrated by the Global Financial Crisis (2008) and the European Sovereign Debt Crisis (2010), it is possible to draw an analogy between the economy and epidemics, referring to the rapid transmission of financial shocks as "contagion" (Atasoy et al. 2024; Brunetti et al. 2019).

In addition to contagion, another concept that came to the fore after the Global Financial Crisis is systemic risk. The bankruptcy of Lehman Brothers fueled fears of a systemic collapse, shifting attention from the individual risks of financial institutions to a systemic risk (Markose et al. 2012; Bougheas and Kirman 2016; Riccetti 2020; Paltalidis et al. 2015).

As with contagion, there is not a clear consensus regarding the definition of systemic risk, though different definitions share some common elements (Di Clemente 2016; Ellis et al. 2022). In the present approach, systemic risk refers to the risk of widespread financial disruptions or failures within a financial system, often caused by the interconnections and dependencies among various entities.

Complex networks play a significant role in the modern financial system, and understanding their dynamics is crucial for assessing and managing systemic risk. Complex networks emerge from the relationships among financial institutions, markets, and other economic agents. These networks are characterized by nodes (representing individual entities) and edges (representing relationships or connections between entities). The connections can be direct, such as contractual obligations or direct financial exposures, or indirect, such as shared exposures to common assets or market dynamics (Covi et al. 2021).

Understanding systemic risk in complex financial networks involves analysing the network structure, identifying key nodes (entities) and edges (connections), and assessing vulnerabilities and potential contagion paths. Network analysis techniques, such as centrality measures, clustering algorithms, and stress testing methodologies, are used to quantify systemic risk and inform risk management strategies, in the consciousness that the measurement of systemic risk is an ongoing and evolving problem actually discussed in literature (Riccetti 2022a, b; Brogi et al. 2021; Avdjiev et al. 2019).

At the same time, it is important to point out that regulators and policymakers can play a crucial role in mitigating systemic risk (Capponi and Chen 2015). They implement measures like stress testing, capital adequacy requirements, and regulations on interconnectedness to enhance the resilience of the financial system and reduce the likelihood and impact of systemic crises (Brunetti et al. 2019).

It is worth considering the European system adopted by the European regulators to tackle systemic risk, considering the strengths and weaknesses elements of the *European Systemic Risk Board* (ESRB) that, with the *European System of Financial*

Supervisors (ESFS), constitute our "safety net" for the supervision of European financial markets, a supervisory system adopted in 2010 on the basis of the de Larosiére Report in response to the deep international financial crisis that spread from 2007 to 2008 (Quirici 2010; ESRB 2023a; Ellis et al. 2022).

Analysing both how complex financial networks can give rise to systemic risk (Paltalidis et al. 2015) and if this kind of risk can be considered effectively controlled/mitigated by the existing dedicated regulator systems, the authors have the aim to enhance the understanding of the dynamics of interconnected financial systems in order to improve adequate risk management practices and policy-makers/regulators actions to tackle systemic risk in a better way, in the consciousness that the study of the relation between this kind of risk and Multilayer Financial Networks in the modern financial system is an ongoing and evolving field.

Moreover, after an analysis of the possible interconnections between these concepts, the authors want also to give an insight into the possible impact on systemic risk and complex networks of the recent sustainable financial context, where the UN Sustainable Development Goals and ESG criteria are more and more new financial paradigms.

The chapter is organized as follows: first an analysis of the structure and role of complex networks in the financial system; then how these networks can give rise to systemic risk, considering possible mechanisms of transmissions; successively, policy-makers and regulators' actions that can tackle the systemic risk, considering in particular the relative EU asset, underlining points of strength and weakness; finally, briefly, the possible impact on systemic risk of the new financial sustainable framework. Some conclusions will close the chapter. An analysis of the extant literature is disseminated in each paragraph.

The Role of Complex Financial Networks

Complex networks actually play a vital role in shaping the dynamics of financial interconnections and vulnerabilities. These networks can encompass a wide range of relationships and interactions, including direct financial exposures, common counterparties, shared assets, and more (D'Arcangelis and Rotundo 2016).

Understanding the structure and behaviour of these complex networks is crucial for identifying potential sources of systemic risk and developing effective risk management strategies. Network analysis techniques can help uncover hidden interdependencies, assess the potential impact of disruptions in one part of the network on other parts, and identify critical nodes or entities that could amplify the propagation of risks (Bougheas and Kirman 2016; Brunetti et al. 2019).

Moreover, as financial systems become increasingly interconnected and globalized, the potential for systemic risk becomes more pronounced. A disturbance in one part of the world can quickly spread across the network, potentially leading to widespread financial crises. This emphasizes the importance of not only understanding the structure of these networks but also implementing robust risk management practices and policy measures to mitigate the impact of systemic risks.

Complex financial networks are a breeding ground for systemic risk due to the intricate interconnections and interdependencies among various entities. Here's a bit more detail on how complex financial networks can give rise to systemic risk:

(a) Contagion and Interconnectedness: The interconnectedness among financial institutions and markets creates pathways for the rapid spread of distress and failures. When one entity experiences financial trouble, it can transmit its problems to others through direct exposures, shared assets, and common counterparties. This contagion effect can lead to a domino effect, where the failure of one entity triggers a cascade of failures in others.[1]

(b) Feedback Loops: Complex networks can amplify shocks through feedback loops. For example, when asset prices start declining due to the distress of a few entities, it can lead to further asset sales and additional price declines. This feedback loop can intensify market stress and contribute to a downward spiral.

(c) Herd Behaviour: In interconnected networks, participants might imitate the actions of others due to uncertainty or a lack of information. Herd behaviour can lead to exaggerated market movements and even the propagation of false information, which can contribute to systemic instability.

(d) Non-linearity and Network Effects: The behaviour of complex networks is often nonlinear, meaning that small disruptions can lead to disproportionately large impacts. As the number of interconnected entities increases, the potential for network effects grows, making the system more vulnerable to systemic risk.

(e) Lack of Transparency: Complex networks can obscure the true extent of interdependencies and risks. This lack of transparency can hinder the accurate assessment of potential vulnerabilities and the formulation of effective risk management strategies.

(f) Emergent Risks: The interactions within complex networks can lead to the emergence of risks that are not readily apparent when considering individual entities in isolation. These emergent risks can catch regulators, policymakers, and market participants off guard.

Given these factors, understanding the structure, dynamics, and vulnerabilities of complex financial networks is crucial for identifying and managing systemic risk. Regulatory measures, stress testing, risk assessment tools, and coordinated policies all contribute to enhancing the resilience of the financial system and reducing the potential for systemic crises arising from the inherent complexities of these networks.

Representing financial systems as networks of interconnected nodes and edges offers a valuable framework for understanding the relationships and dependencies

---

[1] *"The propagation of financial contagion in networks with dense clustering which reflects high concentration or localization of exposures between few participants will be identified as one that is TITF"* (Markose et al. 2012). TITF is an acronym meaning "Too interconnected to fail", to consider those financial networks that dominate in terms of network centrality and connectivity, called also *"super-spreaders"*. Regarding TITF see also Hüser (2015).

that exist among various entities.[2] Considering nodes and edges work within financial networks, it is possible to underline that nodes represent individual entities within the financial system—these entities can include banks, financial institutions, investment firms, corporations, governments, and even specific markets or market segments; each node in the network can be thought of as a distinct player with its own characteristics, assets, liabilities, and interactions—while edges, also known as links or connections, represent the relationships between nodes.

These relationships can be direct or indirect (Riccetti 2020) and can take various forms:

(a)   Direct Connections: These represent direct interactions or obligations between entities. For instance, if Bank A has provided a loan to Bank B, a direct edge between these two banks is present;

(b)   Indirect Connections: These indicate shared dependencies, common exposures, or interactions that might not be direct but still link entities. For example, if both Bank A and Bank B have invested in a particular asset or market, there's an indirect connection between them through that shared exposure.

Analysing these nodes and edges can reveal important insights into systemic risks, vulnerabilities, and potential contagion pathways. A disturbance or failure in one node could propagate through its connections to affect other nodes, potentially leading to systemic repercussions.

Network analysis techniques, such as centrality measures, clustering algorithms, and stress testing simulations, help to identify critical nodes (entities with a disproportionate impact), to detect hidden interdependencies, and to assess the overall resilience of the financial network.

By understanding the structure and dynamics of these complex networks, regulators, policymakers, and financial institutions can better anticipate and manage risks. Additionally, insights gained from network analysis can inform the design of more effective regulatory measures, risk management strategies, and contingency plans to enhance the stability of the financial system.

Complex Financial Networks and Systemic Risk: Possible Mechanism of Transmission

The presence of complex networks in the financial system can give rise to systemic risk through two primary mechanisms (Gai et al. 2011):

---

[2] According to Brunetti et al. (2019), there are two types of financial networks: "correlation networks", where edges are based on asset return correlations, and "physical networks", where links result from agent choices. This study is able to demonstrate that the two types of networks capture related but differing information sets, with correlation networks capturing both direct and indirect linkages, while physical networks capturing more specific direct linkages among banks.

1. Contagion Risk: Contagion is a critical aspect of systemic risk in interconnected financial networks. It occurs due to the interdependencies among entities, causing distress or failure in one entity to quickly spread to others. There are a few key types of contagion (Gai and Kapadia 2010):

   – Direct Exposures: When one entity defaults on its obligations to another entity, the latter's financial stability can be compromised. For instance, if Bank A defaults on a loan to Bank B, Bank B's ability to meet its obligations might be affected.
   – Indirect Exposures: Entities might not have direct relationships but might be linked through common counterparties or shared exposures. If these common elements are affected, the distress can be transmitted through the network.
   – Feedback Loops: Contagion can create self-reinforcing feedback loops. As one entity's distress leads to asset sales, it can trigger a decline in asset prices, affecting the health of other entities that hold those assets, leading to further distress and more asset sales.

2. Amplification Mechanisms: Complex networks can amplify shocks and disturbances, exacerbating the initial impact. This amplification can occur through various mechanisms:

   – Fire Sales: Distressed entities might be forced to sell assets at discounted prices to raise funds. These fire sales can lead to a rapid decline in asset prices, affecting the value of similar assets held by other entities and potentially causing a cascade of selling.
   – Liquidity Spirals: Funding liquidity spirals occur when institutions face difficulties in obtaining funding. This can lead to deleveraging, as entities sell assets to raise cash, driving down asset prices and further reducing their collateral value. This, in turn, tightens funding conditions for others.

Liquidity spirals are a phenomenon in financial markets where a shortage of liquidity, or the ability to quickly convert assets into cash, leads to a self-reinforcing cycle of declining asset prices and worsening market conditions. This can result in a feedback loop that intensifies market stress and can have a broader impact on financial stability. Here's how a liquidity spiral typically unfolds:

(a) Initial Trigger: The spiral often begins with a shock or a perceived negative event in the financial system. This could be a sudden increase in uncertainty, the default of a major counterparty, or economic news that creates uncertainty.
(b) Market Stress: As uncertainty increases, market participants become more cautious and risk-averse. They might start to demand higher returns for holding risky assets, which can lead to a decline in asset prices.
(c) Asset Price Decline: Falling asset prices reduce the value of collateral held by financial institutions. This can make it difficult for these institutions to borrow or raise funds using these assets as collateral.

(d) Tightening Liquidity: As institutions face difficulties in obtaining funds, they might be forced to sell assets to raise cash. This selling pressure further depresses asset prices.

(e) Collateral Devaluation: The decline in asset prices reduces the perceived value of collateral in the financial system. Lenders become more cautious about accepting these devalued assets as collateral, making it even harder for institutions to secure funding.

(f) Deleveraging: Institutions, facing liquidity pressures and concerns about collateral quality, might engage in deleveraging—reducing their overall exposure by selling assets or reducing their trading activities. This can further drive down asset prices.

(g) Continued Feedback Loop: The cycle continues as asset sales lead to further price declines, worsening collateral quality, and increasingly constrained access to funding. This can create a self-reinforcing loop that intensifies market stress.

Liquidity spirals can impact both individual institutions and the broader financial system. As asset prices decline and funding conditions tighten, the risk of insolvency for some institutions increases. Furthermore, if multiple institutions are simultaneously trying to sell assets to raise cash, it can exacerbate the decline in asset prices and spread distress across the financial network.

Mitigating liquidity spirals often involves a combination of monetary policy actions, lender-of-last-resort interventions, and measures to restore market confidence. Central banks might inject liquidity into the system, provide emergency funding to distressed institutions, and offer assurances to market participants to help break the cycle of panic-driven selling and restore stability.

Understanding liquidity spirals is crucial for policymakers, regulators, and financial institutions to design effective strategies to prevent and manage their occurrence.

- Herd Behaviour: In interconnected networks, entities might react to the distress of others by taking similar actions. For example, if one bank experiences run-on deposits, other banks might face similar withdrawals as customers fear contagion.

Herd behaviour refers to the tendency of market participants, including financial institutions, investors, and consumers, to imitate the actions of others in response to uncertainty or perceived information. This behaviour can lead to amplified market movements, increased volatility, and potential systemic risks.

Here's how herd behaviour works and its implications. Herd behaviour can arise due to several factors, such as:

(a) Information Asymmetry: When individuals have limited information about the market or a specific asset, they might look to the actions of others for guidance. If many participants start selling a particular asset, others might follow suit, assuming those participants have superior information.

(b) Social Proof: People tend to believe that if a large number of others are doing something, it must be the correct course of action. This can lead to a self-reinforcing cycle as more participants follow the initial trendsetter.

(c) Risk Aversion: In times of uncertainty or crisis, individuals may choose to adopt the actions of others to avoid standing out or making decisions that might lead to losses.

Considering the implications of Herd Behaviour, it is possible to underline that Herd behaviour can have significant effects on financial markets and systemic stability:

(a) Market Volatility: Herd behaviour can amplify market movements, leading to abrupt price swings and increased volatility.
(b) Contagion: In interconnected financial systems, one institution's troubles can trigger a herd response among other similar institutions. For example, if one bank faces run-on deposits, customers of other banks might fear similar problems and withdraw their funds as well.
(c) Systemic Risk: Herd behaviour can contribute to systemic risks, as simultaneous actions by numerous participants can lead to self-fulfilling prophecies. For instance, if many banks sell assets at the same time due to perceived risks, it can cause asset prices to plummet and negatively impact their capital positions.
(d) Liquidity Crunch: If many participants rush to liquidate assets simultaneously, it can strain market liquidity and make it difficult to sell assets at reasonable prices.

Understanding and managing herd behaviour is crucial for maintaining the stability of financial systems, especially in times of uncertainty or stress. By addressing the psychological factors that contribute to herd behaviour and implementing measures to counteract its negative effects, regulators can help mitigate potential systemic risks.

Understanding these mechanisms is crucial both for risk management and for designing effective measures to mitigate systemic risk by financial regulators and policy-makers.

Regulators and Policy-Makers Measures to Mitigate Systemic Risk in the Presence of Financial Networks

Financial institutions are connected to each other through sophisticated networks of multilateral exposures. As a consequence of these linkages, distress or failure of a financial institution, triggering large unexpected losses on its trades, can seriously affect the financial status of its counterparties. The intricate structure of linkages can be "captured" via a network representation of the financial system, in the consciousness that such network models can assist in detecting important shock transmission mechanisms. (Capponi and Chen 2015; Riccetti 2020).

Consequently, regulators and financial institutions need to consider not only the individual health of entities, but also the potential ripple effects that can propagate through the complex network, leading to broader financial instability (Ellis et al. 2022; Markose et al. 2012; Cont et al. 2013; Galizia 2015).

Systemic risk so represents another dimension that financial regulators have to include in their analysis: when bank failures are contagious regulators, in assigning priority rights, need also to take into account how the bankruptcy resolution of one institution might affect the survival of other institutions that have acted as its creditors. In other terms, if the choice of policy can affect the structure of the financial network, policy design becomes surely more complex (Bougheas and Kirman 2016).

If managing herd behaviour is challenging, there are some steps that regulators and policy-makers can take to mitigate its impact:

(a) Communication and Transparency: Providing clear and timely information to market participants can reduce uncertainty and the potential for panic-driven reactions.
(b) Strengthening Confidence: Implementing measures to enhance the confidence of market participants can reduce the likelihood of mass withdrawals or panic selling.
(c) Macroprudential Measures: Regulatory authorities can introduce measures that encourage more prudent behaviour among market participants, such as implementing circuit breakers or capital requirements.
(d) Monitoring and Intervention: Regulatory bodies can closely monitor market trends and intervene if they detect signs of irrational herd behaviour that could threaten financial stability.

Regulators and policy-makers play a vital role in safeguarding financial stability and mitigating systemic risks (ECB 2011; Hollo et al. 2012). There are various measures indeed that represent key tools in their arsenal for achieving these goals:

a. Stress Testing: Stress testing involves subjecting financial institutions to simulated adverse scenarios to assess their resilience. By evaluating how institutions' capital positions, liquidity, and overall financial health would hold up under severe economic and market conditions, regulators can identify vulnerabilities and take corrective actions if necessary. Stress testing enhances transparency, helps institutions prepare for potential shocks, and contributes to better risk management.
b. Capital Adequacy Requirements: Requiring financial institutions to maintain sufficient capital buffers is a fundamental aspect of preventing systemic risks. Adequate capital cushions protect institutions from unexpected losses and provide a buffer that can absorb shocks. Capital requirements are set by regulators to ensure that banks and other financial entities have the resources to cover potential losses and maintain their operations during times of stress.
c. Regulations on Interconnectedness: Regulators often impose regulations to manage the degree of interconnectedness within the financial system. These regulations might include limits on exposures to specific counterparties, concentration limits for asset holdings, and requirements to hold additional capital for exposures to systemically important institutions. These measures reduce the risk of contagion and limit the potential for a single entity's failure to spread across the network.

d. Macroprudential Policies: Macroprudential policies focus on the stability of the financial system as a whole rather than individual institutions. These policies might involve setting limits on loan-to-value ratios for mortgages, countercyclical capital buffers, and other measures that aim to prevent excessive risk-taking during periods of economic boom and to provide support during downturns.
e. Regulatory Coordination and Cooperation: Global financial systems are interconnected, and systemic risks can transcend national borders. Therefore, effective regulation often requires coordination and cooperation among different regulatory bodies and across jurisdictions. International standards and agreements can help harmonize regulatory approaches and enhance the effectiveness of measures to mitigate systemic risks.
f. Early Intervention and Resolution Frameworks: Regulators often establish frameworks for early identification and intervention in troubled financial institutions. This helps prevent the deterioration of institutions' financial conditions and reduces the likelihood of contagious failures. Resolution frameworks provide a structured process for handling the failure of systemically important institutions while minimizing disruption to the financial system.

By combining these measures and regularly assessing the evolving financial landscape, regulators and policy-makers aim to create a resilient financial system that can withstand shocks, protect consumers and investors, and reduce the potential for systemic crises. The goal is to strike a balance between promoting financial innovation and growth while ensuring stability and risk mitigation.

The European Systemic Risk Board (ESRB)

Reasons of the ESRB Establishment and Its Main Characters

After the collapse of Lehman Brothers in mid-September 2008, global financial market participants were directly impacted by its default, resulting from numerous cross-border and cross-entity interdependencies (Acharya et al. 2012). The shock was rapidly spread in Europe, the effects of both interconnectedness and contagion manifested themselves and systemic risk emerged not only as one of the most challenging aspects but also as a risk that had been enormously underestimated. While the euro area banking system was fundamentally solvent, according to several stress tests (Acharya and Steffen 2015),[3] the intensity and speed with which shocks propagated

---

[3] As the 2008 financial crisis evolved and with the sovereign debt crisis in the euro area in 2010–2011, it became necessary to further integrate the system of surveillance on euro area's banking system: this led to the EU's Banking Union initiative, with a Single Supervisory Mechanism, placing the ECB as the central supervisors for euro-area banks, and a Single Resolution Mechanism, having the aim to ensure the orderly resolution of failing banks covered by the Single Supervisory Mechanism. It is necessary to point out that the rules on reducing risks in the European banking sector are constantly evolving over time.

in the entire financial system, put at risk the same stability of the whole European system, highlighted in a clear way the need to identify, measure and understand the nature and the source of systemic risk (Covi et al. 2021).

The European Systemic Risk Board (ESRB) is the European regulatory answer to this need. It represents a key component of the new European Union's framework for addressing systemic risk in the financial system with the European System of Financial Supervision (ESFS). The ESRB[4] was established on 1st January 2011 just in response to the Global Financial Crisis (2007–2008) representing an innovative element of the new European "safety net". The ESFB introduction was based on recommendations from the de Larosiére Report, which aimed to enhance financial supervision and stability within the EU (de Larosiére Group 2009).

The ESRB consists of the National Central Bank Governors and representatives from the European Central Bank, the European Commission, the European Supervisory Authorities (ESAs), and the chairman and two vice chairmen from the Advisory Scientific Committee (which comprises external experts) with voting rights. Representatives from the National Supervisory Authorities and the chairman of the Economic and Financial Committee attend without voting rights. The President of the European Central Bank is the chairman of the ESRB.

The ESRB is to prevent and reduce systemic risks in the EU. The ESRB is tasked with identifying risks and—if necessary—make recommendations and warnings that may reduce these risks. Recommendations and warnings can be directed towards the entire EU, individual or groups of countries' governments or supervisory authorities, the Commission or the European supervisory authorities. When the ESRB makes a recommendation, the recipient can choose to follow it or not. As such, it is not a binding recommendation for the member countries. However, if a recipient chooses not to follow a recommendation from the ESRB, said recipient must explain why. The recipient of the recommendation shall inform the ESRB and ECOFIN about the status. There is not a similar obligation to comply-or-explain to warning from the ESRB.

In other terms, the European Systemic Risk Board (ESRB) has to provide the macro-prudential oversight of the European Union's financial system and to contribute to preventing or mitigating systemic risks in the EU, while the European System of Financial Supervision has to guarantee surveillance at a microeconomic level, thanks of an active collaboration between the national supervisory authorities and the three new European Supervisory Authorities, represented by the European

---

[4] Considering the elements regarding organization and governance, the ESRB, based in Frankfurt am Main (Germany), has a General Board, chaired by the President of the ECB, that ensures the performance of tasks by taking the necessary decisions, a Steering Committee, that assists in the decision- making processes, an Advisory Scientific Committee and an Advisory Technical Committee, beyond a Secretariat. The voting members of the General Board include the President and Vice-President of the European Central Bank, heads of the central banks of the member states, a representative of the European Commission, the Chairs respectively of the three European Supervisory Authority (or ESAs) represented by EBA, ESMA and EIOPA, the Chair of the Advisory Scientific Committee and the Chair of the Advisory Technical Committee. Non-voting member is the Chair of the Economic and Financial Committee (EFC) (Sciascia 2021; Cafaro 2021).

Banking Authority (EBA), the European Insurance Authority (EIA) and the European Securities Authority (ESA). The new European surveillance system establishes also mutual interconnections between the ESRB and ESFS: the ESFS must provide the ESRB with information on micro-prudential developments, while the ESRB must provide warnings on imminent systemic risks and, when necessary, it will recommend measures in response to the identified risks, but having no binding powers (Quirici 2010).

ESRB also coordinates itself with international financial organizations (such as the International Monetary Fund and the Financial Stability Board).

As underlined in ESRB Annual Report 2022 (ESRB 2023a), in September 2022 the European Systemic Risk Board issued, for the first time, a general warning on vulnerabilities in the EU financial system, calling for heightened awareness of the risks to financial stability and the need of greater resilience in the EU financial sector (ESRB 2022). The economic impact of Russia's war against Ukraine had increased the likelihood of a tail-risk scenario. In the months after the general warming, near-term tail risk to the economic outlook receded to some extent, but banking sector vulnerabilities intensified in March 2023, following the collapse of two mid-sized banks in the USA and the takeover of Credit Suisse by USB. The ESRB continued its regular monitoring activities and contributed to the stress-testing exercise of the ESAs.

At its meeting on 30 November 2023, the General Board of the European Systemic Risk Board concluded that financial stability risks in the EU remained elevated. The ESRB released also the 46th issue of its risk dashboard, which represents a set of quantitative and qualitative indicators measuring systemic risk in the EU financial system (ESRB 2023b).[5] Finally, the ESRB concluded its exploratory work on measuring and modelling the systemic dimension of climate risks and on possible macro-prudential policy options.

Main Elements of Strength and Weakness of the ESRB

The main strengths of the ESRC and ESFS can be considered the following:

(a) Enhanced Coordination: The ESRC brings together representatives from different regulatory and supervisory authorities across the EU, facilitating coordinated actions and information sharing. This collaborative approach improves the understanding and assessment of systemic risks that might affect the entire financial system.

---

[5] The ESRB risk dashboard is published quarterly, one week after its adoption by the General Board, and it is accompanied by two annexes that explain the methodology and describe the indicators (both quantitative and qualitative) measuring systemic risk in the EU financial system. Additional indicators that support the systemic risk assessment by ESRB are available in the macro-prudential database that is present in the ECB (ESRB 2023b).

(b) Holistic Risk Assessment: The ESRC is focused on macro-prudential supervision, which means it assesses risks and vulnerabilities across the entire financial system rather than just individual institutions. This broader perspective helps in identifying and addressing risks that might arise from interconnectedness and network effects.

(c) Early Warning System: The ESRC's primary goal is to provide early warnings about potential systemic risks. Monitoring various indicators and market developments, it can signal potential threats to financial stability, allowing regulators and policymakers to take pre-emptive measures.

(d) Recommendations and Policy Tools: The ESRC can make recommendations to national authorities and the broader EU regarding policies and measures to mitigate systemic risks. These recommendations might include adjusting capital requirements, imposing lending restrictions, or implementing other macro-prudential tools.

But there are also some elements of weakness in the ESRB that it is necessary to point out, including:

(a) Limited Enforcement Power: the ESRB can make recommendations, but it doesn't have direct enforcement power. It relies on national authorities to implement its suggestions, which might lead to variations in how different countries respond to the same systemic risks. The ESRB has to monitor the measures taken in response to its warnings and recommendations, considering that addressees of the ESRB recommendations have to provide an explanation for any inaction.

(b) Interplay with National Interests: Balancing the interests of individual member states with the broader EU objectives can be challenging. National regulators might prioritize local concerns over pan-European systemic risks.

(c) Data Availability and Quality: Effective risk assessment requires accurate and timely data. Ensuring consistent and high-quality data across all member states can be difficult, impacting the accuracy of risk assessments.

(d) Changing Landscape: Financial markets and instruments are constantly evolving. Adapting the regulatory framework to address new and innovative financial products and practices is an ongoing challenge.

(e) Communication and Decision-Making: Coordinating among different regulatory bodies and national authorities can sometimes lead to delays in decision-making and responses to emerging risks.

The European Systemic Risk Board, along with the European System of Financial Supervision, represents a significant effort by the European Union to address systemic risks in the financial system. Its strengths lie in its coordinated approach to risk assessment, early warning capabilities, and recommendations for policy measures. However, challenges related to enforcement, national interests, data quality, and the dynamic nature of financial markets need to be managed for the framework to be truly effective in maintaining financial stability across the EU. So further improvements might be realized by the European Commission to tackle in a better way systemic risks in the EU financial context, considering that the European Systemic Risk Board represents only a "voice" that has no direct enforcement powers regarding this risk.

Some Implications of Systemic Risk and Complex Network in the New Sustainable Financial Context

Environmental, Social, and Governance (ESG) criteria and their relative consider-ations are becoming increasingly important in the financial sector as they address broader sustainability and ethical concerns. These considerations are closely linked to the Sustainable Development Goals (SDGs) established by the United Nations.

The main implications of systemic risk and complex networks in the context of ESG and the SDGs are tentatively the following:

a. Environmental Implications: Complex financial networks can have environ-mental implications, particularly related to sectors with significant environmental impacts such as energy, natural resources, and waste management. Systemic risk in these sectors can lead to widespread disruptions that impact not only financial institutions but also companies involved in environmentally sensitive activities. From an ESG perspective, financial institutions that invest in or lend to envi-ronmentally unsustainable businesses could face reputation risks and financial losses if these businesses fail due to systemic risks.

b. Social Implications: Systemic risk can have significant social implications, espe-cially when it affects sectors that have a direct impact on people's livelihoods, jobs, and access to basic needs. For instance, disruptions in the housing market or labour-intensive industries due to systemic risks can lead to job losses and economic instability. From an ESG perspective, financial institutions have a responsibility to consider the social impact of their investments and lending activities. Investments in companies that prioritize fair labour practices, diver-sity and inclusion, and community development align with social goals and the SDGs.

c. Governance Implications: Complex networks and systemic risks can also high-light governance-related issues within financial institutions. Weak governance structures, inadequate risk management practices, and lack of transparency can exacerbate the potential for systemic risk. Financial institutions with strong governance frameworks are better equipped to identify, assess, and mitigate systemic risks effectively. Ensuring ethical conduct, transparency, and account-ability within financial institutions contributes to long-term sustainability and supports SDGs related to good governance.

Considering the alignment with Sustainable Development Goals (SDGs), it is possible to notice that several SDGs can be directly or indirectly impacted by systemic risk and complex financial networks can play a significant role in their pursuit. It is possible also to do some examples of this:

• SDG 8 (Decent Work and Economic Growth): Systemic risks affecting employ-ment, job stability, and economic growth can impact progress toward this goal.
• SDG 9 (Industry, Innovation, and Infrastructure): Financial disruptions can hinder investments in critical infrastructure and innovation initiatives.

- SDG 10 (Reduced Inequality): Systemic risks that disproportionately affect vulnerable populations can hinder efforts to reduce inequalities.
- SDG 12 (Responsible Consumption and Production): Addressing systemic risk can encourage responsible consumption, production, and sustainable supply chain practices.
- SDG 17 (Partnerships for the Goals): Coordinated efforts among financial institutions, regulators, and governments are vital for managing systemic risk and achieving the SDGs.

Understanding systemic risk and complex networks through an ESG and SDG lens emphasizes the need for responsible and sustainable financial practices. Financial institutions that consider environmental, social, and governance factors in their decision-making can contribute to the broader goals of sustainability, inclusivity, and ethical conduct while also mitigating the potential negative impacts of systemic risk on society and the environment.

An example of this is given by the General Board of the European Systemic Risk Board (ESRB) that, at its 52nd regular meeting held on 30 November 2023, discussed the systemic dimension of climate risk, jointly with the ECB, considering three aspects:

(1) The General Board took stock of ways in which climate risk could trigger systemic risk, noting that EU banks should properly assess climate-related financial risks of lending portfolios that are clearly titled towards higher emitting parts of the economy;
(2) The board discussed the macro-prudential policy toolbox and the possibility of applying instruments provided for by current legislation to address risks to banks, borrowers, and no-bank financial intermediaries;
(3) The board had an initial exchange of views about the risk of natural degradation which could exacerbate the effects of climate change, with repercussions for financial stability too (ESRB 2023c).

Conclusions

This chapter has mainly shown the role of complex networks in the financial context and how these networks can give rise to rapid diffusion of systemic risk through different possible mechanisms of transmission, in the consciousness that understanding these mechanisms is crucial both for risk management and for designing effective measures necessary to mitigate the diffusion of systemic risk.

After the Great Financial Crisis (2007–2008) financial supervisors all over the world understood that the systemic risk had been underestimated and that there was a need to better monitor and tackle it, considering relative ways and channels of contagion.

This is in presence of a modern financial system characterized by multiple layers or levels of interconnectedness. Several key aspects contribute to the complexity of financial networks, such as Interconnectedness (that can create channels for the transmission of shocks and vulnerabilities across the system), Interdependencies (when financial institutions and markets rely on each other for various services, such as payment processing, liquidity provision, and risk hedging. So, if one institution fails or experiences severe stress, it can impact other entities that depend on its services, leading to a cascading effect of failures or disruptions) or Feedback loops (that can amplify shocks and disturbances within the financial system, creating a cycle of contagion and instability).

This multilayer structure (Multilayer Financial Networks) can amplify the spread of risks and, in particular, can amplify the diffusion, or contagion, of the systemic risk. In fact, if this risk can lead to the collapse or severe disruption of an entire financial system or economy, it often arises due to the interconnectedness and interdependencies within the financial system. One small shock in one part of the system can potentially propagate and impact other parts, leading to a chain reaction of failures.

In the presence of complex financial networks, the Contagion, or the widespread transmission of financial distress or negative events from one institution or sector to others, is more rapid. We saw that it can occur through direct exposures (such as loans between banks) or indirect exposures (via common counterparties or interconnected markets). Several studies in literature underline the need for a new concept of "too interconnected to fail" (or TITF), besides the concept of "too big to fail", concerning those financial institutions particularly relevant in the financial network context.

Consequently, a better understanding of these forms of transmission is necessary for policy-makers and financial supervisors to identify mitigation strategies, to reduce the impact of systemic risks and prevent or limit contagion. This involves implementing measures and policies that enhance the resilience of financial institutions and the overall system.

In the last decade significant progress has been made in studying the growing interconnectedness of the global financial system and how shocks are amplified or mitigated depending on the network structure and the heterogeneity of the various financial agents. However, the present analysis shows that many issues need also to be solved.

In literature are presented more and more studies regarding these concepts, but despite this and notwithstanding their growing importance, we have not yet reached a univocal definition of systemic risk nor a shared methodology for measuring this risk. Also, in relation to the concept of contagion, there is no univocal definition and studies analysing the determinants of systemic risk contagion are relatively few (Atasoy et al. 2024). For these reasons it is difficult to find a shared methodology to determine what are the "Systemically Important Financial Institutions" (SIFIs), which should be controlled in a more significant way because able to trigger a systemic crisis (Riccetti 2020, 2022a, b; Brogi et al. 2021; Avdjiev et al. 2019).

Also considering the financial regulators' methods to capture factors that can propagate systemic risk, it is possible to point out that a more holistic regulatory approach, incorporating a range of risk factors simultaneously and utilizing high-frequency data, would be more suitable than the actual ones.

The same analysis of the European Systemic Risk Board, representing the European answer to the need to better understand and tackle the systemic risk in our financial system, shows the presence of elements of improvement, to the extent that it could be transformed from the actual role of "voice"—that can issue general warming—into a European Authority capable of giving effective indications on how to quickly address certain situations of potential systemic risk.

In this context, the presence of ESG risks, such as the risks deriving from climate change, can only represent a further element capable of worsening the existing situation, making it even more necessary for academic authors, financial supervisors, and policy-makers to deepen their studies on the causes, remedies and methods of measuring of the systemic risk in the presence of Multilayer Financial Networks.

References

Acharya, V., Steffen, S.: The "greatest" carry trade ever? Understanding Eurozone bank risk. J. Financ. Econ. **115**(2), 215–236 (2015)

Acharya, V., Engle, R., Richardson, M.: Capital shortfall: a new approach to ranking and regulating systemic risks. Am. Econ. Rev. **102**(3), 59–64 (2012)

Atasoy, B.S., Özkan, I., Erden, L.: The determinants of systemic risk contagion. Econ. Model. **130**, 106596 (2024)

Avdjiev, S., Giudici, P., Spelta, A.: Measuring contagion risk in international banking. BIS Working Papers N° 796, Monetary and Economic Department of the Bank for International Settlements, at https://www.bis.org/ (2019)

Bougheas, S., Kirman, A.: Bank insolvencies, priority claims and systemic risk. In: Commendatore, P., Matilla-Garcìa, M., Varela, L.M., Cànovas, J.L. (eds.) Complex Networks and Dynamics, Lecture Notes in Economics and Mathematical Systems, pp. 195–208. Springer International Publishing, Switzerland (2016). https://doi.org/10.1007/978-3-319-40803-3_8

Brogi, M., Lagasio, V., Riccetti, L.: Systemic risk measurement: bucketing global systemically important banks. Ann. Finance **17**, 319–351 (2021)

Brunetti, C., Harris, J.H., Mankad, S., Michailidis, G.: Interconnectedness in the interbank market. J. Financ. Econ. **133**(2), 520–538 (2019)

Cafaro, S.: Article 134 [Economic and Financial Committee]. In: Springer Commentaries on International and European Law. Springer, Cham (2021)

Capponi, A., Chen, P.C.: Systemic risk mitigation in financial networks. J. Econ. Dyn. Control **58**, 152–166 (2015)

Cont, R., Moussa, A., Santos, E.B.: Network structure and systemic risk in banking system. In: Fouque, J.P., Langsam, J. (eds.) Handbook of Systemic Risk, pp. 327–368. Cambridge University Press (2013)

Covi, G., Gorpe, M.Z., Kok, C.: CoMap: mapping contagion in the euro area banking sector. J. Financ. Stab. **53**, 100814 (2021)

D'Arcangelis, A.M., Rotundo, G.: Complex networks in finance. In: Commendatore, P., Matilla-Garcìa, M., Varela, L.M., Cànovas, J.L. (eds.) Complex Networks and Dynamics. Lecture Notes in Economics and Mathematical Systems, pp. 209–235. Springer International Publishing, Switzerland (2016)

De Larosiére Group: Report (2009)

Di Clemente, A.: Rischio sistemico e intermediari bancari. In: Di Clemente, A. (a cura di), Stabilità finanziaria e rischio sistemico. Problemi di stima e di regolazione, ARACNE, Roma, 45–62 (Novembre) (2016)

ECB (European Central Bank): Systemic risk methodologies. Financ. Stabi. Rev. (2011)

Ellis, S., Sharma, S., Brzeszczynski, J.: Systemic risk measures and regulatory challenges. J. Financ. Stab. **61**, 100960 (2022)

ESRB (European Systemic Risk Board): Warning on vulnerabilities in the financial system of the European Union (EU), 22 September 2022. https://www.ecb.europa.eu/ (2022)

ESRB (European Systemic Risk Board): Annual Report 2022, 30 March 2023. https://www.esrb. europa.eu/ (2023a)

ESRB (European Systemic Risk Board): ESRB Risk Dashboard (with Annex I and Annex II), November 2023 (Issue 46). https://www.esrb.europa.eu/ (2023b)

ESRB (European Systemic Risk Board): The General Board of the European Systemic Risk Board held its 52nd regular meeting on 30 November 2023, PRESS RELEASE, 7 December 2023. https://www.esrb.europa.eu/ (2023c)

Foglia, M., Di Tommaso, C., Wang, G., Pacelli, V.: Interconnectedness between stock and credit markets: the role of European G-SIBs in a multilayer perspective. J. Int. Financ. Markets Inst. Money **91** (2024)

Gai, P., Haldane, A., Kapadia, S.: Complexity, concentration and contagion. J. Monet. Econ. **58**, 453–470 (2011)

Gai, P., Kapadia, S.: Contagion in financial networks, Bank of England Working Papers n. 393, Bank of England. https://bankofengland.co.uk/ (2010)

Galizia, F.: Should SIFIs protect themselves from systemic risk? J. Risk Manage. Financ. Inst. **8**(1), 27–33 (2015)

Hollo, D., Kremer, M., Lo Duca, M.: CISS: a composite indicator of systemic stress in the financial system. Working Paper Series, N. 1426, ECB (2012)

Hüser, A.C.: Too interconnected to fail: a survey of the interbank networks literature. J. Netw. Theory Finance **1**(3), 1–50 (2015)

Markose, S., Giansante, S., Shaghaghi, A.R.: Too interconnected to fail" financial network of US CDS market: topological fragility and systemic risk. J. Econ. Behav. Organ. **83**(3), 627–646 (2012)

Paltalidis, N., Gounopoulos, D., Kizys, R., Koutelidakis Y.: Transmission channels of systemic risk and contagion in the European financial network. J. Banking Finance **61**(1), S36–S52 (2015)

Quirici, M.C.: Dalla crisi finanziaria internazionale a nuove forme di vigilanza integrata sulla base del Rapporto de Larosiére. In: Quirici, M.C. (ed.) Il mercato mobiliare. L'evoluzione strutturale e normativa, pp. 376–395. FrancoAngeli, Milano (2010)

Riccetti, L.: Agent-based multi-layer network simulations for financial systemic risk measurements: a proposal for future developments. Int. J. Microsimul. **15**(2), 44–61 (2022a)

Riccetti, L.: Gestire il rischio sistemico con la teoria delle reti. In: Busilacchi, G., Cedrola, E. (a cura di). La forza delle reti, pp. 85–103. Gioacchino Onorati Ed., Roma (2020)

Riccetti, L.: Systemic risk analysis and SIFI detection: mechanisms and measurement. J. Risk Manage. Financ. Inst. **15**(3), 245–259 (2022b)

Sciascia, G.: La regolazione giuridica del rischio sistemico. Stabilità finanziaria e politiche macroprudenziali, Giuffré, Torino (2021)

Chapter 6
The Impact of Inflation and Financial Stability on the European Financial System: A Network Approach

Javier Sánchez-García and Salvador Cruz-Rambaud

Abstract Inflation and financial stability are pivotal elements in the fields of economics and finance, exerting a profound influence on economic performance and the overall stability of financial systems. The intricate interplay between these factors has garnered significant attention from researchers, policymakers, and market participants due to its far-reaching implications, especially since recent inflationary shocks have put many economies around the world under pressure. This chapter builds an econometric design to estimate a network of volatility connectedness, and an Exponential Random Graph Model (ERGM) is proposed to analyse the structure, capturing both endogenous and exogenous effects on the network. The results show no significant relationship between inflation and financial stress for this set of European countries, shedding light on potential macro-financial vulnerabilities and systemic risks within the European financial system.

Keywords Inflation · Systemic risk · Financial contagion · Financial crisis · Macro-financial links

Introduction and Overview

Inflation and financial stability are key elements in the fields of economics and finance, exerting a profound influence on economic performance and the overall stability of financial systems. The intricate interplay between these factors has garnered significant attention from researchers, policymakers, and market participants due to its far-reaching implications, especially since recent inflationary shocks have put many economies around the globe under pressure.

J. Sánchez-García (✉) · S. Cruz-Rambaud
Mediterranean Research Center on Economics and Sustainable Development, CIMEDES, 04120 Almería, Spain
e-mail: scruz@ual.es

Department of Economics and Business, University of Almería, 04120 Almería, Spain
e-mail: jsg608@ual.es

© The Author(s) 2025   113
V. Pacelli (ed.), *Systemic Risk and Complex Networks in Modern Financial Systems*, New Economic Windows, https://doi.org/10.1007/978-3-031-64916-5_6

Building on the seminal work of Diebold and Yilmaz (2012, 2014, 2015) on network connectedness, this study builds an econometric design to estimate a network of volatility connectedness. Afterward, an Exponential Random Graph Model (ERGM) (Cranmer and Desmarais 2011; Hunter et al. 2008; Butts et al. 2014; Ghafouri and Khasteh 2020) is proposed to analyse the structure, capturing both endogenous and exogenous effects on the network. This methodological framework enriches the analysis, enabling a comprehensive exploration of the multifaceted interactions between inflation and financial stress.

The chapter is based on a comprehensive analysis of the interconnections between inflation and financial stress, using data from 11 European countries. These countries, characterized by significant commercial and financial ties facilitated by the free movement of goods, capital, and labour, are especially suitable for analysing network phenomena. The Composite Indicator of Systemic Stress (CISS) provided by the European Central Bank (ECB) is employed to gauge financial stress, while the Headline Consumer Price Index (CPI) from the World Bank serves as a key metric for measuring inflation. The 10 year yield of the public bond of each country is used to capture interest rates, which are considered to have a crucial moderator effect in the bidirectional relationship. Therefore, this dataset enables a thorough examination of the intricate relationship between inflation and financial stress, shedding light on potential vulnerabilities and systemic risks within the financial system.

The results show no significant relationship between inflation and financial stress (and vice versa) for this set of European countries. A hierarchical approach sheds light on the robustness of the coefficients, which confirms the moderator potential of interest rates.

This approach allows for a nuanced understanding of the intricate web of relationships between inflation and financial stress, providing valuable insights for policymakers and market participants. In particular, the methodology applied in this research chapter is poised to offer valuable implications for policymakers, financial regulators, and market participants. By unravelling the intricate relationship between inflation and financial stress interconnectedness, the chapter provides a nuanced understanding of the potential risks and vulnerabilities within the European financial system.

However, significant limitations are to be considered for the interpretation of this chapter. First, only cross-sectional data is used in the econometric analysis, as the temporal evolution of the random variables enters the model through their intertemporal mean. This is an important limitation, since real-world financial networks are dynamic. Second, only 11 elements are used in the network model. This is a small sample from which to derive inferences. Therefore, we consider our contribution to be mainly methodological, since we expose an empirical approach to derive inferences about the effect of a factor on the probability of formation of new links in financial networks.

The remainder of the chapter is structured as follows. Section "Econometric Design" presents the econometric design. Section "Results" shows the results. Eventually, Sect. "Conclusions" concludes.

Econometric Design

Section "Econometric Design" exposes the econometric framework employed in this study. In particular, section "Estimation of the Network" shows the methodology employed for the estimation of the network of volatility connectedness. Section "Empirical Multipliers" presents the methodology used to estimate the empirical multipliers of connectedness. Finally, section "Data" shows the data employed.

Estimation of the Network

In order to calculate the volatility networks, we begin by analysing the variance decompositions of VAR models as outlined by Diebold and Yilmaz (2012, 2014, 2015). Take into account the subsequent VAR model:

$$y_t = A_1 y_{t-1} + A_2 y_{t-2} + \cdots + A_p y_{t-p} + \epsilon_t, \tag{6.1}$$

In this context, we have a vector y_t consisting of k endogenous variables, represented as $(y_{1t}, y_{2t}, \ldots, y_{kt})$. The matrices A_i, where i ranges from 1 to p, are $k \times k$ coefficient matrices. The vector ϵ_t represents the residuals, which are assumed to be white noise. Specifically, we have $\epsilon_t \sim (0, \Sigma_\epsilon)$, where Σ_ϵ is the covariance matrix of ϵ_t given by $E(\epsilon_t \epsilon_t')$.

The Forecasted Error Variance Decompositions (FEVD) are derived from the moving average representation of the model.

$$y_t = \sum_{i=0}^{\infty} B_i \epsilon_{t-i},$$

The $k \times k$ parameter matrices B_i are obtained through recursion. Consequently, the forecast for the h step is given by:

$$y_{T+h} - y_{T+h|T} = \Phi_0 \epsilon_{T+h} + \Phi_1 \epsilon_{T+h-1} + \cdots + \Phi_{h-1} \epsilon_{T+1} \tag{6.2}$$

and when we set $\Sigma = I_K$, the Forecast Error Variance Decomposition (FEVD) of the kth element of y_{T+h} is defined as

$$\sigma_k^2 = \sum_{j=0}^{K} (\phi_{kj,0}^2 + \cdots + \phi_{kj,h-1}^2), \tag{6.3}$$

where ϕ_{nm} represents the nm element of Φ_i (Lütkepohl 2013).

Using a Cholesky decomposition, it is possible to obtain a matrix $\beta = \beta_{ij}$ $(i, j = 1, \ldots, k)$ of size $k \times k$, where each element represents the influence of variable j on variable i. The elements of β can be normalized by dividing them by the sum of the remaining elements in their respective row.

$$\bar{\beta}_{ij} = \frac{\beta_{ij}}{\sum_{j=1}^{k} \beta_{ij}}, \tag{6.4}$$

The elements $\bar{\beta}_{ij}$ are normalized. Nevertheless, the Cholesky decomposition considers the ordering of the variables important. To address this issue, the generalized VAR proposed by Koop et al. (1996), Pesaran and Shin (1998) can be utilized (Diebold and Yilmaz 2012).

Once the estimation of the FEVD is completed, the Directional Spillover Index can be computed using the following metric:

$$DSI_{i \leftarrow j} = \frac{\sum_{\substack{j=1 \\ i \neq n}}^{k} \bar{\beta}_{ij}}{n} \times 100, \tag{6.5}$$

The calculation can be performed for pairs when $j = 1$, or for all the variables $j = 1, \ldots, k$ with $i \neq j$. In this case, n represents the decomposition horizon.

Additionally, let us consider a weighted adjacency matrix A, which can be represented as:

$$A = \begin{bmatrix} \alpha_{1,1} & \alpha_{1,2} & \cdots & \alpha_{1,m} \\ \alpha_{2,1} & \alpha_{2,2} & \cdots & \alpha_{2,m} \\ \vdots & \vdots & \ddots & \vdots \\ \alpha_{m,1} & \alpha_{m,2} & \cdots & \alpha_{n,m} \end{bmatrix}, \tag{6.6}$$

Each element $\alpha_{n,m}$ in the matrix A represents a weighted edge from element n to element m in a network. The matrix A forms a directed volatility network by incorporating the volatility interconnections (DSI). Each $DSI_{i \leftarrow j}$ estimate corresponds to the volatility edge (n, m) in matrix A. For instance, $DSI_{1 \leftarrow 2} = \alpha_{2,1}$, indicating the directed volatility edge from component 2 to component 1 in the network. Since the volatility network is directed, $DSI_{i \leftarrow j} \neq DSI_{j \leftarrow i}$.

Empirical Multipliers

Exponential Random Graph Models (ERGMs) belong to the exponential family, which includes probabilistic models used to analyse networked data (Ghafouri and Khasteh 2020). ERGMs offer the advantage of being able to capture both endogenous (structural) and exogenous (covariate) effects on the network, without making any assumptions about the independence of the data (Cranmer and Desmarais 2011; Hunter et al. 2008; Butts et al. 2014; Ghafouri and Khasteh 2020). By estimating networks as described in section "Estimation of the Network", it becomes possible to infer the factors responsible for volatility transmissions.

ERGM models aim to calculate the likelihood of the network generated by a model (Y) being equal to an observed network (y), taking into account a set of random variables and parameters. This can be expressed mathematically as:

$$\log P(Y = y) \propto \theta \cdot v(G) \tag{6.7}$$

or, equivalently,

$$P(Y = y) \propto \exp\{\theta \cdot v(G)\}, \tag{6.8}$$

where θ represents a vector containing the relevant parameters, G denotes a network, and $v(G)$ represents a vector containing the variables associated with that network.

In order to obtain a probability distribution, it is important to normalize all potential networks that have the same number of nodes. Consequently,

$$P(Y = y) = \frac{\exp\{\theta \cdot v(G)\}}{\sum_{G \in \mathcal{G}} \exp\{\theta \cdot v(G)\}}, \tag{6.9}$$

G belongs to the set $\mathcal{G}$, which represents all possible networks with the same number of nodes.

The overall structure of an Exponential Random Graph Model (ERGM) is

$$X = \frac{1}{\psi} \exp\{\theta \cdot v(G)\}, \tag{6.10}$$

The equation $\psi = \sum_{G \in \mathcal{G}} \exp\{\theta \cdot v(G)\}$ represents the normalization constant, where $\mathcal{G}$ is the set of all possible graphs, θ is a parameter vector, and $v(G)$ is a function that maps a graph G to a real number. The variable X represents the probability $P(Y = y)$. Equation (6.10) is the canonical representation of the Exponential Random Graph Model (ERGM).

In order to obtain the model parameters, consider a network G_{ij} with nodes i and j. If there is a link between i and j, then G_{ij} is equal to 1. Otherwise, G_{ij} is equal to 0. Using this notation, we can represent the likelihood of a link between nodes i and j as:

$$\text{odds}(G_{ij} = 1) = \frac{P(G_{ij} = 1)}{P(G_{ij} = 0)} \tag{6.11}$$

and the Logit, which guarantees a probability space ranging from 0 to 1, is:

$$\text{Logit}[G_{ij} = 1] = \log[\text{odds}(G_{ij} = 1)] = \log \frac{P(G_{ij} = 1)}{P(G_{ij} = 0)}. \tag{6.12}$$

Assume that the logarithm of the probability of the connection between nodes i and j is influenced by a collection of n explanatory variables and associated parameters, as follows:

$$\log P(G_{ij} = 1) = \sum_{k=1}^{n} \theta_n \qquad (6.13)$$

Given that $\theta := (\theta_1, \theta_2, \ldots, \theta_n)$ represents the relevant parameters and $X := (X_1, X_2, \ldots, X_n)$ represents the explanatory variables, the following relationship holds:

$$P(G_{ij} = 1) = \exp\left\{\sum_{k=1}^{n} \theta_k X_k\right\} \qquad (6.14)$$

The Eq. (6.12) represents a Logit model that assumes the data to be independently and identically distributed (i.i.d.). However, in the case of network data, there is a dependency among the observations. Therefore, the Logit equations need to consider this dependency. To account for this, we introduce G_{ij}^o, which represents a network without the connection between nodes i and j. Consequently, we have the following expression:

$$\text{odds}(G_{ij} = 1|G_{ij}^o)) = \frac{P(G_{ij} = 1|G_{ij}^o)}{P(G_{ij} = 0|G_{ij}^o)}, \qquad (6.15)$$

In other words, a new expression is proposed for the probabilities of an edge in the network. These probabilities are now conditioned on the structure of the network without the link ij, denoted as G_{ij}^o.

Since some characteristics of the model can be subgraphs and are included in the model through counts, the counts of these features differ when the link ij is present or absent (Van der Pol 2019). Let $v(G_{ij}^+)$ represent the vector of features when the link ij is present, and $v(G_{ij}^-)$ represent the vector of features when the link is not present. Using Eqs. (6.14) and (6.15), where θ' denotes the transpose of θ and θ_k' denotes the transpose of θ_k for $k = 1, 2, \ldots, n$, we can express this relationship.

$$
\begin{aligned}
\text{odds}(G_{ij} = 1|G_{ij}^o) &= \frac{P(G_{ij} = 1|G_{ij}^o)}{P(G_{ij} = 0|G_{ij}^o)} \\
&= \frac{\exp\{\theta' \cdot v(G_{ij}^+)\}}{\exp\{\theta' \cdot v(G_{ij}^-)\}} \\
&= \exp\{\theta'(v(G_{ij}^+) - v(G_{ij}^-))\} \\
&= \exp\{\theta_1'(v_1(G_{ij}^+) - v_1(G_{ij}^-))\} + \cdots \\
&+ \exp\{\theta_n'(v_n(G_{ij}^+) - v_n(G_{ij}^-))\},
\end{aligned}
\qquad (6.16)
$$

The essential components of the ERGM are the relevant parameters that quantify the extent of linear dependence between the differences in counts within the subgraphs of the model. It is important to note that $v_k(G_{ij}^+) - v_k(G_{ij}^-)$ ($k = 1, 2, \ldots, n$) represents the discrepancy in the measurement of statistic k for the network when an additional edge is added. In this manner, v_k represents the k-th change statistic.

The change statistic of a ij tie of the feature k is denoted as $v_k(\Delta_k G_{ij})$, which can be calculated as the difference between $v_k(G_{ij}^+)$ and $v_k(G_{ij}^-)$. To eliminate the exponential term in Eq. (6.16), natural logarithms are used.

$$
\begin{aligned}
\text{Logit}[G_{ij} = 1|G_{ij}^o] &= \log[\text{odds}(G_{ij} = 1|G_{ij}^o)] \\
&= \log \frac{\exp\{\theta' \cdot v(G_{ij}^+)\}}{\exp\{\theta' \cdot v(G_{ij}^-)\}} \\
&= \theta_1'(v_1(G_{ij}^+) - v_1(G_{ij}^-)) + \cdots + \theta_n'(v_n(G_{ij}^+) - v_n(G_{ij}^-)) \\
&= \theta_1'v_1(\Delta_1 G_{ij}) + \cdots + \theta_n'v_n(\Delta_n G_{ij}).
\end{aligned}
$$

Therefore, the model predicts the Logit of a link by considering the impact of a variation in the count of the network statistics, which is weighted by the parameters θ' (Sánchez-García and Cruz-Rambaud 2023a, b, c, d).

Data

The datasets utilised in this chapter are openly accessible on the websites of the ECB, World Bank, OCDE, FRED, and BIS. To measure financial stress, the Composite Indicator of Systemic Stress (CISS) provided by the ECB has been utilised (for more information, refer to Hollo et al. 2012). Instead of focusing on individual countries, this research concentrates on the systemic risk of the entire financial network, specifically the interconnectedness of financial stress. This approach effectively approximates systemic risk, as an escalation in the interconnectedness of financial stress among individual countries corresponds to an increase in stress within the entire network.

The network consists of eleven European countries that are anticipated to have significant commercial and financial connections because of the free movement of goods, capital, and workers as outlined in the Treaty on the Functioning of the European Union. This shared behaviour is indeed confirmed by the uniform behaviour of the CISS for the different countries, as Fig. 6.1 shows.

This chapter uses the Headline Consumer Price Index (CPI) of the World Bank to measure inflation. The yields of 10-year government debt are used to measure interest rates. The intertemporal mean is calculated for all variables. The dataset includes twelve countries as shown in Table 6.1. To make the ERGM estimation feasible and focus on significant systemic linkages, only the edges beyond the median are considered.

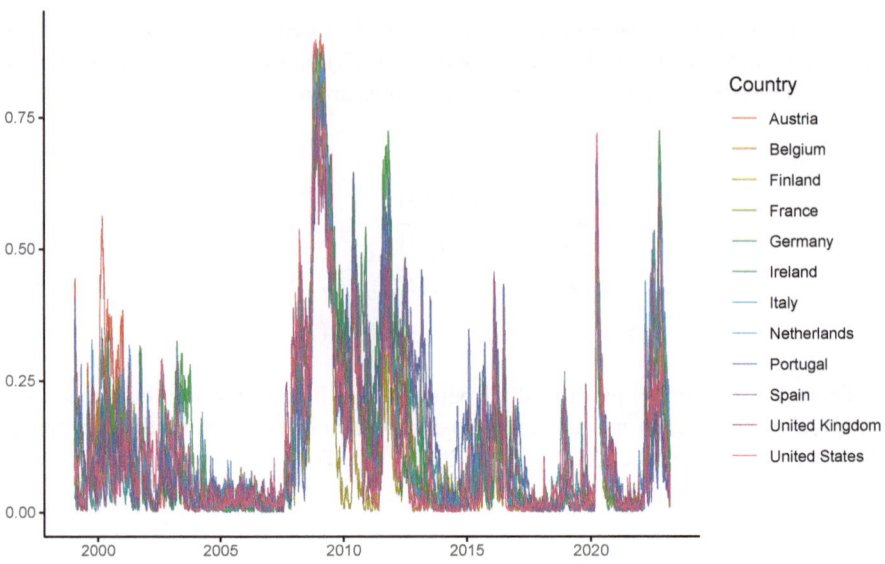

Fig. 6.1 Financial Stress Index by country. The recession probabilities are represented in the y axis and range from 0 (less probability) to 1 (most probability)

Table 6.1 Summary statistics of the inflation data

Statistic	N	Mean	St. Dev.	Min	Max
Austria	41	2.497	1.467	0.506	6.803
Belgium	41	2.694	2.062	−0.053	8.727
Germany	41	2.057	1.507	−0.129	6.344
Spain	41	4.544	4.117	−0.500	15.562
Finland	41	3.027	3.039	−0.208	11.595
France	41	2.971	3.380	0.038	13.563
United Kingdom	41	3.635	3.291	0.402	17.966
Ireland	41	3.738	4.926	−4.478	20.374
Italy	41	4.586	4.984	−0.138	21.064
Netherlands	41	2.238	1.524	−0.691	6.739
Portugal	41	6.652	7.396	−0.836	28.385

Fig. 6.2 Dynamic connectedness of the financial stress of the countries. The calculations were made for a VAR(1) model, for windows of 526 observations and a generalized VAR

Results

Figure 6.2 shows the dynamic connectedness of the financial stress index for the countries of the sample. The results are for a VAR(1) model and a calculation of FEVD $h = 16$, which is used for all estimates of the interconnectedness of volatility. Volatility connectedness exhibits procyclical behaviour, since it increases in periods of recession (Great Recession (2007–2009), European Debt Crisis (2010–2012), Energy Prices Recession (2014–2016), COVID-19 (2020–2022)). This variation is of more of 20%. Interestingly enough, this connectedness, a crucial component of systemic risk, increases not only when the crises are of financial nature, such as the first two, but when they are generated due to economic or health-related circumstances, such as the latter two.

In second place, Fig. 6.3 shows the directional volatility spillovers that each country transmits and receives. The countries that exhibit high or low values do not coincide when talking about inflation and financial stress volatility spillovers. Take the case of Finland and the Netherlands. Regarding inflation, their volatility spillovers are of the lowest amount, while for the case of financial stress, these are of the highest magnitude. Similar situations happen with the UK and Ireland, the UK and Austria, Italy and Ireland, etc.

Eventually, Tables 6.2 and 6.3 show the point estimates for the empirical multipliers of volatility connectedness from financial stress to inflation and from inflation to financial stress, respectively. The estimates are for the regular variables, the variables in absolute differences, and for the effect of the variables on incoming and outgoing links. The hierarchical approach shreds light about the robustness of the coefficients, and interest rates are included due to their moderator potential.

Table 6.2 Empirical volatility connectedness multipliers from financial stress to inflation

	Model 1	Model 2	Model 3	Model 4	Model 5	Model 6	Model 7	Model 8	Model 9	Model 10
Edges	5.71*	−0.06	6.91**	5.57*	6.04**	8.13**	0.57	10.04**	9.96*	11.66**
	(2.40)	(0.48)	(2.15)	(2.29)	(2.31)	(2.93)	(0.61)	(3.60)	(4.06)	(4.39)
Mutual	1.23*	1.48*	0.97	2.57***	2.27*	1.19	1.31*	0.97	3.42***	3.07**
	(0.60)	(0.60)	(0.63)	(0.75)	(0.92)	(0.62)	(0.62)	(0.66)	(1.01)	(1.09)
nodecov.Stress	−25.25**		−27.44***			−26.49**		−26.29**		
	(9.42)		(8.17)			(9.53)		(9.76)		
absdiff.Stress		−20.37	−25.22*		−24.54		−20.43	−32.68*		−35.22
		(11.52)	(12.56)		(15.54)		(12.62)	(14.96)		(17.99)
nodeicov.Stress				24.13	20.23				29.98	26.40
				(15.93)	(17.43)				(17.48)	(20.04)
nodeocov.Stress				−79.02***	−73.89***				−95.12***	−88.45***
				(17.45)	(17.58)				(22.94)	(21.54)
nodecov.Interest rates						−0.35		−0.53		
						(0.25)		(0.35)		
absdiff.Interest rates							−0.80*	−0.06	−0.05	−0.05
							(0.38)	(0.53)	(0.59)	(0.59)
nodeicov.Interest rates									0.97	0.67
									(0.51)	(0.58)
nodeocov.Interest rates									−2.16***	−2.26***
									(0.64)	(0.64)
AIC	138.06	145.33	136.05	122.10	121.14	138.02	142.75	135.55	110.28	109.83
BIC	146.16	153.43	146.85	132.90	134.64	148.82	153.55	151.75	126.48	131.43
Log Likelihood	−66.03	−69.67	−64.02	−57.05	−55.57	−65.01	−67.38	−61.77	−49.14	−46.92

*** $p < 0.001$; ** $p < 0.01$; * $p < 0.05$

Table 6.3 Empirical volatility connectedness multipliers from inflation to financial stress

	Model 1	Model 2	Model 3	Model 4	Model 5	Model 6	Model 7	Model 8	Model 9	Model 10
Edges	2.12	−0.47	2.35	2.07	2.23	1.51	−0.55	1.79	1.32	1.69
	(1.12)	(0.43)	(1.22)	(1.16)	(1.25)	(2.75)	(0.46)	(2.66)	(2.67)	(2.62)
Mutual	1.56*	1.81**	1.57*	1.81*	1.74*	1.51*	1.88**	1.61*	1.75*	1.81**
	(0.64)	(0.59)	(0.64)	(0.73)	(0.70)	(0.62)	(0.64)	(0.71)	(0.72)	(0.70)
nodecov.Inflation	−0.42**		−0.48*			−0.52		−0.58		
	(0.14)		(0.19)			(0.35)		(0.37)		
absdiff.Inflation		−0.28	0.13		0.11		−0.41	−0.03		−0.01
		(0.16)	(0.23)		(0.23)		(0.31)	(0.37)		(0.37)
nodeicov.Inflation				−0.17	−0.23				−0.06	−0.01
				(0.21)	(0.24)				(0.61)	(0.63)
nodeocov.Inflation				−0.70***	−0.72**				−1.09	−1.10
				(0.20)	(0.24)				(0.57)	(0.57)
nodecov.Interest rates						0.22		0.20		
						(0.80)		(0.76)		
absdiff.Interest rates							0.33	0.44		
							(0.74)	(0.84)		
nodeicov.Interest rates									−0.34	−0.54
									(1.32)	(1.33)
nodeocov.Interest rates									0.92	0.89
									(1.25)	(1.19)
AIC	132.58	142.99	133.73	131.33	133.21	133.92	144.81	137.48	134.71	138.32
BIC	140.68	151.09	144.53	142.14	146.71	144.72	155.61	153.68	150.91	159.93
Log Likelihood	−63.29	−68.49	−62.87	−61.67	−61.61	−62.96	−68.40	−62.74	−61.35	−61.16

*** $p < 0.001$; ** $p < 0.01$; * $p < 0.05$

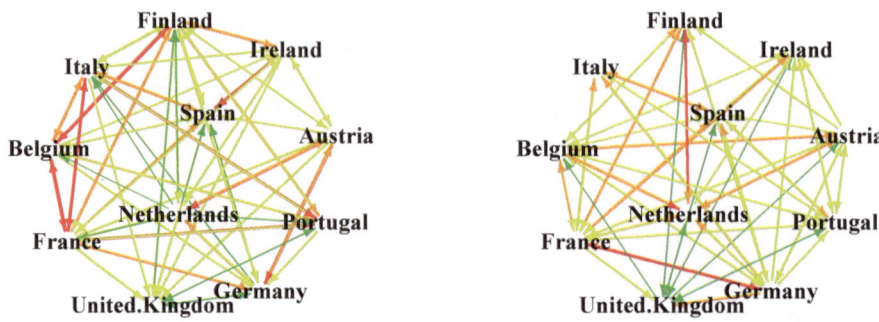

Fig. 6.3 Volatility spillovers between the countries of the sample. Left figure corresponds to inflation, right edges correspond to financial stress. Cutoff values are of = 0.5, 0.7, 0.9 and 1 for inflation and 0.5, 0.7, 0.8, 0.9 for financial stress

As Table 6.2 shows, there is an inverse relationship between the countries with the highest financial stress and the probability of increasing the interconnectedness of inflation volatility. The sign of the coefficient is robust to multiple model choices .However, the marginal effect of this relationship is practically inexistent, since $\exp -88.45 = 0.0000$. Therefore, no economically significant relationship was found. For the case of inflation, no evidence is found of that it increases financial stress interconnectedness. In effect, no coefficient is statistically significant after including interest rates.

Conclusions

In conclusion, this chapter provides a comprehensive analysis of the intricate relationship between inflation and financial stress, leveraging data from eleven European countries. The proposed econometric design estimates a network of volatility connectedness, and the Exponential Random Graph Model captures both endogenous and exogenous effects on the network.

The results show no significant relationship between inflation and financial stress for this set of European countries, shedding light on potential macro-financial vulnerabilities and systemic risks within the European financial system. However, the empirical application of the study has limitations, such as the use of only cross-sectional data or the small size of the sample of the econometric analysis.

Overall, this chapter contributes to the ongoing discussion on the interplay between inflation and financial stability, highlighting the need for further research in this area. Furthermore, the methodology applied in this study offers valuable implications for policymakers, financial regulators, and market participants, providing a nuanced understanding of the intricate web of relationships between real-economy and financial factors.

References

Butts, C.T., Morris, M., Krivitsky, P.N., Almquist, Z., Handcock, M.S., Hunter, D.R., Goodreau, S.M., de Moll, S.B.: Introduction to Exponential-Family Random Graph (erg or p*) Modeling with ERGM. European University Institute, Florence (2014)

Cranmer, S.J., Desmarais, B.A.: Inferential network analysis with exponential random graph models. Polit. Anal. **19**(1), 66–86 (2011)

Diebold, F.X., Yilmaz, K.: Better to give than to receive: predictive directional measurement of volatility spillovers. Int. J. Forecasting **28**(1), 57–66 (2012)

Diebold, F.X., Yılmaz, K.: On the network topology of variance decompositions: measuring the connectedness of financial firms. J. Econometrics **182**(1), 119–134 (2014)

Diebold, F.X., Yılmaz, K.: Financial and Macroeconomic Connectedness: A Network Approach to Measurement and Monitoring. Oxford University Press, USA (2015)

Ghafouri, S., Khasteh, S.H.: A survey on exponential random graph models: an application perspective. Peer. J. Comput. Sci. **6**, e269 (2020)

Hollo, D., Kremer, M., Lo Duca, M.: Ciss-a composite indicator of systemic stress in the financial system (2012)

Hunter, D.R., Handcock, M.S., Butts, C.T., Goodreau, S.M., Morris, M.: Ergm: a package to fit, simulate and diagnose exponential-family models for networks. J. Stat. Softw. **24**(3), 54860 (2008)

Koop, G., Pesaran, M.H., Potter, S.M.: Impulse response analysis in nonlinear multivariate models. J. Econometrics **74**(1), 119–147 (1996)

Lütkepohl, H.: Vector autoregressive models. In: Handbook of Research Methods and Applications in Empirical Macroeconomics, pp. 139–164. Edward Elgar Publishing (2013)

Pesaran, H.H., Shin, Y.: Generalized impulse response analysis in linear multivariate models. Econ. Lett. **58**(1), 17–29 (1998)

Sánchez-García, J., Cruz-Rambaud, S.: Estimation and inference in financial volatility networks. In: Data Analytics for Management, Banking and Finance: Theories and Application, pp. 95–111. Springer (2023a)

Sánchez-García, J., Cruz-Rambaud, S.: Inflation and systemic risk: a network econometric model. Finance Res. Lett. 104104 (2023b)

Sánchez-García, J., Cruz-Rambaud, S.: The network econometrics of financial concentration. Rev. Manager. Sci. 1–39 (2023c)

Sánchez-García, J., Cruz-Rambaud, S.: Macro financial determinants of volatility transmission in a network of European sovereign debt markets. Finance Res. Lett. **53**, 103635 (2023d)

Van der Pol, J.: Introduction to network modeling using Exponential Random Graph Models (ERGM): theory and an application using R-project. Comput. Econ. **54**(3), 845–875 (2019)

Chapter 7
Credit Risk Transfer and Systemic Risk

Francesco Moliterni

Abstract This chapter investigates the relationship between the banking and insurance industry by focusing on systemic risk. The concept of credit risk transfer stems from banks' inclination to offload credit risks. Insurance companies, particularly those specializing in risk transfer services, emerge as natural recipients for these risks. Notably, credit insurance firms are equipped with specialized expertise in risk assessment and selection. Banks seek to mitigate their exposure to credit risks by transferring them to insurance companies. This transfer occurs because insurance companies, particularly those offering risk transfer services, possess the necessary expertise to assess and manage these risks effectively.

Keywords Insurance sector · Banking sector · Credit default swap · Credit risk transfer

The Reasons Behind Research on the Systemic Relevance of Credit Risk Transfer from Banks to Insurance Companies

The principle of "same activities, same risks, same rules" is the explicit guideline of the European Union's legislative policy in financial matters (see whereas 9 of the Mi.Ca Regulation). The "European legislator is concentrating its efforts in this direction" (Siani 2022, p. 3, www.bancaditalia.it).

Moreover, this "regulatory approach" is fundamental for the certainty and efficiency of rules protecting the financial system from systemic risks. "It is therefore important to ensure that (…) the intermediaries currently under supervision follow the same rules, in accordance with the usual approach 'same activity, same risk, same rules'" (Siani 2022, p. 3) and in accordance with the principle of reasonableness.

In particular, this argument serves as a strategic premise for legal reasoning (U. Breccia) concerning the transfer of credit risks, specifically credit risk transfer from

F. Moliterni (✉)
University of Bari Aldo Moro, Bari, Italy
e-mail: francesco.moliterni@uniba.it

V. Pacelli (ed.), *Systemic Risk and Complex Networks in Modern Financial Systems*, New Economic Windows, https://doi.org/10.1007/978-3-031-64916-5_7

banks to insurers. This phenomenon shifts the circulation of credit risks from the banking to the insurance system.

The systemic relevance of risks belongs to the reality of complex systems, of which legal rules are a part, and especially to their inherent nature and the factual rules governing them. General network theory and general systems theory have highlighted this (Higgins 2012, p. 53; Barabàsi 2004, p. 45 ff., especially p. 56).

Therefore, it is a phenomenon that must be observed primarily in its reality, as Richard Feynman might say: "Science as a method of inquiry (…) is based on the principle that observation is the ultimate judge of how things are" (Feynman 1999, p. 25).

Circulation of Credit Risk, Systemic Risks, and Insurance Supervision

To extract a lesson from the 2007/2008 financial crisis, we must focus on the exponential growth of widespread and multidirectional credit risks assumed by banks. This phenomenon serves as a "model" and vehicle for the systemic expansion of risks well beyond the boundaries of the banking system (see Merusi 2009, I, p. 254 ff.; Merusi 2013, p. 4 ff.).

For this purpose, let us revisit some considerations previously discussed in part in an earlier contribution (Moliterni 2016, Treccani–Diritto on line).

The phenomenon of credit risk transfer naturally arises from banks' interest in transferring credit risks. Among the natural recipients of these credit risks are insurance companies, particularly those offering "risk transfer" services. Notably, credit insurance companies (as per Article 2 of the insurance code) possess specific expertise in risk selection (see Donati 1955, p. 37 ff.; Donati p. 289 ff.; also refer to Moliterni 2016, p. 15 ff., especially p. 30 ff.).

The phenomenon of the "marked expansion of the market" for Credit Risk Transfers attracts non-specialized investors who sometimes lack the necessary skills to manage the complexity of such products (see Banca d'Italia–Isvap 2004, p. 9 ff., www.bancaditalia.it).

Regarding the extension of systemic risk toward insurance companies, it assumes a category specific to the sector.

In the context of banking and, in particular, payment systems, the "systemic importance" lies in the transfer of credit risks from banks to insurance companies. This phenomenon, observed over time, has gradually expanded with the proliferation of global credit risk insurance (Pauscht and Welzel 2012, n. 5, available at www.bun desbank.de/Redaktion/eu; Duffie 2008, in BIS, Working Paper, n. 255).

However, Italy presents a different story concerning the exponential growth of the credit risk transfer market from the banking sector to the insurance sector (and other sectors). This divergence is due to legal constraints set by Italy's national regulations, particularly in secondary regulation. The "use of credit derivatives by insurance companies is entirely negligible, especially considering that, based on current sector provisions, they can only be employed for the purpose of investment risk reduction or effective portfolio management" (see Banca d'italia – ISVAP 2004, p. 5; also refer to the Letter to the Market from IVASS, Rome, March 15, 2013, p. 2 ff., specifically point 3, 2nd paragraph, available at www.ivass.it).

On a broader European Union level, with reference to the Solvency II framework and the resulting "restricted use of derivatives and transferable financial instruments," the European Commission states the following: "Derivatives can only be used for hedging currency and interest rate risk. This also excludes synthetic securitizations (…). The pool of underlying exposures must not include transferable financial instruments (effectively excluding CDOs), except for financial instruments issued by the securitization special purpose entity itself, in order to accommodate master trust structures" (Solvency II Overview, January 12, 2015, paragraph 15.1.3, available at.

Conversely, in the absence of legal constraints, one must consider whether—especially in certain areas of the Anglo-Saxon "world" (USA)—we can still speak of the banking system and the insurance system as distinct sectoral systems or whether it is more accurate to view them as a single banking-insurance system (see IAI).

The "strictly personal" nature of credit obligations (as per Article 1260, paragraph 1 of the Civil Code) is derived from their connection to "money." Among other characteristics, credit obligations borrow their fungibility and suitability for circulation from "money" (highlighting the "problem of whether money determines the peculiarity of the obligation that it is the object of or, conversely, whether the obligation assigns a particular 'legal form' to money," Di Majo 1979, p. 223).

Regarding the fungibility of "monetary units," Ascarelli (1959, p. 17 ff.) and others have discussed this concept. The broader concept of money or currency, including legal tender, has been explored extensively (see pages 9 ff., especially page 11). Notably, the fungibility of "monetary pieces" is relevant.

Furthermore, the original predisposition of credit and its associated risks (credit risks) to circulate is enhanced by the "ease of transit" among participants in the payment system. This ease is facilitated by the network of connections within the system itself. The case of Herstatt Bank illustrates this phenomenon (for the payment system and the so-called "Herstatt risk," see Padoa Schioppa 1992, p. 45 and p. 28).

Indeed, it's essential to recognize that Herstatt Bank was not considered part of the category of "major banks," and as far as records show, it was not. This is a "history." Another "history" is the systemic risk triggered by the Herstatt crisis at the time. When combined, these "histories" lead to an inevitable conclusion: systemic risk does not reside solely within "major banks."

Returning to the wisdom of the masters (Vivante) and their illustrious students (Ascarelli), the "nature of facts" or the "legal reality in effect" asserts itself. Therefore, reversing the order of factors, just as a bank (by its nature) is relevant due to the credit risks it holds, each of these credit risks contributes to the systemic relevance of the bank. The size and "concentration" of risks (see Article 53 of the Banking Consolidation Act; BANCA D'ITALIA, Supervisory Instructions for Banks, Title IV, Chap. 1, Sect. 7.1, available at www.bancaditalia.it) can provide a measure of the resulting "systemic importance." However, it is the nature of these risks, specifically their nature as credit risks, combined with their ownership by a bank (part of the banking system), that gives them "systemic" relevance (for the definition of "credit risk," see BANCA D'ITALIA, Governor's Report 2012, Appendix, Rome, 2013, Glossary, ad vocem; also T. Padoa Schioppa, cited, p. 285, Glossary, ad vocem; for credit risk calculation methods, see BANCA D'ITALIA, Supervisory Instructions for Banks, Title II, Chap. 1, and Title I, Chap. 1, p. 4 ff., which also refers to the "use of ratings expressed by export credit agencies (ECA)"; on this last point, I refer back to my previous observations in Moliterni 2016, p. 43 ff.).

Indeed, the transfer of credit risks from banks to insurance companies does not automatically strip them of their systemic significance. Instead, it inevitably extends this significance to the credit insurance sector. Moreover, the mechanisms of credit reinsurance further create systemic links between insurers and reinsurers (see Boglione 2012, p. 85 ff.). Specific concerns expressed by the International Association of Insurance Supervisors (IAIS) on a global scale have been echoed in BANCA D'ITALIA—ISVAP (2004, p. 9, text and note no. 10).

An empirical confirmation of this can be found in the case of AIG, which played a crucial role in the tragedy of the 2007/2008 financial crisis and its subsequent evolution. Conversely, its rescue also played an equally crucial role in containing its systemic expansion (albeit in a different dimension, referring to the systemic relevance of "credit risk transfer through insurance products," Merusi, Per un divieto di cartolarizzazione del rischio di credito, p. 261; also see Boglione, p. 85 ff.).

While the "shocking cost of Solvency II" is burdensome for insurers (Bailey 2013, p. 5), the "shock" of extending another potential banking crisis to the insurance sector could create systemically difficult-to-calculate and containable damages.

The recent financial crisis has highlighted the commonality and/or transfer of "systemic risk factors" between the banking and insurance systems, prompting the adoption of not only appropriate supervisory tools but also a different "institutional" supervisory model focused on factors such "common risks - in an activity of control effectively integrated in the banking-insurance sectors with 'groups of truly expert insurance and banking supervisors' (Bailey 2013, p. 2). In other words, the principle of adequacy with respect to the assigned institutional objectives (Article 3 of the Insurance Code; Article 5 of the Banking Consolidation Act) must be applied to administrative authorities and their organizational model. This principle is essentially analogous to the concept of 'proportionality,' understood as the 'coherence between means and ends' (see Guarraccino 2010, p. 246). The adequacy principle applies not only to the structure and rules of the 'organizational model' in a static dimension but also to its governance in a dynamic dimension. To borrow the words of the Financial Services Act 2012 (Section 2E(1)), the strategy of the Ivass (or any other supervisory administrative authority) must be determined in relation to its objectives: 'the Prudential Regulation Authority (PRA) must—(a) determine its strategy in relation to its objective, and (b) from time to time review, and if necessary revise, the strategy,' as stated in Section 2E(1) of the Financial Services Market Act 2023 (also cited in the Bank of England PRA, 29 August 2023, p. 2 ff., www.bankof england.co.uk). This ensures the efficiency or smooth functioning of the resulting activity (Article 97, paragraph 2 of the Constitution, see M. S. Giannini, Istituzioni, p. 262 ff.; M. D'Alberti, Lezioni, p. 23 and p. 40).

References

Ascarelli, T.: Obbligazioni pecuniarie. Giuffrè, Milano, p. 17 (1959)

Bailey, A.: The Evolution of Insurance Regulation; a Shifting Scope and New Frontiers. www.bis.org (2013)

Barabàsi, A.: Link. La scienza delle reti. Einaudi, Torino (2004)

Boglione, A.: La Riassicurazione, Giuffrè (2012)

D'alberti, M.: Lezioni di diritto amministrativo. G Giappichelli Editore (2017)

Donati, A.: L'assicurazione del credito, in Riv. trim. dir. proc. civ., p. 37 (1955)

Duffie, D.: Innovation in Credit Risk Transfer: Implications for Financial Stability, BIS. Working Paper, n. 255 (2008)

Feynman, R.P.: Il senso delle cose, Adelphi, Milano (1999)

Giannini, M.S.: Istituzioni di diritto amministrativo. Giuffrè, Milano, p. 262 (1981)

Guarraccino: L'adozione degli atti di regolazione delle autorità del mercato bancario, finanziario, assicurativo e previdenziale, in Studi in onore di Francesco Capriglione, Padova, I (2010)

Higgins, P.: La matematica dei social network. Una introduzione alla teoria dei grafi, Dedalo editore, Bari (2012)

Merusi, F.: Per un divieto di cartolarizzazione del rischio di credito, Banca e borsa e titoli di credito, I, p. 254 s (2009)

Moliterni, F.: Ivass, Treccani – Diritto on line (2016)

Padoa Schioppa, T.: La moneta e il sistema dei pagamenti, Bologna, p. 285 (1992)

Pausch, T., Welzel, P.: Regulation, credit risk transfer with CDS, and Bank Lending, Dia n. 5. www.bundesbank.de/Redaktion/eu (2012)

Siani, G.: Regulating New Distributed Ledger Technologies (DLT): Market Protection and Systemic Risks, p. 3. www.bancaditalia.it (2022)

Chapter 8
Systemic Cyber Risk in the Financial Sector: Can Network Analysis Assist in Identifying Vulnerabilities and Improving Resilience?

Ida Claudia Panetta and Sabrina Leo

Abstract The increasing interconnectedness and digitalisation of the financial sector have exposed it to a new and pervasive threat: systemic cyber risk. Systemic cyber risk in finance refers to the potential for a cyber-attack or breach to cause widespread disruption and instability across financial systems and markets. This type of risk can arise from various sources, including hackers, insider threats, and technological failures. Financial institutions and policymakers can help safeguard the global economy and protect against potential disruptions and instability by addressing systemic cyber risk. To effectively mitigate systemic cyber risk, it is important to have a deep understanding of the potential threats and vulnerabilities within their systems. This requires ongoing analysis and study of the evolving nature of cyber threats and the latest technological advancements in cybersecurity. Ongoing analysis and study of cyber threats and advancements in cybersecurity are crucial to staying ahead of evolving risks and ensuring the financial system's stability. In this context, Network analysis can be a valuable tool in studying systemic cyber risk in the financial domain since it is a powerful tool for understanding the interconnectedness of financial institutions and markets and the potential pathways for cyber risk to spread throughout the system. By mapping out these networks and identifying key nodes and vulnerabilities, institutions can better prepare for and respond to cyber-attacks.

Keywords Systemic Cyber risk · Network science · Financial crisis · Financial system

I. C. Panetta · S. Leo (✉)
Management Department, Sapienza University of Rome, Rome, Italy
e-mail: sabrina.leo@uniroma1.it

I. C. Panetta
e-mail: ida.panetta@uniroma1.it

© The Author(s) 2025
V. Pacelli (ed.), *Systemic Risk and Complex Networks in Modern Financial Systems*,
New Economic Windows, https://doi.org/10.1007/978-3-031-64916-5_8

Introduction

The increasing interconnectedness and digitalisation of the financial sector have exposed it to a new and pervasive threat, such as systemic cyber risk, which in finance refers to the potential for a cyber-attack or breach to cause widespread disruption and instability across financial systems and markets. This type of risk can arise from various sources, including hackers, insider threats, and technological failures. According to the European Systemic Risk Board (ESRB), cyber risk is one of the sources of systemic risk to the financial system, which could have serious negative consequences for the real economy. The Financial Stability Board (FSB) also warned that "a major cyber incident, if not properly contained, could seriously disrupt financial systems, including critical financial infrastructure, leading to broader financial stability implications". As such, it is essential for financial institutions and policymakers to take proactive measures to mitigate this risk and ensure the stability and security of financial systems. This can involve implementing robust cybersecurity measures, conducting regular risk assessments, and collaborating with other stakeholders to share information and best practices. Financial institutions and policymakers can help safeguard the global economy and protect against potential disruptions and instability by addressing systemic cyber risk. To effectively mitigate systemic cyber risk, it is important to have a deep understanding of the potential threats and vulnerabilities within their systems. This requires continuous analysis and study of the evolving nature of cyber threats and the latest technological advances in cybersecurity. These are critical to keeping pace with evolving risks and stabilising the financial system.

In this context, Network analysis can be a valuable tool in studying systemic cyber risk in the financial domain. Network analysis is a powerful tool for understanding the interconnectedness of financial institutions and the potential pathways for cyber risk to spread throughout the system. By mapping out these networks and identifying key nodes and vulnerabilities, institutions can better prepare for and respond to cyber-attacks. However, it is important to recognise that network analysis is just one component of a comprehensive finance approach to managing cyber risk.

Considering the potential for cyber incidents to disrupt the stability and functioning of the entire financial system, leading to cascading effects and significant economic consequences, this chapter proposes a network analysis approach to understanding and mitigating systemic cyber risk in the financial sector. Network analysis has already proven to be a powerful tool for understanding the interconnectedness of financial institutions and the potential pathways for other risks to spread throughout the system. Considering the nature of cyber risk and its attitude towards widespread disruption and instability across financial systems, Network analysis allows us to gain some new insights into the potential propagation and systemic consequences of cyber incidents, aiding in developing effective risk management and resilience-enhancing measures. By leveraging network methods, the study wants to identify the possible steps to construct a network model that captures and uncovers the complex relationships, vulnerabilities, information flows, and cascading effects of cyber incidents within the financial system. This chapter aims to theoretically deepen the different methodologies to forecast and measure systemic risk and financial crises, apply various network analysis techniques to identify key nodes, evaluate their centrality and criticality, and simulate cascading effects under different cyber-attack scenarios. The resulting theoretical framework will contribute to understanding systemic cyber risk and provide insights into potential policy interventions and resilience-enhancing strategies for the financial sector.

To support this framework, in the next paragraph we present an overview of the systemic cyber risk in the financial system, while in the third paragraph we deepen the logical-conceptual paradigms that make network science a useful tool for analysing systemic risk. Finally, the chapter ends with some summary reflections.

Systemic Cyber Risk: Overview in the Financial System

The Systemic Nature of Cyber Risk in the Financial System

> Systemic cyber risk is defined as the combination of the probability of cyber incidents occurring and their impact on financial stability. (ESRB 2020)

It is commonly accepted that there are two key dimensions to systemic risk (Borio 2003) easily verifiable in the case of cyber risk:

1. The **cross-sectional dimension** refers to the way risks are spread across the financial system and how specific shocks might spread to become systemic. It relates to the dimensions and interlinkages of financial institutions and markets and the potential for spreading financial distress through factors such as direct or perceived connections to unstable institutions. In this context, assume relevance to the concept of **substitutability,** related to the possibility that the risk affects a critical infrastructure (e.g., the payment system) that is not readily substitutable.
2. The **time-related dimension** pertains to the internal evolution of threats to financial stability across time. This encompasses the gradual and cyclical buildup of financial vulnerability during periods of economic expansion, as well as the heightened caution and even panic that can occur during economic downturns. Cybersecurity threats exhibit a temporal aspect, intensifying during periods of increased political and economic instability. The evidence suggests that advanced economies experience cyberattacks more frequently than their developing counterparts, a trend that likely mirrors the varying degrees of digitalisation across these economies.

In addition, the previous dimension has to be underlying the **risk correlation and level of interconnectedness** pertain to the likelihood that a breach or cyber incident will result in extensive and destabilising consequences throughout interconnected systems (Fig. 8.1).

Fig. 8.1 Interconnections between entities operating in the financial system. *Source* DALL.E 3. (2024). Generated by OpenAI

Considering cyber risk, it is undeniable that, under the dimension considered, it can be relevant at a systemic level.

The growing prevalence of innovation and digitalisation in the financial sector and other industries heightens the vulnerability of both financial and non-financial sectors to cyberattacks. Furthermore, the increasing interconnectedness between the real and financial sectors amplifies the potential for cyberattacks originating in the real economy to spread and disrupt the operations of financial institutions. Cyber threats emerging from the real economy can quickly extend their impact on the financial sector, harming financial intermediation and causing a broader ripple effect on the real economy and vice versa.

According to main reports on cyberattacks related to the financial system, we notice that, as for other industries, the main part of attacks is due to financial motivations (75% in 2021, quoting ECB 2021) and caused by criminals; a few parts of attacks are nation-state-nature, then oriented to cause a disruption for the financial system directly. Despite the objective not intending to provoke instability, the resulting distress for the financial system can be systemic in nature because affects a large number of less significant institutions or because the target is a systemically important institution or market infrastructure.

The cross-sectional nature of this risk is also evident in the possibility of the financial system suffering the consequences of a sector interrelated with or many financed entities in other sectors.

Of course, if cyber attacked, systemic entities *di-per-sè* could become a source of risk for the financial system's stability. Systemic vulnerabilities may emerge if an attack successfully compromises a financially vulnerable entity yet crucial to the system's overall stability. Consequently, risks are most pronounced among entities characterised by a high reliance on technology, limited financial robustness, and a lack of awareness regarding such threats. Conversely, for those firms that do not have systemic relevance, the cyber-attack could mine financial stability proportionally with the level of connection between the entities in the financial system and the ability to have a domino effect (ECB 2016) through different channels of propagation. Channels can amplify operators' financial, operational, and confidence relationships (ESBR 2020; Koditis and Schreft 2022).

A cyberattack on a critical infrastructure that cannot be replaced (substitutability dimension) or on one service that reveals vulnerabilities in another (risk correlation) might quickly cause system-wide repercussions. In addition, the increased reliance on some information technologies, such as cloud ones, increases the potential for cyber-attacks that might disrupt, even temporarily, financial and economic activities on a broad scale.

Because cyber risk could affect the whole system and cause big losses in the financial world (Bouveret 2019), policymakers need to improve infrastructure for surveillance and analysis, make macroprudential tools more useful, and encourage cooperation and sharing of information at both the operational and strategic levels. This concerted effort is crucial to bolstering the financial system's resilience and reducing the systemic repercussions of cyberattacks.

The need for a deep understanding of systemic cyber risk for cybersecurity in the financial industry is gaining greater relevance in policy-making interest in preserving the stability of the system, as the growing initiative at the international level demonstrates:

- The *G7 Cyber Expert Group* focuses on addressing third-party entities' risks, conducting threat-led penetration tests, and implementing cyber exercise programmes.
- The *Financial Stability Board* (FSB) has created a *cyber lexicon* to standardise terminology and has recommended a toolkit for responding to and recovering from cyber incidents.
- The Committee on Payments and Market Infrastructures at the *Bank for International Settlements* (BIS) has released guidelines for bolstering the cyber resilience of financial market infrastructures (BIS 2022).
- The *European Systemic Risk Board* (ESRB), at the macroprudential level, has advocated for creating a continent-wide framework to coordinate responses to systemic cyber incidents, aiming to reduce failure in coordinating responses (ESRB 2022a). The magnitude, rapidity, and spread of a cyber incident necessitate an immediate reaction from both corporations and financial regulators to maintain financial stability.
- European supervisors, at the micro-prudential level, have prioritised cyber risks within their oversight agendas and have established a reporting framework for cyber incidents to monitor the challenges key financial institutions face closely.
- The *EU's Digital Operational Resilience Act* (DORA) is set to introduce a thorough regulatory framework for digital operational resilience among financial entities in the EU. This act is intended to significantly enhance cybersecurity practices

across the financial services sector, comprehensively addressing an expansive range of operational risks.

The ESRB (2020, 2022a, b) has developed an analytical framework to assess how cyber risk can become a source of systemic risk to the financial system. The four stages of this conceptual model (context, shock, amplification, systemic event) facilitate a systematic analysis of how a cyber incident can grow from operational disruption into a systemic crisis. More in particular:

- Context: This initial stage sets the background, detailing the operational and cyber-security posture of the financial system, including existing vulnerabilities, the level of interconnectedness among institutions, and the regulatory environment.
- Shock: This stage represents the occurrence of a cyberattack or cybersecurity incident that directly impacts one or more entities within the financial system. The nature, scale, and target(s) of the cyberattack are defined here.
- Amplification: Following the initial shock, this stage outlines how the impact of the cyber incident can escalate due to the interconnectedness of the financial system, lack of timely information sharing, and other systemic vulnerabilities. Factors that contribute to the amplification of the initial shock, such as panic, misinformation, or the failure of critical operational functions, are explored.
- Systemic Event: The culmination of the model, where the amplified shock reaches a threshold that causes significant disruption or destabilization of the financial system as a whole. This stage assesses the broader economic and financial conse-quences of the cyber incident, including loss of confidence, liquidity crises, or significant operational disruptions across the financial sector.

To simplify and graphically represent the ESRB framework assessing how cyber risk can become a source of systemic risk to the financial system refer to Fig. 8.2.

Analysis of Financial Cyber Risk in Light of Other Systemic Risks

A cyberattack can have devastating consequences for a financial institution, causing operational disruptions, reputational damage, and financial losses. Additionally, the attack's impact can create a ripple effect through operational or financial conta-gion, eroding confidence in the entire financial system. This escalation process can potentially spread the disturbance throughout the financial system, even affecting institutions not directly targeted by the initial attack. Although the later phases of a systemic cyber crisis may mirror those of a conventional financial crisis, the disrup-tion to the financial system's functionality introduces an additional layer to crisis management, encompassing the activation of systemic safeguards.

The propagation of cyber risk at a systemic level, especially in parallel to other financial systemic risks, may involve distinctive channels and mechanisms.

Fig. 8.2 Infographic synthesising ESRB framework. *Source* Own elaboration on DALL.E 3. (2024). Generated by OpenAI

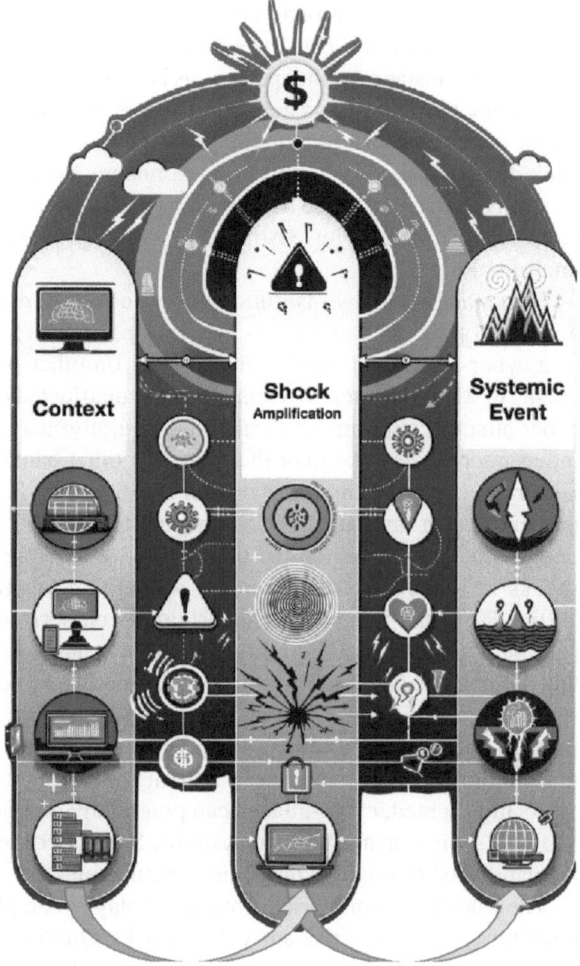

Grasping these routes is essential for crafting efficient risk management and mitigation approaches. It also aids in comprehending how network analysis might be applied to examine this issue, either in a manner similar to or distinct from the analysis of other financial risks. In this synthetic section, we briefly recall these similarities and discontinuities, focusing on the channels and mechanisms of propagation for cyber risk versus other financial systemic risks. Similarly to financial contagion, cyber risk can propagate through the networked financial system, but the vector is technological rather than purely financial. As a distinct feature, unlike traditional financial risks, cyber risks can directly impair the operational capabilities of financial institutions, affecting their ability to conduct transactions, process payments, or maintain liquidity. It follows that an attack on critical infrastructure (e.g., payment systems) can immediately and widely affect multiple entities.

A cyber-attack on a bank can lead to a sudden withdrawal of funds and raise concerns about the ability to meet financial obligations, like any other operational risk event. In this instance, either panic—as Diamond and Dybvig described it in (1983)—or fundamentals—as Goldstein and Pauzner explained in 2005—can be responsible for the occurrence of a run. Inefficient asset disposal at a bank incurs expenses for the bank itself and negative effects on the financial sector (counterparties) and the real economy (borrowers). The occurrence of cyber assaults has the potential to hinder the bank's ability to serve its existing debtors if payments or access to accounts become unavailable.

Banks and financial institutions operate in different marketplaces and carry out a variety of services for clients. Therefore, if there is uncertainty about the extent of a cyber-attack and the possible impact on other parts of the firm, it may cause clients to withdraw their investments from unaffected operations. In a broader sense, cyber-attacks could render capital and liquidity inactive, resulting in costs that are independent of a bank run or the actions of other banks and creditors/counterparties. Cyber-attacks can immobilise capital and liquidity, leading to expenses that are not influenced by a bank run or the behaviour of other banks and creditors/counterparties.

As Allen and Gale pointed out in 2000, a network's topology can also transmit shocks in addition to the so-called *accounting-based effects* (Eisenberg and Noe 2001). These traditional models, originally elaborated to illustrate the propagation pattern of solvency and liquidity shocks, also apply to cyber shocks. A cyber shock that disrupts the distribution or access to liquidity could lead to contagion, as demonstrated in the study by Allen and Gale (2000). A virus or technical flaw can spread through data and communication networks, shared service providers, or similarities in technology, which can affect payments or lending between banks.

As highlighted, cyber-attacks can potentially produce interconnected disruptions and pose a significant threat to systems. Similar consequences to bank defaults can be manifested through technological connections.

The role that asymmetric information plays in amplifying risk is what causes the other type of systemic risk and cyber shocks. Quoting Caballero and Simsek (2013), it is possible to demonstrate how vague data regarding the exact location of a solvency shock in a financial network significantly increases the motivation for institutions to adopt cautious yet highly detrimental actions to the overall system. It works in the same way for the cyber-attack. Cyber shocks prompt banks to implement their most effective financial network solutions. The impact of cyber shocks can be significantly greater when their origin is undisclosed. News or rumours of impairment may lead to pre-emptive withdrawals from banks or financial institutions that are not affected. The disclosure tactics employed by institutions to engage with clients are being criticised for generating worries about asymmetric knowledge.

If we compare the channel and mechanism of systemic cyber risk to other financial risks, we note that cyber risk:

– can spread faster than traditional financial risks due to the instantaneous nature of digital networks and operations;

- is often less predictable and more difficult to detect in advance compared to traditional risks, which can often be modelled and anticipated based on economic indicators and past trends;
- is not limited to financial losses but extends to operational disruption, data integrity, and loss of consumer confidence, adding layers of complexity to crisis management and recovery.
- requires a different set of tools focused on technological resilience, cybersecurity measures, and information sharing protocols, while traditional financial systemic risks are mitigated through financial policies, capital adequacy requirements, and liquidity provisions.

Consequently, while cyber risk shares some propagation channels and mechanisms with traditional financial systemic risks, its distinct nature necessitates a unique approach to risk management. The digital and interconnected landscape of modern finance introduces new vulnerabilities and requires a blend of cybersecurity, operational resilience, and traditional financial risk management to protect the stability of the financial system.

The Utility of Network Science for Systemic Cyber Risk Analysis in the Financial System: The Proposed Theoretical Framework

If we assume that cyber risk could be relevant at the systemic level for the financial system and for other systemic risks in this domain, it is beneficial to use a promising method to study systemic cyber risk, which could be constituted by network analysis.

Drawing upon this premise of systemic cyber risk, numerous scholars have ventured into network analysis to devise robust models and methods for cyber risk assessment.

A range of studies have explored network analysis and cyber risk, offering various methods and models for risk assessment. Huiying and Yuanda (2008) and Lv (2009) both propose quantitative evaluation algorithms based on threat analysis and attack probability, respectively. Mixia et al. (2007) emphasise the importance of visualisation in network security situation analysis, while Lamichhane et al. (2018) introduce a quantitative risk analysis model for enterprise networks. Adams and Heard (2014) and Wang and Wang (2010) both focus on data analysis, with Adams and Heard discussing the use of graph analysis and Wang introducing nonlinear system analysis and forecasting techniques. Kalinin et al. (2021) and Aktayeva et al. (2023) discuss more specific settings. Kalinin et al. look at assessing cybersecurity risks in smart city infrastructures, and Aktayeva et al. talk about how cognitive modelling can be used in social networks for critical infrastructure.

Numerous studies have applied network analysis methods to financial systems analysis, focusing on understanding interdependencies and systemic risk. In particular, network analysis has been widely applied to studying systemic risk in the financial system. Chen et al. (2016) and Tsankov (2021) provide comprehensive overviews of the development and application of network theory in this context, while Saltoğlu and Yenilmez (2010) and Gai and Kapadia (2019) emphasise the importance of this approach in understanding systemic risk and identifying systemically important institutions.

Ahelegbey (2016) and Chapman et al. (2011) specifically explore the use of network models in analysing contagion, spillover effects, and risk propagation channels. Jo (2012) and Furfaro et al. (2016) suggest frameworks that connect liquidity loss, solvency ability, and macroeconomic shocks. They also use a goal-oriented approach to think about how to stop and respond to cyber systemic risk in banking systems.

Allen and Babus (2008) further discuss the role of network theories in explaining economic interactions and modelling systemic risk in banking systems. Chen and wang (2013) and Cont et al. (2013) present network models for the Credit Default Swap market and interlinked financial institutions, respectively, highlighting the role of network parameters and counterparty exposures in systemic risk.

Iori and Mantegna (2018) review empirical work on the fragility and resilience of financial and credit markets, focusing on systemic risk and evaluating systemically important institutions. Gong et al. (2019) and Bougheas and Kirman (2014) both review the application of network analysis to systemic risk, with Gong et al. (2019) specifically focusing on the Chinese financial market and the identification of risky financial firms. To the best of our knowledge, no theoretical or empirical contribution to the application of network science to cyber risk in the financial system is currently available in the literature.

Bridging the gap between these two areas of study, we aim to apply the insights gained from network analysis in financial systems to the relatively unexplored field of cyber risk in the financial sector. Trying to make the most of both strands of literature, in Table 8.1, we propose the possible main steps for analysing the impact of a cyberattack on the financial system using network science. This table outlines a structured approach to utilising network science for cybersecurity analysis in the financial sector, emphasising the importance of each step in building a comprehensive understanding and response strategy to cyber threats.

Cyber Mapping Challenges

Once the valuable steps for analysing the impacts of a cyberattack on the financial system have been identified, we now focus on step 3 (Network Modelling, Table 8.1). This step aims to retrace the links (the edges) between banks, financial institutions, and other interested parties (the nodes) by looking at data flows, financial relationships, and communications. This helps find weaknesses and possible risks by looking

Table 8.1 The main steps for analysing the impact of a cyberattack on the financial system

Step	Description	Key activities
1. Define the scope and objectives	Determine the analysis boundaries and aims	− Identify financial institutions and systems to be analysed − Define analysis goals (vulnerability assessment, impact analysis, etc.)
2. Data collection and preparation	Gather and prepare the necessary data for network analysis	− Collect data on entities and their connections − Clean and preprocess data for analysis
3. Network modelling	Model the financial system as a network of nodes and edges	− Construct a network with entities as nodes and relationships as edges − Consider multilayer models for different interaction types
4. Network analysis	Apply network science methodologies to analyse the structure and dynamics	− Calculate centrality measures to identify key nodes − Detect communities within the network − Assess node and edge vulnerabilities
5. Simulate cyberattack scenarios	Develop and simulate various cyberattack scenarios	− Create scenarios based on different attack vectors and methods − Simulate attacks to study propagation and impact
6. Impact assessment	Evaluate the potential impact of cyberattacks on the network	− Quantify the financial, operational, and confidence impacts − Highlight critical nodes for systemic risk
7. Mitigation and resilience building	Develop strategies to mitigate risks and enhance resilience	− Propose cybersecurity enhancements and system redundancies − Plan for rapid response and recovery
8. Policy recommendations and regulatory compliance	Make policy and regulatory recommendations based on the findings	− Suggest policy changes or regulatory measures − Ensure strategies comply with existing cybersecurity regulations
9. Continuous monitoring and updating	Establish ongoing monitoring and update analyses regularly	− Monitor the network for new vulnerabilities and threats − Update the analysis to reflect changes in the network and threat landscape

at the links between nodes and following up on any strange activities or threats to the financial system from inside and outside the system.

Even the European Systemic Risk Board (ESRB) in 2022 emphasised the importance and usefulness of a network analysis approach, particularly for cyber mapping and identifying systemic nodes. This methodology could be crucial for understanding the complexities and interdependencies within financial networks, enabling the pinpointing of critical points that, if compromised, can proliferate and pose systemic risks. Systemically important nodes[1] are essential to the financial system or provide irreplaceable services. A cyber-attack at these nodes could cause a much greater financial system disruption than at less essential nodes. Thus, monitoring and analysing systemic cyber risk requires identifying these nodes in financial and operational networks. As underscored in the preceding sections, the system is not just through a solitary, irreplaceable entity (systemic node) but also across a cluster of institutions within the financial network.

Within the realm of cybersecurity, financial networks designed to manage market risks and ensure liquidity flow exhibit susceptibility to cyber-attacks and may become targets for cyber threats. Such attacks have the potential to impact numerous interconnected organisations across various networks. Particularly vulnerable are networks with a central-peripheral configuration that lacks alternative routes for data flow when their core is breached. This can lead to a rapid attack spread, complicating the network's ability to identify and understand its interconnections. Such a scenario could temporarily disrupt access to their services, underscoring the intricate nature of systemic risks.

The financial system is inherently complex, with intricate interconnections in financial transactions. This complexity is further amplified in the cyber domain as new actors and potential channels for contagion are introduced. Therefore, it is essential to perform an examination of the interrelated network of third-party entities. Due to their ability to cause business disruptions or systemic failures, these entities require a comprehensive analysis to identify and reduce risks that could have widespread systemic consequences.

In light of what has been said, the most formidable yet critical task lies in the initial stages of mapping the network: accurately identifying each node and precisely delineating the edges representing the relationships and dependencies between these nodes. This foundational step is essential for understanding the network's vulnerability to systemic cyber threats.

In network analysis, identifying key financial and operational nodes means shedding light on the structural intricacies of the financial network by pinpointing the nodes that play a crucial financial or operational role within the sector. Recognising these nodes helps comprehend the network's topology, which the cyber risk conceptual model suggests may magnify risk. This process of identification gives us a first

---

[1] See ESRB (2020), p. 17 for the definition: "Systemic nodes are any agents fulfilling a critical financial or operational role in the financial sector. Such systemic nodes are often characterised by the importance or lack of substitutability of the financial or operational services they provide to the financial system."

look at possible ways for the infection to spread, which helps us find institutions and third-party service providers for stress tests on systemic cyber resilience scenarios.

While it is quite easy to identify financial systemic relevant institutions and infrastructure, only recently could a pattern to identify systemic relevant entities at the operational level be found at the EU level. Thanks to the new prescription of DORA, EU-critical ICT third-party service providers are commonly defined for operational services. Additionally, DORA asks financial institutions (starting from January 2025) to document all individual, sub-consolidated, and consolidated ICT service agreements, contributing to the design of a centralised EU overview of essential financial system nodes. A *cyber map* helps detect systemic nodes and monitors and analyses the main technologies, services, and linkages of financial institutions, service providers, and in-house and third-party systems.

So, a good mapping should show the technological and financial connections between banks and other tech and service providers. It should use operational and financial data to find systemic nodes based on their importance, connectivity, and dependencies. By using these mappings and identifying important systemic nodes, we can better comprehend the interconnectedness within the EU financial ecosystem, as summarised by Figs. 8.3, 8.4, 8.5, 8.6. In particular, the figures are an example, for the payment system and capital market, of how the interconnections between types of actors develop and how the interconnections are transformed with the entry into the scene of IT service providers.

Specifically, the figures illustrate how the relationships between different types of actors evolve and change when IT service providers enter the market, as seen in the case of the payment system and capital market.

As for the payments system, a conceptual network analysis method can represent the connection between actors involved in the EU's payment systems. This framework (Table 8.2) can aid in comprehending the interlinkages, potential weaknesses, and consequences of cyber threats on the payment infrastructure.

Figures 8.3 and 8.4 provide a simplified depiction of the actors involved in the EU payment system. These figures highlight the important infrastructure, the movement of financial transactions, and the regulatory framework crucial for efficient financial activities throughout the European Union. The diagram uses circles to represent the nodes, while lines of various colours and styles indicate their interactions or relationships.

The diagram in Fig. 8.4 encapsulates the comprehensive network of entities involved in the EU payment system, from the foundational infrastructure and regulatory oversight to the service provision to end-users and the essential support provided by IT Service Providers. This visualisation underscores the complexity and interdependence of the payment ecosystem, highlighting how technology underpins the seamless operation and security of financial transactions. The graphical analysis would visually demonstrate the foundational roles of traditional actors and how the integration of IT Service Providers introduces new dynamics, enhances capabilities, and potentially introduces new vulnerabilities.

The absence of IT Service Providers highlights traditional vulnerabilities and operational inefficiencies, while their presence underscores the critical role of technology

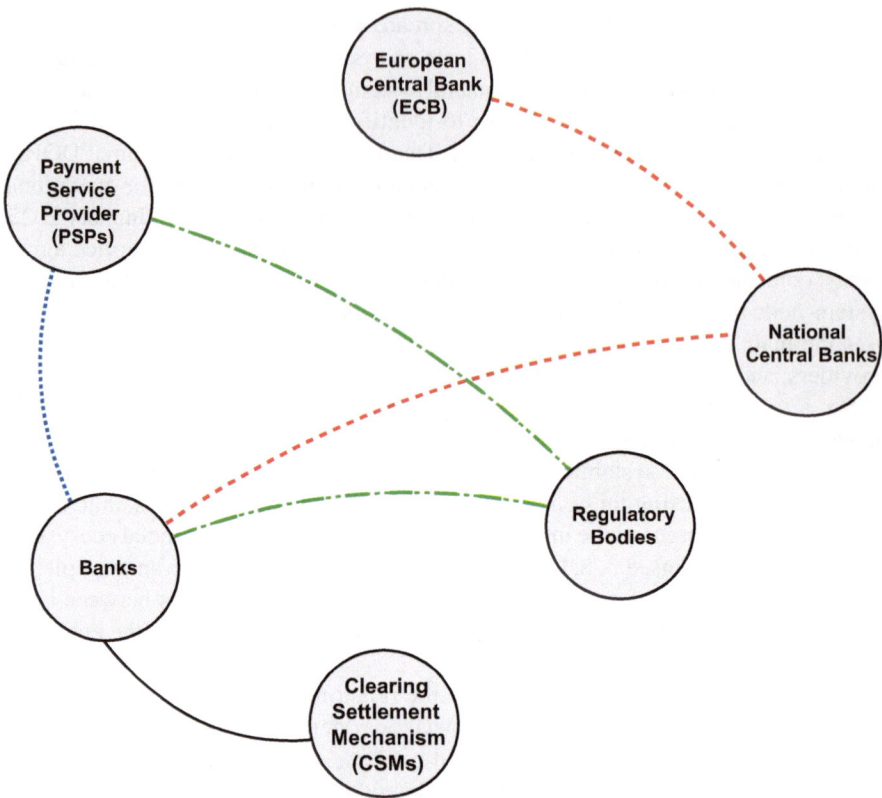

Fig. 8.3 EU payment system network analysis (without IT service provider). *Note* (i) solid black lines connect Banks and Clearing and Settlement Mechanisms (CSMs) representing the primary transactional flows essential for daily financial operations; (ii) blue dotted lines show how Payment Service Providers (PSPs) are linked to banks; (iii) green dash-dot lines represent compliance interactions between regulatory bodies and both banks and PSPs, emphasising the importance of adhering to established regulations and standards to maintain system integrity. *Source* Generated by OpenAI (2024).

in modernising and securing these systems. Then, this comprehensive approach illustrates the interconnectedness of financial systems and the pivotal role of technology in ensuring their resilience, efficiency, and adaptability to new challenges.

Even with regard to the capital market, it is possible to use a conceptual network analysis method to represent the actors involved. This framework (Table 8.3) can help understand the interconnections, potential weaknesses, and consequences of cyber threats on the capital market.

The diagram displayed in Fig. 8.5 now features differentiated edges to represent various types of relationships within the European Union's capital markets network. Each colour and style of the edges convey a distinct aspect of the market interactions. This differentiation helps to highlight the complexity of the capital markets,

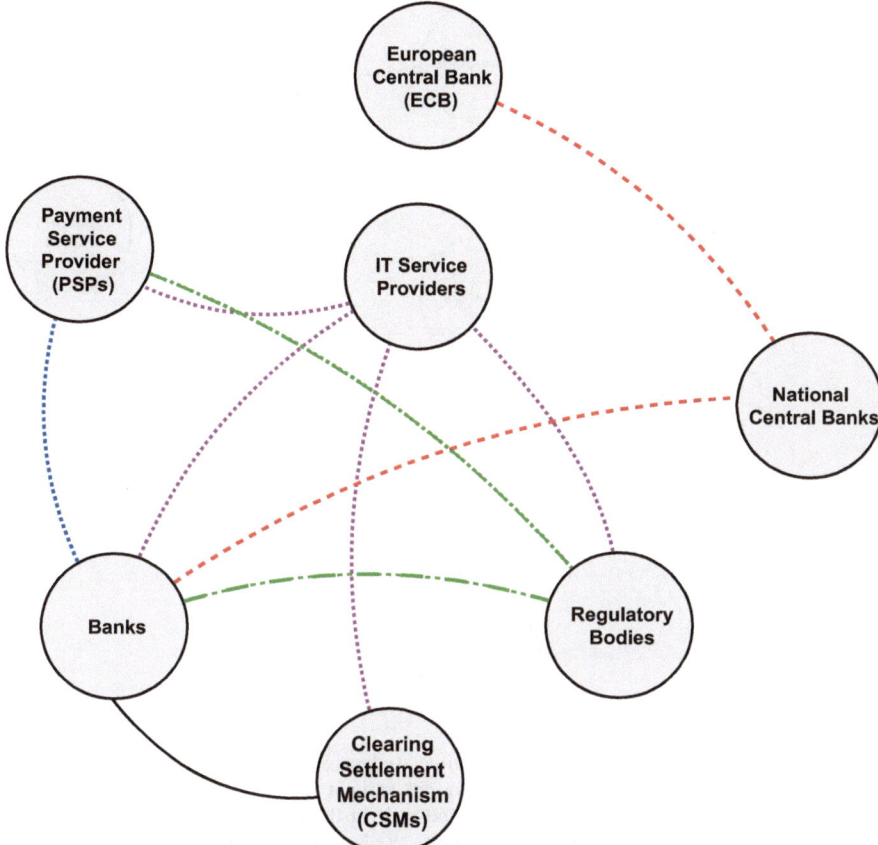

Fig. 8.4 EU payment system network analysis (with IT service provider). *Source* Generated by OpenAI (2024). *Note* Different from Fig. 8.3, in this one, purple dotted lines mark the support provided by IT service providers to banks, PSPs, CSMs, and regulatory bodies, underlining the critical role of technology and cybersecurity in the payment ecosystem

illustrating not just who the participants are but how they interact and the nature of their interdependencies, which is crucial for understanding potential vulnerabilities and areas needing robust cybersecurity measures. For example, the central roles of clearing houses and custodians highlight how critical these entities are to market stability and security. Similarly, the interconnectedness of exchanges and intermediaries with issuers and investors underscores the potential for cyberattacks to impact a wide range of market participants.

The simplified network diagram in Fig. 8.6 is updated to include IT Service Providers, highlighting their integral support role within the European Union's capital markets. The magenta dashed lines represent these new connections, underscoring the importance of IT services in the functioning, security, and resilience of capital

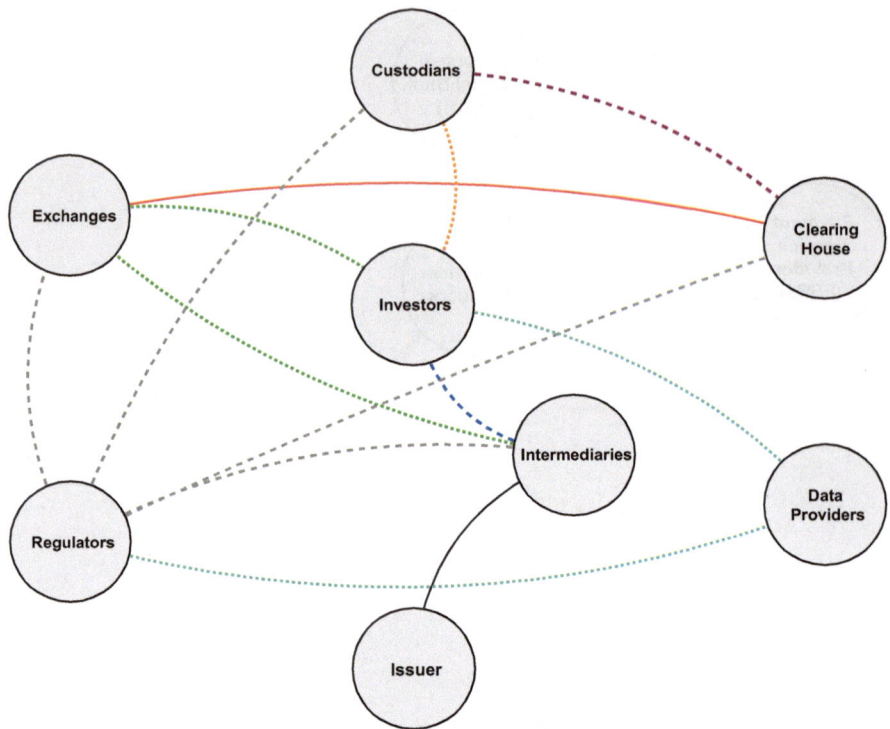

Fig. 8.5 EU capital market network with differentiated edges. *Source* Generated by OpenAI (2024). *Note* The diagram differentiates edges to represent various relationships within the European Union's capital markets network. Each colour and style of the edges convey a distinct aspect of the market interactions: (i) Solid black lines signify the issuance and intermediary relationships, highlighting the direct path of securities from issuers to the market through intermediaries; (ii) Blue dashed lines represent the flow of investments between intermediaries and investors, indicating the advisory and transactional services; (iii) Green dotted lines show trading activities between investors and exchanges, as well as the role of intermediaries in facilitating these trades; (iv) Red solid lines emphasise the critical clearing and settlement process between exchanges and clearing houses; (v) Purple dashed lines denote the safeguarding of assets by custodians post-clearance; (vi) Orange dotted lines illustrate the custodial services provided to investors, ensuring the safekeeping of their investments; (vii) Grey dashed lines indicate regulatory oversight exerted by regulators to maintain market integrity; (viii) Cyan dotted lines represent the flow of information from data providers to investors and regulators, supporting decision-making and regulatory compliance

markets. This addition emphasises how critical IT service providers are in maintaining the integrity and efficiency of market operations, especially in cybersecurity. This visualisation helps identify potential points of vulnerability within the capital markets, especially concerning cyber threats.

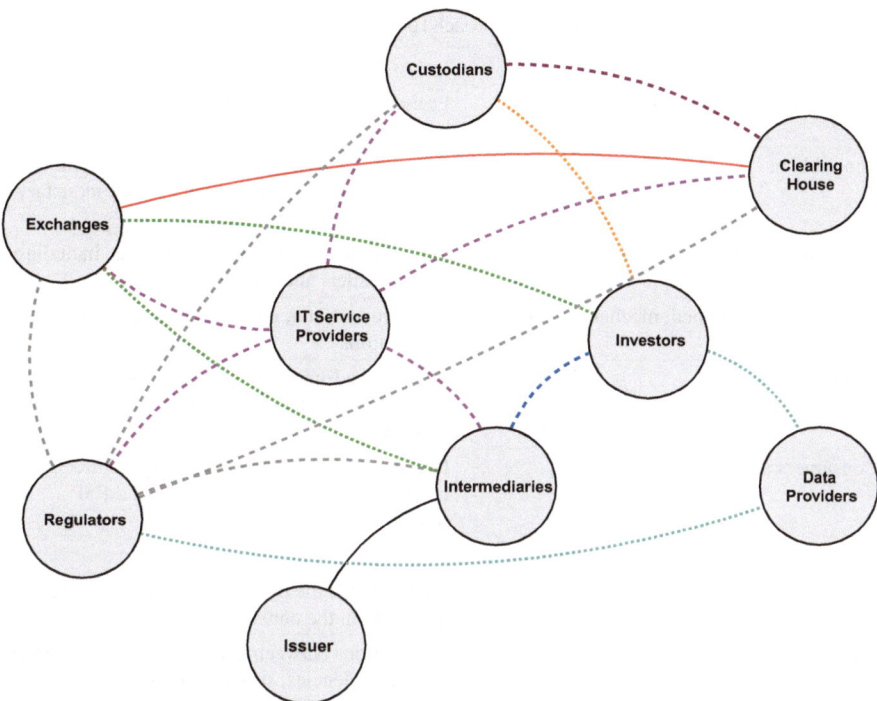

Fig. 8.6 EU capital market with IT service providers. *Source* Generated by OpenAI (2024). *Note* compared to Fig. 8.4, in addition to the previous lines, the magenta dashed lines represent new connections, underscoring the importance of IT services in the functioning, security, and resilience of capital markets

Concluding Remarks

Our world is complex and interdependent, with various systems and phenomena influencing each other, even when they may seem unrelated. This interconnected reality exposes societies to systemic vulnerabilities that demand a holistic approach to analyse complex phenomena. In this interconnected digital ecosystem, nothing operates in isolation. Modern financial markets are no exception; they are part of this intricate network, susceptible to cyber threats that can cascade through the system. In today's globalized and highly interconnected financial world, we face not only traditional systemic risks but also an emerging threat in the form of systemic cyber risk.

Systemic cyber risk arises from the growing reliance on technology and the internet even in the financial industry. Each entity, whether it's a corporation, government, financial institution, infrastructure or individual, is a node in this vast digital web. The connections, or edges, represent the various digital interactions that bind them together additional to the financial transaction ones.

Table 8.2 Conceptual network analysis approach (payment system)

Actors	
Banks and financial institutions	Primary actors that facilitate and process payments
Central banks	Including the European Central Bank (ECB) and national central banks, which oversee monetary policy and payment systems
Payment service providers (PSPs)	Offer payment services and solutions, including digital wallets and payment gateways
Clearing and settlement mechanisms (CSMs)	Infrastructure that processes and settles payment transactions
Regulatory bodies	Such as the European Banking Authority (EBA), which sets standards and regulations for payment services
Consumers and businesses	End-users of the payment systems, indirectly connected through their banking and PSP relationships
Network analysis approach	
Nodes	Each actor in the payment system is represented as a node in the network
Edges	Connections between nodes signify relationships and dependencies, such as transaction flows, regulatory oversight, and information exchange

Constructing an accurate model of these networks is a complex task that requires collaboration between experts from different fields. However, once established effectively, these digital networks can provide invaluable insights into how cyber shocks propagate through the system, potentially triggering a systemic crisis.

Expanding our analysis, we can create networks of interconnected economic agents and establish links between these different networks. This approach allows us to predict the potential spread of a cyber crisis across various sectors and domains. In this context, a well-constructed network can serve as a powerful tool for identifying early warning signals of system-wide vulnerabilities.

The correct and efficient construction of these networks relies on both methodological expertise, including data analysis models, and a deep understanding of economic and financial systems.

To enhance the capacity of the current macroprudential policy toolkit to tackle cyber risks, the increasing significance of these threats necessitates that macroprudential overseers pre-emptively engage with them (ECB 2022). This involves assessing the financial system's resilience against such threats and issuing risk warnings as necessary.

Crucially, econometric evidence indicates that cyberattacks follow discernible patterns, influenced by factors like, among others economic robustness, the extent of financial globalization, and the prevailing policy and political uncertainty. This

Table 8.3 Conceptual network analysis approach (capital market)

Actors	
Issuers	Corporations, governments, and other entities that issue securities (stocks, bonds)
Investors	Institutional (pension funds, insurance companies) and retail investors
Intermediaries	Investment banks, brokers, dealers, and market makers
Exchanges and trading venues	Stock exchanges, electronic trading platforms
Clearing houses	Entities that facilitate the clearing and settlement of trades
Custodians	Institutions that hold securities for safekeeping
Regulators	European Securities and Markets Authority (ESMA), national regulators
Data providers	Agencies that provide market data, analysis, and ratings
IT providers	Company or organization that provides services or products related to Information Technology
Relationships and vulnerabilities	
Issuers and investors	Connected through intermediaries; vulnerable to misinformation and market manipulation
Intermediaries and trading venues	Facilitate trading; susceptible to cyber-attacks on trading platforms
Clearing houses	Central to the settlement process; a single point of failure could disrupt market operations
Custodians	Hold assets; targeted for cyber theft and data breaches
Regulators	Oversee market operations; depend on accurate data for effective regulation

evidence highlights the pivotal role of policymakers, who uniquely possess or have access to the necessary data, in mapping the cyber landscape for the financial system. Their exclusive position underscores the imperative for authorities to bridge the data gaps concerning cyberattacks, enabling a more robust defence mechanism against potential cyber threats.

In conclusion, systemic cyber risk is an emerging threat in growing interconnected financial world. Scholars, policy makers and economic agents must adapt, collaborate, and invest wisely to mitigate this risk, just as we strive to address environmental challenges. The balance between economic recovery and risk mitigation is delicate, but it is essential for our economic system's continued survival. So, in the light of what has been argued above, to understand the complex holistic nature of the phenomenon of systemic risk, today it is needed an integrated and prospective analysis, through the logical-conceptual schemes borrowed from the network science, of the multiple and often obscure relationships that bind the various economic, political, social, health and environmental phenomena.

References

Adams, N.M., Heard, N.A.: Data Analysis for Network Cyber-Security (2014)

Ahelegbey, D.F.: The econometrics of Bayesian graphical models: a review with financial application. Risk Manage. Anal. Financ. Inst. eJ. (2016)

Aktayeva, A., Makatov, Y., Tulegenovna, A.K., Dautov, A., Niyazova, R., Zhamankarin, M., Khan, S.: Cybersecurity risk assessments within critical infrastructure social networks. Data **8**, 156 (2023)

Allen, F., Babus, A.: Networks in Finance. Econ. Netw. eJ. (2008)

Allen, F., Gale, D.: Financial contagion. J. Polit. Econ. **108**(1), 1–33 (2000)

BIS: Implementation monitoring of the PFMI: level 3 assessment on financial market infrastructures' cyber resilience. BIS Committee on Payments and Market Infrastructure (2022)

Borio, C.: Towards a macroprudential framework for financial supervision and regulation? BIS Working Papers, No 128, Bank for International Settlements (2003)

Bougheas, S., Kirman, A.: Complex financial networks and systemic risk: a review. CESifo Working Paper Series (2014).

Bouveret, A.: Cyber risk for the financial services sector. Risk Manage. eJ. (2019)

Caballero, R.J., Simsek, A.: Fire sales in a model of complexity. J. Finan. **68**(6), 2549–2587 (2013)

Chapman, J.T., Embree, L., Roberts, T.L., Zhang, N.: Payment networks: a review of recent research. Bank Can. Rev. 21–27 (2011)

Chen, H., Wang, S.S.: A Network Model Approach to Systemic Risk in the Financial System January 2013 (2013)

Chen, Y., Zhang, L., Li, K.: The review of network theory applied to the financial system. In: Wuhan International Conference on E-Business (2016)

Cont, R., Moussa, A.B., Santos, E.B.: Handbook on Systemic Risk: Network Structure and Systemic Risk in Banking Systems (2013)

Diamond, D.W., Dybvig, P.H.: Bank runs, deposit insurance, and liquidity. J. Polit. Econ. **91**(3), 401–419 (1983)

ECB (European Central Bank) (2016)

ECB (European Central Bank): ECB Banking Supervision: SSM Supervisory Priorities for 2022–2024 (2021)

ECB (European Central Bank): Financial Stability Review (2022)

Eisenberg, L., Noe, T.H.: Systemic risk in financial systems. Manag. Sci. **47**(2), 236–249 (2001)

ESRB (European Systemic Risk Board): Systemic Cyber Risk, ESRB, February. Available at: https://www.esrb.europa.eu/pub/pdf/reports/esrb.report200219_systemiccyberrisk~101a09 685e.en.pdf (2020)

ESRB (European Systemic Risk Board): ESRB recommends establishing a systemic cyber incident coordination framework. Press release (2022a)

ESRB (European Systemic Risk Board): Mitigating systemic cyber risk. ESRB (2022b)

Furfaro, A., Gallo, T., Saccà, D.: Modeling cyber systemic risk for the business continuity plan of a bank. In: Availability, Reliability, and Security in Information Systems: IFIP WG 8.4, 8.9, TC 5 International Cross-Domain Conference, CD-ARES 2016, and Workshop on Privacy Aware Machine Learning for Health Data Science, PAML 2016, Salzburg, Austria, August 31–September 2, 2016, Proceedings, pp. 158–174. Springer International Publishing (2016)

Gai, P., Kapadia, S.: Networks and systemic risk in the financial system. Oxf. Rev. Econ. Policy **35**(4), 586–613 (2019)

Gong, X., Liu, X., Xiong, X., Zhang, W.: Financial systemic risk measurement based on causal network connectedness analysis. Int. Rev. Econ. Finan. (2019)

Huiying, L., Yuanda, C.: Research on network risk situation assessment based on threat analysis. In: 2008 International Symposium on Information Science and Engineering, vol. 2, pp. 252–257 (2008)

Iori, G., Mantegna, R.N.: Empirical Analyses of Networks in Finance (2018)

Jo, J.: Managing systemic risk from the perspective of the financial network under macroeconomic distress (2012)

Kalinin, M.O., Krundyshev, V.M., Zegzhda, P.D.: Cybersecurity risk assessment in smart city infrastructures. Machines **9**, 78 (2021)

Kotidis, A., Schreft, S.: Cyberattacks and financial stability: evidence from a natural experiment. Finance and Economics Discussion Series, No 2022-025, Board of Governors of the Federal Reserve System (2022)

Lamichhane, P.B., Hong, L., Shetty, S.S.: A quantitative risk analysis model and simulation of enterprise networks. In: 2018 IEEE 9th Annual Information Technology, Electronics and Mobile Communication Conference (IEMCON), pp. 844–850 (2018)

Mixia, L., Dongmei, Y., Qiuyu, Z., Honglei, Z.: Network security risk assessment and situation analysis. In: 2007 International Workshop on Anti-Counterfeiting, Security and Identification (ASID), 448–452 (2007)

Lv, H.: Research on network risk assessment based on attack probability. In: 2009 Second International Workshop on Computer Science and Engineering, vol. 2, pp 376–381 (2009)

Saltoğlu, B., Yenilmez, T.: Analyzing Systemic Risk with Financial Networks An Application During a Financial Crash (2010)

Tsankov, P.: Overview of network-based methods for analyzing financial markets. Proceedings of the Technical University of Sofia (2021)

Wang, Z., Wang, X.: Research on technologies in quantitative risk assessment and forecast of network security. In: 2010 3rd International Conference on Advanced Computer Theory and Engineering (ICACTE), vol. 6, V6-524–V6-528 (2010)

Chapter 9
Time Sensitive and Oversampling Learning for Systemic Crisis Forecasting

Francesco De Nicolò, Marianna La Rocca, Antonio Marrone, Alfonso Monaco, Sabina Tangaro, Nicola Amoroso, and Roberto Bellotti

Abstract The development of early warning systems for systemic crises has recently received growing interests. Recent studies have proposed possible solutions to address this challenging topic, in particular by means of cutting-edge artificial intelligence (AI) approaches. Financial data are fundamentally characterized by intrinsic temporal dynamics and the presence of both short-/long-term interactions. Hence, it is of paramount importance, when validating the proposed solutions to adopt validation strategies which consider this aspect. To this aim, we show here how Temporal Cross Validation (TCV) deeply affects the models' learning. Moreover, to take into account the data imbalance often characterizing these models, we combine the TCV with a popular solution, which is the SMOTE (Synthetic Minority Oversampling TEchnique) algorithm.

Francesco De Nicolò, Marianna La Rocca, Nicola Amoroso and Roberto Bellotti these authors contributed equally to this work.

F. De Nicolò · M. La Rocca · A. Marrone · A. Monaco (✉) · S. Tangaro · N. Amoroso · R. Bellotti
University of Bari Aldo Moro, Bari, Italy
e-mail: alfonso.monaco@uniba.it

F. De Nicolò
e-mail: francesco.denicolo@poliba.it

M. La Rocca
e-mail: marianna.larocca@uniba.it

A. Marrone
e-mail: antonio.marrone@uniba.it

S. Tangaro
e-mail: sonia.tangaro@ba.infn.it

N. Amoroso
e-mail: nicola.amoroso@uniba.it

R. Bellotti
e-mail: roberto.bellotti@uniba.it

F. De Nicolò
Politecnico of Bari, Bari, Italy

© The Author(s) 2025
V. Pacelli (ed.), *Systemic Risk and Complex Networks in Modern Financial Systems*,
New Economic Windows, https://doi.org/10.1007/978-3-031-64916-5_9

Keywords Early warning systems · Systemic crises · Temporal cross validation · SMOTE

Introduction

In the last decades, the national financial systems worldwide have witnessed the onset of a series of systemic crises that have shown systems' vulnerability and inter-dependence (Wilson et al. 2010; Anginer and Demirguc-Kunt 2014; Bongini et al. 2015; Ellis et al. 2014; Silva et al. 2017; Rizwan et al. 2020).

Early warning models (EWMs) are mathematical models, based on suitable economic hypothesis, that aim to forecast these crises. EWMs are of crucial importance since they help in timing policies by providing information about the likelihood of a crisis. In fact, pioneering works on this subject date back to late 1990s (Kaminsky and Reinhart 1999; Casabianca et al. 2019). These works used the time series data of various economic indicators of countries as input and "precrisis dummy variables" as outputs of different Machine Learning models trained to forecast the onset of systemic crises. Precrisis dummy variables are defined as follows: for every country, every time step is labelled as "1" or "0" if the corresponding date represents the onset of a systemic crisis or not, respectively. Then, these values are shifted backward in time according to the considered forecast horizon. The corresponding value in each time-point is called "precrisis dummy variable".

Nonetheless, these studies rarely consider the features characterizing time series and adopt two approaches (Tölö 2020): (1) a fixed train-test splitting that biases the evaluation of the generalization power of the EWMs; (2) a classical cross-validation framework that discards the temporal ordering of time series data.

Moreover, even though systemic crises have deep impacts on countries' economic systems making them radically change to respond to such shocks, they represent rare events in the economic history of countries. Accordingly, there are few time points that represent the onset of a systemic crisis with respect to those that represent "non-crisis" periods. This imbalance between crisis and non-crisis events makes the learning phase of the EWMs hard and affects Machine Learning systems ability of recognizing future crises (Kim et al. 2020). In order to counterbalance this effect, some works (see for example Tölö 2020) consider a one-century or two-centuries time span in order to deal with more crises, but on one hand this does not augment data to delete the unbalancing effect and, on the other hand, there is the risk to train model on old data that do not influence at all the onset of future crises.

In this work, for the first time, at the best of our knowledge, we propose to face both these problems by the combined use of the following techniques: (1) Temporal Cross Validation, aiming at extending the classical cross-validation framework taking into account the chronological ordering of time series data; (2) the SMOTE (Synthetic Minority Oversampling TEchnique) algorithm (Gosain and Sardana 2017), used to augment the minority class in datasets, to balance crisis and non-crisis time points. In particular, the SMOTE algorithm creates "artificial" crisis events based on the input variables characterizing the "natural" crises that are present in the training dataset.

This work is organized as follows. First, the data used for describing the overall crisis forecasting pipeline will be thoroughly described. Then, the two previously introduced techniques (Temporal Cross Validation and SMOTE) will be explained. Afterwards, the results of applying these techniques to data will be shown and discussed. Finally, the limitations and future perspectives of this work will be exposed.

Economic-Financial Indicators for EU Countries and Crisis Events

This chapter is concerned with forecasting the onset of a systemic crisis in an EU country. Accordingly, first, the data used to flag a time period as a *crisis period* in a European country will be shown. Then, the economic indicators used to forecast these crises will be described.

Dependent Variable

To identify the systemic events, we use the database on financial crises provided by the European Systemic Risk Board for EU countries, which gives precise chronological definitions of crisis periods (ESRB 2022). The crises dataset covers all EU Member States and Norway for the period 1970–2020 and consists of a core set of 50 systemic crises, which fulfil a number of conditions including (i) the financial system acting as a shock originator or amplifier and/or (ii) systemic financial intermediaries experiencing distress or going bankrupt and/or (iii) substantial crisis management policy interventions (lo Duca et al. 2017). However, to ensure that the analysis is based on reliable and consistent information, we exclude those countries not reporting important data on a country-level. This process brought to focus on analysing 16 EU countries over the period 2001–2020 in which we identified 12 financial systemic crises. These countries are: Belgium, Czech Republic, Denmark, Finland, France, Germany, Greece, Ireland, Italy, Netherlands, Poland, Portugal, Spain, Sweden, Switzerland, UK.

For our purpose, we define the dependent variable as a binary variable, also denoted as a *dummy variable*, taking a value of "1" in case of a systemic crisis and "0" otherwise. Table 9.1 contains the dates of the registered systemic crises.

Table 9.1 Systemic events in EU countries over the period 2001–2020

Country	Start date
Belgium	November 2007
Denmark	January 2008
France	April 2008
Germany	August 2007
Greece	May 2010
Ireland	September 2008
Italy	August 2011
Netherlands	January 2008
Portugal	October 2008
Spain	March 2009
Sweden	September 2008
UK	August 2007

Source European systemic risk board

Independent Variables

To measure the key variables able to predict systemic events, we study several monetary, economic and financial indicators. We divide 17 indicators into 5 groups, which hereafter we call *classes*. The classes are as follows; **monetary**: monetary aggregate (m1), three-month interbank rates of loans (ir3), five-year treasury rate (tr5); **bond**: government bond (gov_bond10); **real economy**: unemployment rate (Unemployment), consumer price index (cpi), trade balance (trade_balance), industrial production index (ind_prod), economic policy uncertainty index (econ_pol), inflation rate (inflation_rate), productivity rate (prod_rate); **stock**: Morgan Stanley Capital International index (msci_action), Morgan Stanley Capital International banks index (msci_bank); **financial**: total assets (total_asset), stock market capitalization (stock_capitalization), domestic credit to the private sector as a percentage of the Gross Domestic Product (GDP) (priv_sec_credit) and sovereign loans as a percentage of the GDP (sovereign_loans). All the indicators not providing monthly data are transformed into monthly observations using linear interpolation.

To run our analysis, since our economic-financial indicators do not have the same unit of measure and models are sensitive to the range of the input data, the former should be put in a normalized form before being fed to the latter (Zheng and Casari 2018). Accordingly, we consider the following normalization pipeline:

1. Every variable is normalized in the training period for all the countries: the maximum value of the variable in the training period is 1, while its minimum is 0.

2. For each country, we calculate the first differences of each normalized variable: if $Z_i^C(t)$ is the ith normalized variable at time t of country C, then we consider:

$$X_i^C(t) = Z_i^C(t) - Z_i^C(t-1)$$

The normalized variables X_i^C are all pure numbers and comparable: they can be safely used in the forecasting models to avoid the range bias.

The monetary aggregate has been on an upward trend while interest rates and treasury rates have been declining over the sample period. This trend is also reflected in the yield spread of government bonds, which has decreased by around 0.06%. This can be attributed to the sovereign debt crisis that occurred in 2009–2010, which resulted in a surge in the yield spread of government bonds. In terms of real economic indicators, only the inflation rate declined during the period under consideration. However, other indicators have remained stable with a growth of no more than 0.39% in the case of the consumer price index. The Morgan Stanley Capital International (MSCI) Index has shown a slight increase, while the MSCI bank index has seen a slight decrease, which is likely due to the non-performing loan (NPL) issues that banks in the Eurozone have faced. All financial indicators have shown slight growth from 2001 to 2017, with the average total assets of banks growing by around 0.06%, the stock market capitalization by around 0.02%, loans granted to the private sector as a percentage of GDP by around 0.04%, and sovereign loans by around 0.1%. Overall, the analysis provides insights into the behaviour of various economic and financial indicators, highlighting the impact of the 2009–2010 sovereign debt crisis on government bonds and the banking sector, as well as the overall growth of financial indicators during the sample period.

Forecasting Time Series: Windowing, Temporal Cross Validation and SMOTE

In this section, we focus on the methods used to forecast the onset of a systemic crisis in a country one year in advance. First, we will show how this problem can be properly framed in a binary classification setting, then we will describe the machine learning algorithms used to forecast systemic crises.

Time Series Input Variables and Forecast Horizons

Following the approach of Holopainen and Sarlin (2017) and Ristolainen (2018), we introduce a pre-crisis dummy variable (Detken et al. 2014), which involves shifting the crisis dummy variable backward in time based on the forecast horizon (one year, in our case) and removing all observations from that point forward. The pre-crisis dummy variable is a crucial output variable in the crisis forecasting task. In fact, it allows us to focus the analysis on the period leading up to a potential systemic crisis and identify early warning signals of an impending crisis. This approach is especially relevant for policymakers and regulators who can take proactive measures to prevent or mitigate the impact of a crisis.

Since the inputs are represented by the time series of economic and financial variables, they need to be divided into shifted time-windows before being used in a forecasting model (Granger 2014). This approach is necessary to capture the temporal dynamics and correlations between the variables, as well as the lead time of each variable with the crisis event. In particular, we consider time-windows with a 1-month shift. Nonetheless, as underlined by Jeon and McCurdy (2017), it is not possible to establish a priori the choice of an optimal size for the length of the time-windows. Accordingly, in this work, we use two window-sizes: 12 and 24 months. Since the original time intervals of the variables are not homogeneous and the maximal time interval is 1-year, we found it wise to use a 12-months-step (i.e., 1-year-step) in going from a time-window size to another. Moreover, we chose to consider up to two years of data in order to use the more recent available data (from the crisis date) and, at the same time, avoid the presence of *not-available* values in the data.

Since the input features are the time-windows of the economic-financial variables and the output variables are the pre-crisis dummy variables, we need to pair these two elements to obtain a dataset to train and test our forecasting models.

Suppose we consider a country C and N economic-financial variables depending on time $(X_1^C(t), \ldots, X_N^C(t))$, a pre-crisis dummy variable at time t_0 (say it y_0) and consider a time window with size T. Then, for each $i = 1, \ldots, N$, we consider the set of elements $X_i^C(t_0 - T - 1), \ldots, X_i^C(t_0)$, representing the variable X_i^C in a time window of size T, beginning at time $t_0 - T - 1$ and ending at time t_0 (the time corresponding to the pre-crisis dummy variable). Finally, we associate the sets $(X_i^C(t_0 - T - 1), \ldots, X_i^C(t_0))$, $i = 1, \ldots, N$, to the pre-crisis dummy variable y_0. Varying t_0 in all the time interval we aim to explore (from 1/2001 to 7/2017 in our case), we end up with a dataset pairing the time-windows of the N variables of each country C and the corresponding pre-crisis dummy variable.

Temporal Cross Validation

The usual cross-validation framework is well represented in the left part of Fig. 9.1: the dataset is divided into some *folds* (usually five or ten folds), a Machine Learning model is trained on all folds but one (the *validation fold*) and then it is tested on this validation fold. This approach is repeated until all folds are used for validation. Since this approach does not take into account the temporal ordering of folds in case of time

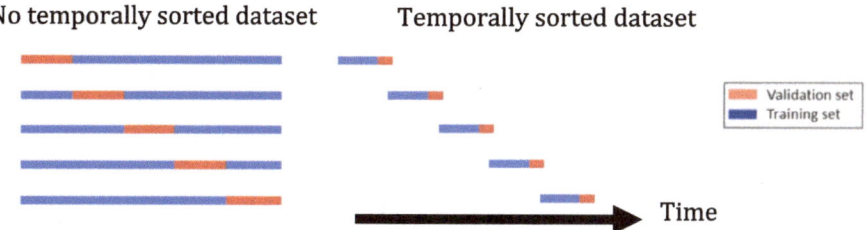

Fig. 9.1 Left: classical cross-validation framework; Right: Temporal cross-validation. The length of training and test folds are chosen by the experimenter

series data, a modification of this approach is needed. Accordingly, this modification is shown in the right part of Fig. 9.1: the temporal length of the training and test folds is a-priori determined by the experimenter, then the partition chosen for the training and test folds is continuously shifted along the time interval considered in the dataset, in accordance with the length of the training and test folds. This approach is known as Temporal Cross Validation (Roberts et al. 2017). In particular, in our case we consider a 1-year long validation-fold, while training-folds' length, i.e. the window-length, is set equal to 12 and 24 months, as explained in the previous section. The threshold date between training and validation folds varies from 8/2007 (the date of the first registered systemic crises in UK and Germany, as reported in Table 9.1) to 8/2010 (1 year before the Italian systemic crisis) with a 1-month shift.

For each training fold and window-length, the windowing approach described in the previous section is applied. As regards data in the validation fold, the pipeline is slightly modified as follows: if a country's crisis happens after the threshold date and within 1 year from it (since our forecast horizon is 1 year), then all dates following this crisis are not discarded and are also labelled as "1" for the considered country. This modification is due to two main reasons: (1) ideally, the experimenter does not know what happens after the threshold date so that it should be impossible for her/him to choose a priori the single time window associated to the onset of the future crisis; (2) if the post-crisis dates for a country are not considered, it is impossible to correctly label the corresponding test time-windows. In fact, the time windows related to post-crisis dates could not be labelled at all, so determining a data leakage phenomenon (if a test time-window cannot be labelled, this means that it belongs to a crisis period, so implicitly labelling that time-window through the knowledge of what happens in the test time period). Moreover, following this framework, it can be determined the onset of a crisis *within* one year in the future.

Figure 9.2 depicts what happens in back-shifting the dummy variable and why the test fold must be treated differently from the training one.

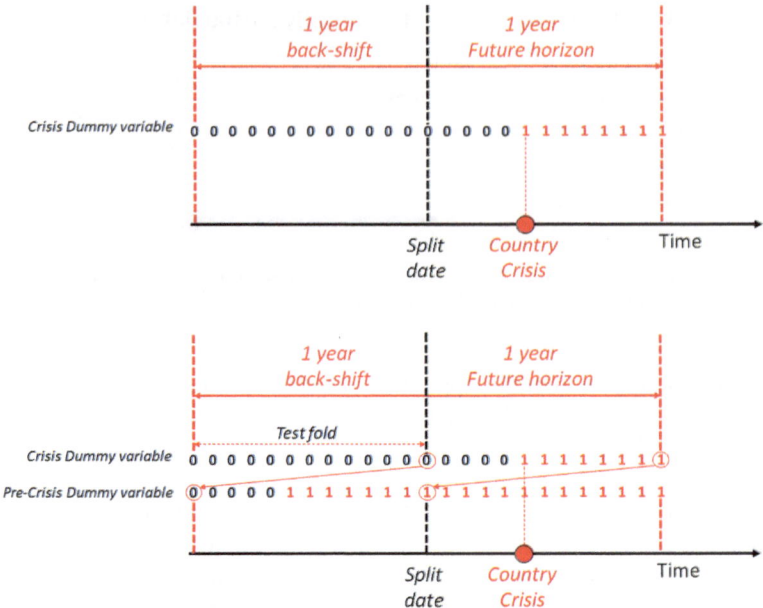

Fig. 9.2 The back-shift operation on the test fold containing one systemic crisis. Top panel: the *split date* divides the train fold from the test one (denoted as *Future horizon* in the figure and having a temporal extent of 1 year). The *crisis dummy variable* marks every single time step as "non-crisis" ("0") or "crisis" ("1"). Bottom panel: the values of the crisis dummy variable is shifted backward by one year (the Future horizon), creating the *pre-crisis dummy variable*. Accordingly, every time step, and every window length, can now be labelled by the crisis label pertaining to the one-year-away time step from it. Without labelling the post-crisis dates, some labels could not be assigned to the dates in the test fold

SMOTE Algorithm

Unbalanced classification is characterized by the fact that there are too few examples of the minority class for a model to effectively learn the decision boundary. One naive way to solve this problem is oversampling the examples in the minority class. This can balance the class distribution but does not provide any additional information to the model. An improvement on simply duplicating examples from the minority class is to synthesize new samples for this class, implementing a data augmentation algorithm.

The most widely used approach to accomplish this task is the Synthetic Minority Oversampling TEchnique, or SMOTE for short. This technique was introduced and described by Nitesh Chawla and collaborators (Chawla et al. 2002).

SMOTE works according to the following steps:

Fig. 9.3 Graphical
representation of the
SMOTE algorithm in a 2-D
feature space

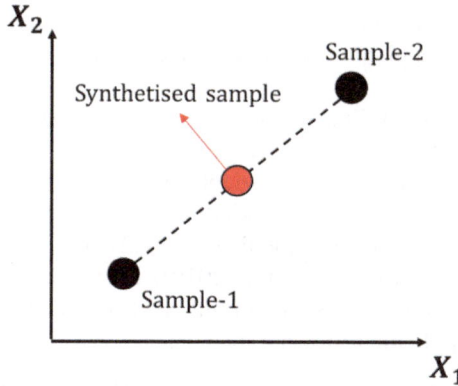

1. select a random sample of the minority class (i.e. the class of the events representing a crisis in our case), denote as Sample-1.
2. Find its k-nearest neighbours in the feature space. In our case, the features characterizing each sample are the time windows of the economic-financial variables used as input of the Machine Learning models. Moreover, even though typically $k = 5$, in this work we use k equal to the number of all the remaining crisis events, because of their small number.
3. Randomly select one of these neighbours, denoted as Sample-2.
4. Randomly select a point in the feature space lying on the line joining Sample-1 and Sample-2. This point represents a synthetic sample.

Figure 9.3 represents the functioning of the SMOTE algorithm for a 2-D feature space, for the sake of representability. The SMOTE algorithm is used in the training folds in order to have the same number of crisis ("1") and non-crisis ("0") events.

Time Series Forecasting: Machine Learning Algorithms and the Crisis Likelihood

Our dataset is heavily unbalanced because we have a set of events corresponding to non-crisis periods (labelled as "0", as described previously) overwhelming the set of crisis periods (labelled as "1"). As underlined by Kim et al. (2020), we face the problem that models trained on unbalanced datasets have poor results when they have to generalize. Many techniques have been developed to solve this problem (Spelmen and Porkodi 2018), such as "undersampling" and "oversampling" techniques. Among the "oversampling" techniques we use, as explained in the previous section, the SMOTE algorithm is based on the input variables characterizing the "natural" crises that are present in the training dataset. Thanks to this technique, we created a number of artificial crises in the training folds, equal to that of non-crisis events. Afterwards, we applied different Machine Learning algorithms to forecast systemic crises in the test folds.

According to the previous discussion, we have input variables composed of windowed economic-financial variables (with different window-sizes) and a binary pre-crisis dummy variable as the output. Then we are faced with a binary classification problem. These kinds of problems are common in applications (credit card fraud detections (Moumeni et al. 2022), medical testing (Buch et al. 2018), and information retrieval (Li et al. 2019)). Many classification algorithms have been developed for these tasks and, in this work, following the economic literature in this field, we apply the following models:

- Logistic regression (Logit) (Kiley 2021): it directly models the probability a systemic crisis begins in a given date $p(y = 1)$ (where y is the pre-crisis dummy label of an instance), given inputs $X_1, \ldots, X_N$.
- Random Forest (RF) (Breiman 2001): it is a generalization of decision trees [88]. In fact, Random Forest is an ensemble learning method that works by constructing a multitude of decision trees at training time. In particular, every tree is trained on a bootstrapped sample of training data (i.e. sampling with replacement from training data) and each tree uses a random subset of predictors to take decisions, in order to overcome the presence of strong predictors. The output of the random forest is the class selected by most trees (majority vote rule). Taking decisions based on an ensemble of trees greatly improves the performance of a single decision tree (Ho 1995).
- Support Vector Machine (SVM) (Noble 2006): it can be used for both regression and binary classification problems and it is based on finding, in the feature-space, the best hyperplane subdividing training points (i.e. data) of one class from those belonging to the other one. In particular, consider a training dataset of N items and with M input features. These items may be represented as $(\vec{x_1}, y_1), \ldots, (\vec{x_N}, y_N)$, where $\vec{x_i}$ is the M-dimensional vector of input variables of i-th data item and y_i is the corresponding binary label (0 or 1, for example). They may be considered as geometrical points in the M-dimensional feature space. The target of SVM algorithm is to find the maximum margin hyperplane: the hyperplane which is defined so that the distance between the hyperplane and the nearest point from either group is maximized. Figure 9.4 clearly explains the result of the SVM algorithm in a dataset with two input-features.
- eXtreme Gradient Boosting (XGB) (Bentéjac et al. 2021): it is a model in the form of an ensemble of decision trees (also called weak learners), but, differently from RF, it is built in an iterative fashion and learns slowly. In fact, trees in RF are trained on different bootstrapped samples taken from the training dataset, independently of each other; XGB, instead, does not involve bootstrap sampling but every tree is grown using information from previously grown trees, being fit on a modified version of the training dataset. The main idea underpinning XGB

Fig. 9.4 The result of the SVM algorithm applied to a dataset with two input features (x_1 and x_2). The two classes are reported in blue and green as colors of the data points. The maximum margin hyperplane is reported in red

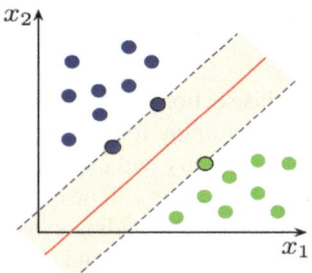

is that, given the current model, we fit a decision tree to the residuals from the current model. Then, we add this new decision tree into the current model in order to update the residuals. By iteratively fitting trees to the residuals, we improve the current model in areas where it does not perform well.

Since the problem we are facing is a classification task, it could seem reasonable to consider the usual classification metrics (e.g. AUC-ROC, sensitivity, specificity) to determine the best model in identifying the onset of a systemic crisis in the test folds. Nonetheless, it should be considered that the events labelled as "1" in the test folds are not only those representing the onset of a crisis, but even those representing immediately following periods, as explained previously. Accordingly, the classical metrics would give a biased picture of the models' performances. As a result, it seems wise to consider the *crisis likelihood* as a parameter for comparing the different models. As a matter of fact, in order to determine if an event represents the onset of a crisis or not, a model outputs a number (the *crisis likelihood*) between 0 and 1 and if this number is over a given threshold (usually 0.5), then it is labelled as "1", otherwise as "0". This likelihood is sometimes denoted also as the *crisis probability*, but it cannot be interpreted directly as a probability in the mathematical sense (for more information on this aspect, please refer to Vaicenavicius, et al. 2019). Moreover, considering how the crisis likelihood of a country varies with time is helpful in determining precisely the date in which it becomes more prone to develop a systemic crisis and what model is more effective in pointing it out. Moreover, following this approach, every crisis will be discussed individually, so identifying the presence of problematic scenarios.

In particular, in the next section, we will show, for every country and every Machine Learning model, the crisis likelihood corresponding to the furthest threshold-date away from the country's crisis date, but no more than one year away from the latter. This will be done for each window-length of the training fold (12 months, 24 months). Accordingly, we will be also able to point out the best training folds' length for forecasting the onset of a systemic crisis.

Dynamics of Crisis Likelihoods

In this section we will show the models' crisis likelihood for every window length of the training fold in the Temporal Cross Validation framework. In particular, for every country and every training folds' length, the furthest threshold date from the country's crisis will be considered (no more than one year away from the crisis date) and the corresponding evolution of the crisis likelihood will be shown. It should be noted that, as UK's and Germany's crisis are the first, they cannot be forecast by any model but they are used to train models and forecast future crises. Accordingly, these two countries do not appear in any test fold and are not reported in the following results. It should be noted that, since models' training phase is implemented through stochastic algorithms (in Python 3.9), the whole pipeline is repeated 100 times and the corresponding standard deviation of the results is reported as vertical error bars in the following graphs.

Figure 9.5 shows the crisis likelihood for Belgium determined by all the considered Machine Learning models. The furthest threshold date from Belgium's crisis (11/2007) is 8/2007, just three months before it. Dates missing in the right-side plot are due to the presence of not-available values of one or more indicators (i.e. independent variables) in the corresponding time windows. This consideration applies also to all the following graphs. It can be readily seen that XGB radically changes its behaviour in going from the 12-months window-length case to the 24-months one. In particular, in the first case, it clearly signals the onset of a crisis one year in the future beginning from 2/2007 (left plot in Fig. 9.5), while it completely rejects this hypothesis in the second case (right plot in Fig. 9.5). This indicates a swinging behaviour an experimenter cannot rely on, especially if compared with all the other models.

In fact, RF has a smoother behaviour in going from the 12-months window-length to the 24-months one. Nonetheless, it is not able to forecast the onset of the crisis in both cases.

In contrast, LR and SVM have both a smooth behaviour and clearly signal the onset of a crisis in Belgium in one year beginning from 6/2007. In particular, in the 12-months case LR has a monotonically growing crisis likelihood that goes over 0.5 at 5/2007; SVM's crisis likelihood is also monotonically growing in time but it overcomes 0.5 at 6/2007. This indicate that both models point out the outbreak of a systemic crisis in Belgium in one year beginning from 6/2007.

Moreover, as the window-length grows from 12-months to 24-months (i.e. older data are fed in the models), both models show that a crisis' outbreak will take place within a year from 5/2007. This means that even older data are useful for the models in order to determine the onset of crisis in Belgium.

This is a remarkable result: the experimenter's knowledge stops at 8/2007, where just two crises have just taken place (in UK and Germany, as reported in Table 9.1), but this limited knowledge is sufficient to forecast the onset of a crisis in Belgium and to determine the date in which the indicators have such values to determine its

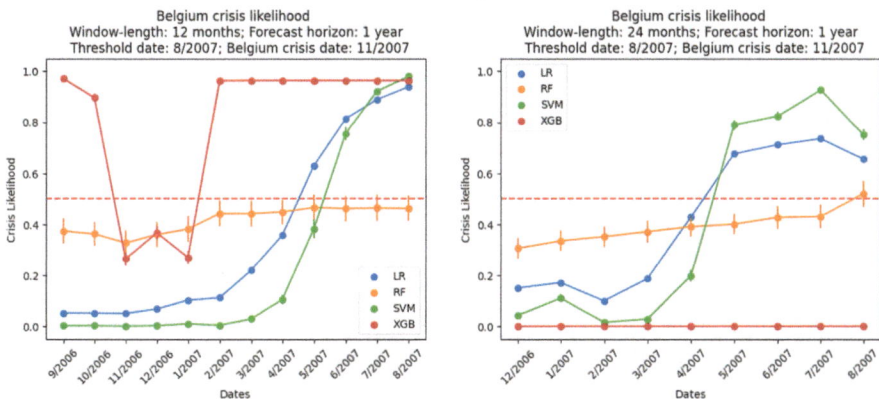

Fig. 9.5 Crisis likelihoods determined by all models for Belgium. The furthest threshold date from Belgium's crisis corresponds to 8/2007. Left: Crisis likelihoods for window-length equal to 12 months. Right: Crisis likelihoods for window-length equal to 24 months. Error bars are due to the stochastic nature of the models' training phase

onset. This could also indicate that the Belgian crisis has the same features as those happened in UK and Germany.

Figure 9.6 depicts models' crisis likelihoods for Denmark. Even in this case, the furthest threshold date from the Danish crisis (1/2008) is 8/2007, just five months before. It is possible to underline the same behaviour of the models reported in the Belgian case: XGB has a too swinging behaviour; RF is not able to forecast the onset of the crisis; LR and SVM have a smoother behaviour and are able to signal the outbreak of a crisis. In particular, for the 12-months case, LR has a growing crisis likelihood that overcomes 0.5 at 7/2007, while SVM outputs a strong crisis' signal at 6/2007. This indicates the onset of a crisis in one year beginning from 7/2007. Nonetheless, in the 24-months case, LR and SVM have both much lower likelihood than in the 12-months case: LR is no more able to output a strong signal crisis (it is always under 0.5), while SVM has a maximum likelihood of 0.72 at 6/2007 and then decreases. This means that adding older data does not help models, in the Danish case, in forecasting the future onset of the crisis. This behaviour is different from what has been seen previously for the Belgian crisis. It seems wise to underline that, even in this case, just the crises in UK and Germany are known to the experimenter, since they are the only systemic crises known at 8/2007.

Figure 9.7 shows models' crisis likelihoods for the Netherlands. Since the date of the Dutch crisis is the same as the Danish one, the furthest threshold date from the crisis (1/2008) is again 8/2007. It is possible to observe the same behaviour for almost all models: LR, SVM and XGB give a strong crisis signal at 8/2007 in the case of a window-length having a 12-months size (left plot in Fig. 9.7). This means that, according to these models, a crisis will outbreak in the Netherland in one year

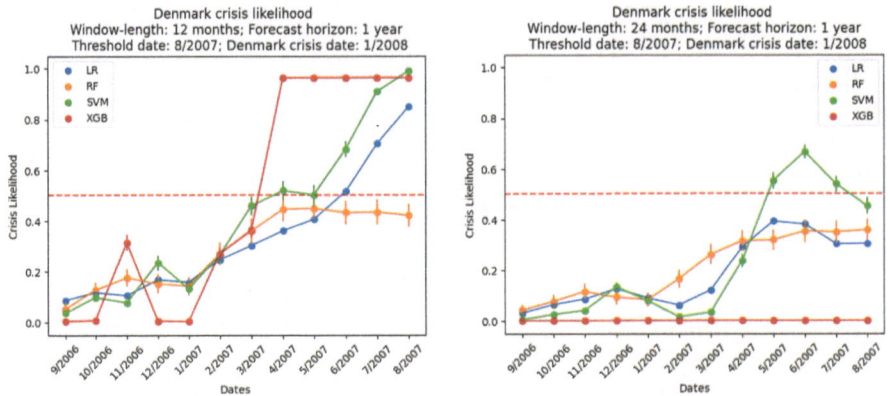

Fig. 9.6 Crisis likelihoods determined by all models for Denmark. The furthest threshold date from Denmark's crisis corresponds to 8/2007. Left: Crisis likelihoods for window-length equal to 12 months. Right: Crisis likelihoods for window-length equal to 24 months. Error bars are due to the stochastic nature of the models' training phase

beginning from 8/2007. Moreover, these signals are completely missed in the 24-months case (right plot in Fig. 9.7). This indicates that older data are confounding and are useless in determining the onset of the systemic crisis in the Netherlands: just more recent indicators are useful in pointing out the outbreak of the Dutch crisis.

Figure 9.8 depicts models' crisis likelihoods for France for a window size of 12-months (left plot) and 24-months (right plot). The furthest threshold date from the French crisis (4/2008) is 8/2007. It can be noted again the swinging behaviour of the XGB crisis likelihood in passing from the 12-months case to the 24-months one

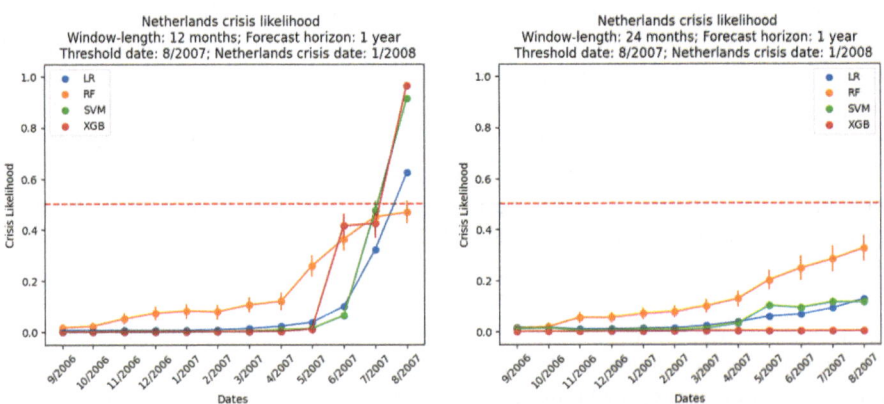

Fig. 9.7 Crisis likelihoods determined by all models for the Netherlands. The furthest threshold date from the Dutch crisis corresponds to 8/2007. Left: Crisis likelihoods for window-length equal to 12 months. Right: Crisis likelihoods for window-length equal to 24 months. Error bars are due to the stochastic nature of the models' training phase

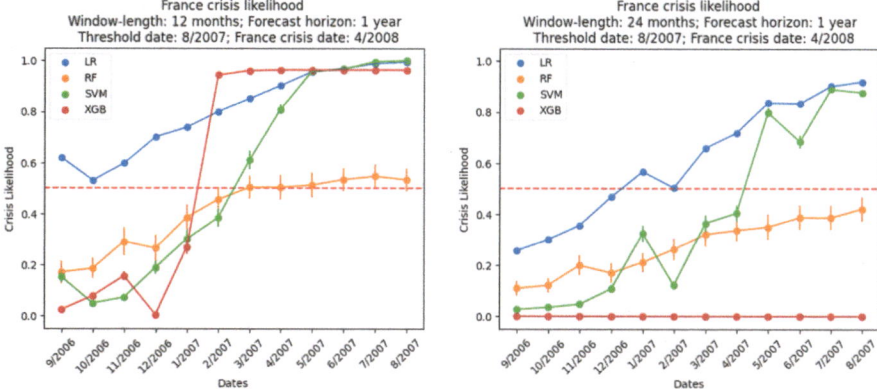

Fig. 9.8 Crisis likelihoods determined by all models for France. The furthest threshold date from the French crisis corresponds to 8/2007. Left: Crisis likelihoods for window-length equal to 12 months. Right: Crisis likelihoods for window-length equal to 24 months. Error bars are due to the stochastic nature of the models' training phase

and the lower crisis signals of the RF model. Moreover, it can be seen that, in the 12-months case, LR's crisis likelihood is always over 0.5, signalling the outbreak of a crisis in one year beginning from 9/2006, while SVM outputs a strong signal at 3/2007. Moreover, in the 24-months cases, the likelihood of both LR and SVM are all lower than in the 12-months one, even though both point out a crisis in one year beginning from 5/2007. This indicates that older data are confounding for the French case.

Figure 9.9 shows models' crisis likelihoods for Ireland for a window size of 12-months (left plot) and 24-months (right plot). The furthest threshold date, at most one year away from the Irish crisis (9/2008), is now 9/2007. It should be underlined that, in the 12-months case, LR and SVM have a monotonically growing likelihood: LR's likelihood overcomes 0.5 at 6/2007, while SVM at 5/2007. Then, it can be stated that, according to these models, a crisis will take place in Ireland in one year beginning from 6/2007. This means that the signals of a systemic crisis in Ireland are present more than one year before its onset. RF's and XGB's likelihoods are both well under 0.5.

In the 24-months case, all models give a likelihood over 0.5 at 9/2007, while they are all lower in all the previous dates. Then, despite the fact that using older data confounds LR and SVM, a strong crisis-signal in Ireland can be observed just one year before its onset (9/2007). It should be noted that, even in this case, the only crises seen before the threshold date are those happened in UK and Germany in 8/2007.

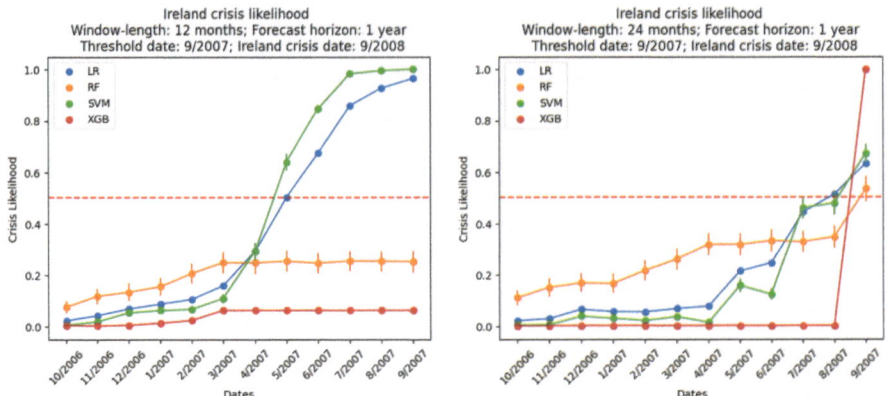

Fig. 9.9 Crisis likelihoods determined by all models for Ireland. The furthest threshold date from the Irish crisis corresponds to 9/2007. Left: Crisis likelihoods for window-length equal to 12 months. Right: Crisis likelihoods for window-length equal to 24 months. Error bars are due to the stochastic nature of the models' training phase

Figure 9.10 illustrates models' crisis likelihoods for Sweden for a window size of 12-months (left plot) and 24-months (right plot). The furthest threshold date is the same as the Irish one: 9/2007. The behaviour of the models' likelihood is similar to that observed in the Irish crisis. In particular, in the 12-months case LR and SVM have likelihood both over 0.5 at 4/2007, signalling the onset of a systemic crisis in Sweden in one year beginning from 4/2007. This means that the Swedish crisis could have been forecast with more than one year in advance and with the knowledge of only the crises happened in UK and Germany in 8/2007.

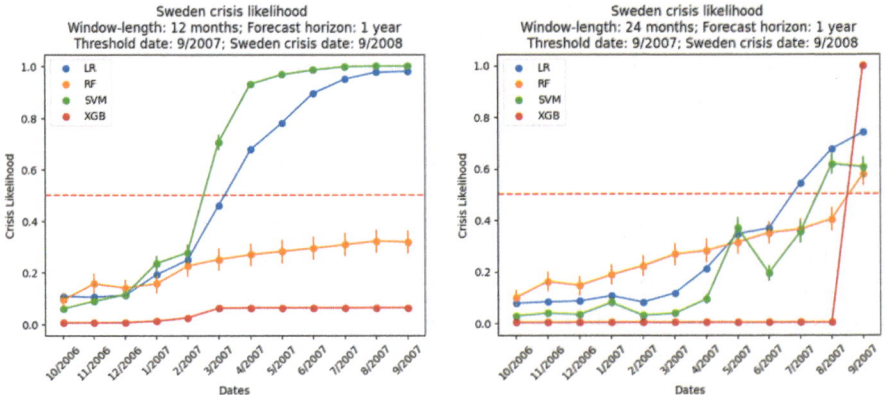

Fig. 9.10 Crisis likelihoods determined by all models for Sweden. The furthest threshold date from the Swedish crisis corresponds to 9/2007. Left: Crisis likelihoods for window-length equal to 12 months. Right: Crisis likelihoods for window-length equal to 24 months. Error bars are due to the stochastic nature of the models' training phase

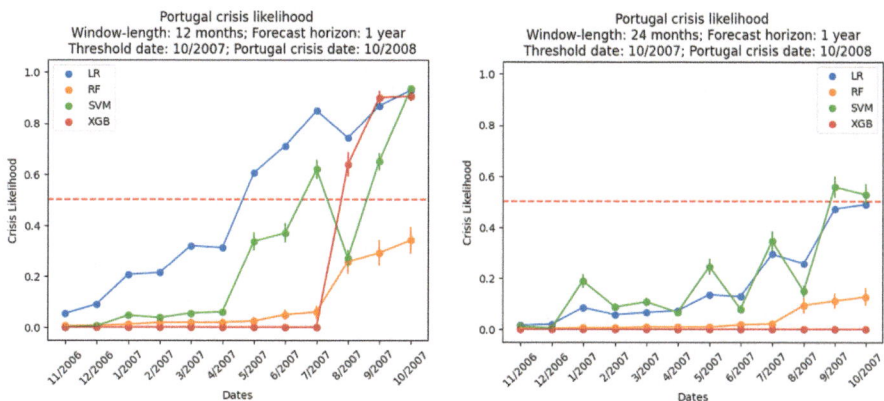

Fig. 9.11 Crisis likelihoods determined by all models for Portugal. The furthest threshold date from the Portuguese crisis corresponds to 10/2007. Left: Crisis likelihoods for window-length equal to 12 months. Right: Crisis likelihoods for window-length equal to 24 months. Error bars are due to the stochastic nature of the models' training phase

In the 24-months case, LR and SVM have both lower signals than before but both point out a crisis in Sweden in one year beginning from 8/2007. RF and XGB both have a crisis likelihood greater than 0.5 at 9/2007.

Figure 9.11 depicts models' crisis likelihoods for Portugal for a window size of 12-months (left plot) and 24-months (right plot). The furthest threshold date now is 10/2007. In this case, LR gives strong crisis signals from 5/2007 onwards in the 12-months case, while SVM does not have a clear behaviour before 9/2007. RF is unable to detect the onset of the crisis, while XGB signals the crisis' outbreak from 8/2007 onwards. Then LR, SVM and XGB all point out the onset of a crisis more than one year before its onset. The situation is completely different in the 24-months case: XGB and RF are unable to forecast the crisis; LR's likelihood shows a growing trend but it never overcomes 0.5. SVM slightly overcomes 0.5 at 9/2007.

Then, even in this case, adding older data do not help in forecasting the systemic crisis in Portugal.

Figure 9.12 shows models' crisis likelihoods for Spain for a window size of 12-months (left plot) and 24-months (right plot). The furthest threshold date now is 3/2008. It can be readily stated that, in the 12-months case, LR and SVM both give strong crisis-signals for Spain at 7/2007. This means that both models forecast the onset of the Spanish crisis in one year from 7/2007, that is 9 months before its effective outbreak. RF and XGB both give a likelihood near to 0 to this crisis. Moreover, the situation slightly changes in the 24-months case: LR and SVM have likelihood overcoming 0.5 from 10/2007 onwards, while RF and XGB are again not able to forecast the onset of the Spanish crisis. Then, in this second case, LR and SVM are able to forecast the outbreak 5 months before its real onset. It should be

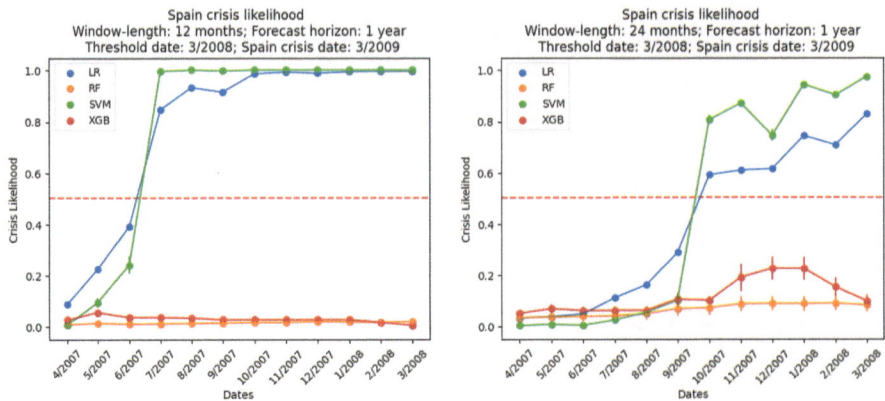

Fig. 9.12 Crisis likelihoods determined by all models for Spain. The furthest threshold date from the Spanish crisis corresponds to 3/2008. Left: Crisis likelihoods for window-length equal to 12 months. Right: Crisis likelihoods for window-length equal to 24 months. Error bars are due to the stochastic nature of the models' training phase

noted that, as the threshold date is 3/2008, the models are trained using all the crises happened up to that date (UK, Germany, Belgium, Denmark, the Netherlands).

Figure 9.13 illustrates models' crisis likelihoods for Greece for a window size of 12-months (left plot) and 24-months (right plot). The furthest threshold date now is set equal to 5/2009, one year before the Greek crisis. It can be readily observed that, in the 12-months case, no model is able to forecast the onset of a systemic crisis in Greece. On the contrary, in the 24-months case, LR and SVM both give strong signals of crisis at 4/2009, pointing out the onset of the systemic crisis in Greece one year before its outbreak in 5/2010. The situation is different from what has been observed previously: 12-months data of the economic-financial indicators are not sufficient to point out the onset of a systemic crisis in Greece; adding older data (window size of 24-months) helps the model in forecasting the crisis. This may indicate that the roots of the Greek crisis are not described by recent (a time-window size of 12-months) values of the indicators, but in their older values (a time-window of size 24-months).

Figure 9.14 shows models' crisis likelihoods for Italy for a window size of 12-months (left plot) and 24-months (right plot). The furthest threshold date now is set equal to 8/2010, one year before the Italian crisis (8/2011). It can be readily seen that the Italian crisis quite different from all those seen before. In fact, no model is able to forecast the onset of this crisis, for both 12-months and 24-months window-length. This may be due to two main reasons: (1) even older data are needed to obtain a crisis likelihood overcoming 0.5; (2) the Italian crisis is different from all the other ones seen before. In fact, since the threshold date is 8/2010, the models are trained on all the other previous crises. Accordingly, the Italian case deserves more attention from an economical point of view: it should be outlined in what aspects the Italian

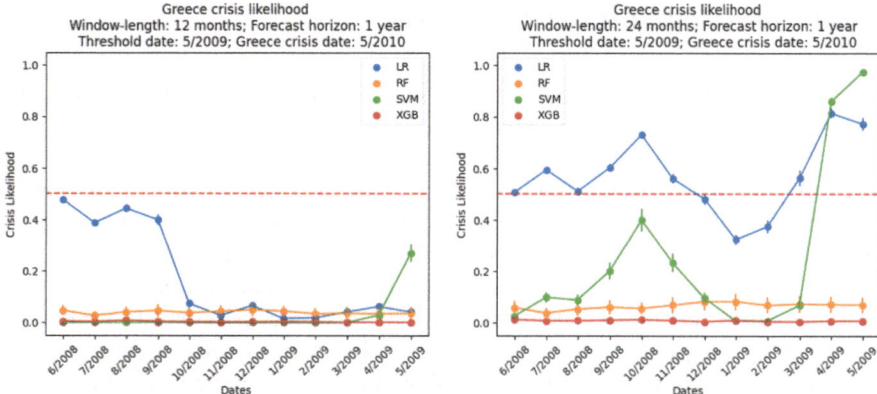

Fig. 9.13 Crisis likelihoods determined by all models for Greece. The furthest threshold date from the Greek crisis corresponds to 5/2009. Left: Crisis likelihoods for window-length equal to 12 months. Right: Crisis likelihoods for window-length equal to 24 months. Error bars are due to the stochastic nature of the models' training phase

crisis is different from all the others and, consequently, which indicators should be considered in order to let the models point out this crisis effectively.

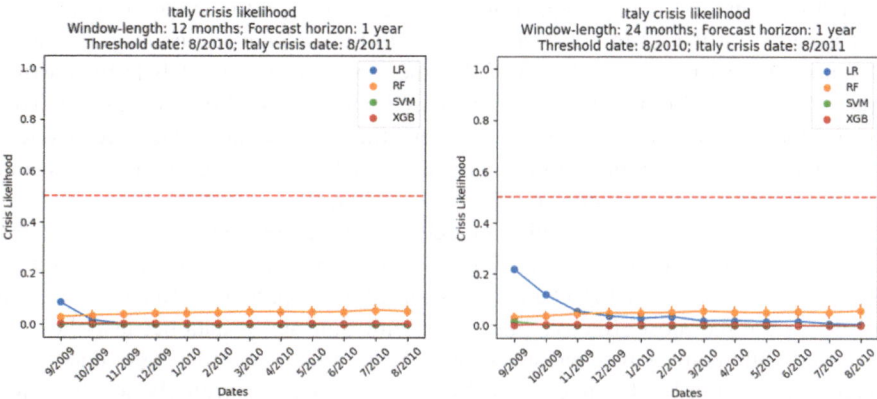

Fig. 9.14 Crisis likelihoods determined by all models for Italy. The furthest threshold date from the Greek crisis corresponds to 8/2010. Left: Crisis likelihoods for window-length equal to 12 months. Right: Crisis likelihoods for window-length equal to 24 months. Error bars are due to the stochastic nature of the models' training phase

Conclusions and Future Perspectives

From the results shown in the previous sections, some conclusions and future perspectives of this work can be drawn.

First, a globally optimal window-length does not exist. In fact, most of the crises (e.g. Ireland, Sweden, Portugal, Spain) can be clearly pointed out using a 12-months window-length, while adding older data, using a 24-months window-length, confounds models and lowers their crisis likelihood. This indicates that the roots of these crises, as described by the considered indicators, should be found in the 12 months before the threshold date. Nonetheless, other crises need an extension of the window-length to 24-months in order to be clearly forecast, like the Greek crisis. In this case, the roots of the crisis should be sought in the values the indicators assume in the 24-months before the threshold date. Greater attention should be paid to the Italian crisis, since all models are not able to forecast its onset. This can be seen as an indication that it is quite different from all the other crises happened before and, accordingly, more effort should be devoted in understanding its origin and which indicators should be used to describe it.

Second, Logistic Regression (LR) and Support Vector Machine with linear kernel (SVM) are the more effective in forecasting the onset of systemic crises. On the other hand, while Random Forest (RF) is unable to point out their outbreak one year in advance, the eXtreme Gradient Boosted model (XGB) is both ineffective in predicting some crises (e.g., Spain and Greece) and has a too swinging behaviour in passing from the 12-months window-size to the 24-months one (e.g. Belgium, Denmark, France). It should be noted that the more effective models in forecasting crises, LR and SVM, are also the more biased ones among those considered. This means that those that are more prone to overfitting (like RF and XGB) are also less able to leverage their training to forecast future crises. This result is well known in the literature (Tölö 2020) and is confirmed by this work.

Only classical Machine Learning models have been considered in this work, but a novel improvement on Early Warning Models has been the introduction of Neural Network architectures (MLP) (Yang and Wang 2019). In particular, Recurrent Neural Networks are the most effective in dealing with time series data because they are endowed with memory mechanisms (Tölö 2020). In fact, the ordering of data is not taken into account by the models used in this analysis. Accordingly, a future step forward of this work will be the use of Neural Network models, both in their MLP and Recurrent form.

In this work just two window-lengths have been considered (12-months and 24-months). Even though considering longer data may be questionable from an economic point of view (Hendry and Richard 1983), it could be useful from a modelling perspective, since it could let us point out the older periods in which the roots of a crisis can be found. Moreover, it is also possible to consider different sizes of the time windows (e.g. 6-months, 8-months, 10-months) in order to be more accurate in identifying the period in which the indicators signal the future onset of a crisis.

References

Anginer, D., Demirguc-Kunt, A.: Has the global banking system become more fragile over time? J. Financ. Stab. **13**, 202–213 (2014)

Bentéjac, C., et al.: A comparative analysis of gradient boosting algorithms. Artif. Intell. Rev. **54**, 1937–1967 (2021)

Bongini, P., Nieri, L., Pelagatti, M.: The importance of being systemically important financial institutions. J. Bank. Finance **50**, 562–574 (2015)

Breiman, L.: Random forests. Mach. Learn. **45**, 5–32 (2001)

Buch, V.H., Ahmed, I., Maruthappu, M.: Artificial intelligence in medicine: current trends and future possibilities. Br. J. Gen. Pract. **68**(668), 143–144 (2018)

Chawla, N.V., Bowyer, K.V., Hall, L.O., Kegelmayer, W.P.: SMOTE: synthetic minority over-sampling technique. J. Artif. Intell. Res. **16**, 321–357 (2002)

Detken, C., Weeken, O., Alessi, L., Bonfim, D., Boucinha, M., Castro, C., Welz, P.: Operationalising the countercyclical capital buffer: indicator selection, threshold identification and calibration options. ESRB: Occasional Paper Series (2014/5) (2014)

Ellis, L., Haldane, A., Moshirian, F.: Systemic risk, governance and global financial stability. J. Bank. Finance **45**, 175–181 (2014)

Gosain, A., Sardana, S.: Handling class imbalance problem using oversampling techniques: a review. In: 2017 International Conference on Advances in Computing, Communications and Informatics (ICACCI), pp. 79–85. IEEE (2017)

Granger, C.W.J.: Forecasting in Business and Economics. Academic Press (2014)

Hendry, D.F., Richard, J.F.: The econometric analysis of economic time series. Int. Stat. Rev. 111–148 (1983)

Ho, T.K.: Random Decision Forests (PDF). In: Proceedings of the 3rd International Conference on Document Analysis and Recognition, Montreal, QC, pp. 278–282 (1995)

Holopainen, M., Sarlin, P.: Toward robust early-warning models: a horse race, ensembles and model uncertainty. Quant. Finance **17**(12), 1933–1963 (2017)

Jeon, Y., McCurdy, T.H.: Time-varying window length for correlation forecasts. Econometrics **5**(4), 54 (2017)

Kaminsky, G.L., Reinhart, C.M.: The twin crises: the causes of banking and balance-of-payments problems. Am. Econ. Rev. **89**(3), 473–500 (1999)

Kiley, M.T.: What macroeconomic conditions lead financial crises? J. Int. Money Finance **111**, 102316 (2021)

Kim, H., Cho, H., Ryu, D.: Corporate default predictions using machine learning: literature review. Sustainability **12**(16), 6325 (2020)

Li, X., Li, K., Qiao, D., Ding, Y., Wei, D.: Application research of machine learning method based on distributed cluster in information retrieval. In: 2019 International Conference on Communications, Information System and Computer Engineering (CISCE), pp. 411–414. IEEE (2019)

Lo Duca, M., Koban, A., Basten, M., Bengtsson, E., Klaus, B., Kusmierczyk, P., Peltonen, T.: A new database for financial crises in European countries. ECB occasional paper (2017194) (2017)

Moumeni, L., Saber, M., Slimani, I., Elfarissi, I., Bougroun, Z.: Machine learning for credit card fraud detection. In: Proceedings of the WITS 2020 Conference, pp. 211–221 (2022)

Noble, W.S.: What is a support vector machine? Nat. Biotechnol. **24**(12), 1565–1567 (2006)

Ristolainen, K.: Predicting banking crises with artificial neural networks: the role of nonlinearity and heterogeneity. Scand. J. Econ. **120**(1), 31–62 (2018)

Rizwan, M.S., Ahmad, G., Ashraf, D.: Systemic risk: the impact of COVID-19. Financ. Res. Lett. **36**, 101682 (2020)

Roberts, D.R., et al.: Cross-validation strategies for data with temporal, spatial, hierarchical, or phylogenetic structure. Ecography **40**(8), 913–929 (2017)

Silva, W., Kimura, H., Sobreiro, V.A.: An analysis of the literature on systemic financial risk: a survey. J. Financ. Stab. **28**, 91–114 (2017)

Spelmen, V.S., Porkodi, R.: A review on handling imbalanced data. In: 2018 International Conference on Current Trends towards Converging Technologies (ICCTCT), pp. 1–11. IEEE (2018)

Tölö, E.: Predicting systemic financial crises with recurrent neural networks. J. Financ. Stab. **49**, 100746 (2020)

Vaicenavicius, J., et al.: Evaluating model calibration in classification. In: The 22nd International Conference on Artificial Intelligence and Statistics, pp. 3459–3467 (2019)

Wilson, J.O., Casu, B., Girardone, C., Molyneux, P.: Emerging themes in banking: recent literature and directions for future research. Br. Acc. Rev. **42**(3), 153–169 (2010)

Yang, Q., Wang, C.: A study on forecast of global stock indices based on deep LSTM neural network. Stat. Res. **36**(6), 65–77 (2019)

Zheng, A., Casari, A.: Feature Engineering for Machine Learning: Principles and Techniques for Data Scientists. O'Reilly Media, Inc. (2018)

Chapter 10
A Fiber Bundle Model of Systemic Risk in Financial Networks

Soumyajyoti Biswas and Bikas K. Chakrabarti

Abstract Failure statistics of banks in the US show that their sizes are highly unequal (ranging from a few tens of thousands to over a billion dollars) and also, they come in "waves" of intermittent activities. This motivates a self-organized critical picture for the interconnected banking network. For such dynamics, recent developments in studying the inequality of the events, measured through the well-known Gini index and the more recently introduced Kolkata index, have been proved to be fruitful in anticipating large catastrophic events. In this chapter we review such developments for catastrophic failures using a simple model called the fiber bundle model. We then analyse the failure data of banks in terms of the inequality indices and study a simple variant of the fiber bundle model to analyse the same. It appears, both from the data and the model, that coincidence of these two indices signal a systemic risk in the network.

Keywords Self-Organized Criticality (SOC) · Fiber Bundle Model (FBM) · Bank failure · Gini & Kolkata indices

Introduction

Systemic Risk in the context of financial markets refers to the risks imposed by the network interlinks (e.g., in an interdependent bank network) in the market, where the failure of a single component or a number of them (a bank or a cluster of banks) can

S. Biswas (✉)
Department of Physics, SRM University-AP, Amaravati, Andhra Pradesh 522240, India
e-mail: soumyajyoti.b@srmap.edu.in

Department of Computer Science and Engineering, SRM University-AP, Amaravati, Andhra Pradesh 522240, India

B. K. Chakrabarti
Saha Institute of Nuclear Physics, Kolkata 70064, India
e-mail: bikask.chakrabarti@saha.ac.in

Economic Research Unit, Indian Statistical Institute, Kolkata 700108, India

© The Author(s) 2025   179
V. Pacelli (ed.), *Systemic Risk and Complex Networks in Modern Financial Systems*,
New Economic Windows, https://doi.org/10.1007/978-3-031-64916-5_10

cause a cascading failure, which could potentially bankrupt or bring down the entire system or market. Although such failures are extensively addressed and studied in the financial literature (see e.g.. Abergel et al. 2013 for a collection of reviews), straight forward modelling of these cascading failures in such networks are still absent.

In physics, however, there are precise and very well studied models, called the fiber bundle models (FBM). Indeed, the model follows an old well known and extensively studied model of materials failures, like fracture, earthquake etc., (see e.g., Pierce 1926; Pradhan et al. 2010; Kawamura et al. 2012; Biswas et al. 2015; Hansen et al. 2015; Chakrabarti 2017), where a bundle of fibers or strings of different strength collectively supports the load on (or hanging from) the bundle. Failure of any individual fiber increases the load share on each of the remaining fibers and that may induce failure of the next weakest fiber and so on until the remaining fibers are strong enough to support the load on the bundle. Or else, all the fibers break and a catastrophe (like fracture of the material) occurs.

Another similar model for traffic jams may also be compared. Here a local jam in a part of the traffic network increases the (diverted) traffic load on the available free roads, some of which in turn gets jammed due to this increased load, leading to further increase in traffic load on the surviving links or roads. These (local failures) may not still lead to a total failure of the traffic network if the surviving roads can sustain the traffic load of the system (city traffic) and the system survives. Otherwise, it will lead eventually to a total jamming in the city lasting for the day and comes back to normalcy in the night when traffic load decreases (see e.g., Chakrabarti 2006 for a simple analytically tractable model of such traffic jams). We intend to propose here a similar FBM for systemic risk of financial networks where the net financial "stress" W is assumed to get equally shared among $N(t)$ banking units, where $N(t = 0) = N$ (initially). Each such unit is assumed to have some threshold of its own and that threshold has uniform distribution. When the distributed load on any one goes beyond its capacity, it fails and the load per surviving units increase from $W/N(t)$ to $W/[N(t) - 1]$ and further failure may occur due to this increased load. Representing the fraction of surviving financial units at any time t by $U(t)$, one can write (see e.g., Pradhan et al. 2010; Biswas et al. 2015) the dynamical equation $-\frac{dU}{dt} = [U^2 - U + \sigma]/U$; with $\sigma = I/N$, giving the surviving fraction $U(t) = U^* + (\sigma_c - \sigma)^{1/2}$, with $\sigma_c = 1/4$ near but before the collapse of the system ($\sigma \leq \sigma_c$), when all the fixed point fraction U^* of the units fail, simultaneously. Beyond the critical load ($\sigma > \sigma_c$), $U(\sigma) = 0$. One can study (Diksha et al. 2023), both analytically and numerically, the inequality indices for the avalanche sizes as the FBM approaches the critical point (tuning the stress σ towards σ_c). It may be mentioned at this point that such an FBM is tuned externally (by changing the stress level σ). One can also consider self-tuned FBM systems, like the sand-piles where the external drive need not be tuned and dynamically stays put in the Self-Organized Critical (SOC) state (see e.g., Bak 1996). In such SOC versions of FBM, the inequality indices for the avalanche distributions show some universal behaviour (Manna et al. 2022).

We will attempt to extract and correlate the above-mentioned universal critical (SOC in particular) behaviour with those from real data. Particularly, the data from the Federal Deposit Insurance Corporations (FDIC) (Federal Deposit Insurance Corporations (FDIC)) for bank failures in the US between 1934 and 2023 reveal that such failures often happened in clusters of events and a large event is preceded by smaller events. This is similar to what is seen in the failure models discussed above. Therefore, we will attempt to find statistical similarities and potential precursory signals, specifically from the point of view of the inequalities of the failure events that has recently been proved to be useful in indicating imminent catastrophic events.

In what follows, we first review avalanche dynamics in the fiber bundle model, discuss the dynamics of the model modified for the purpose of banking network. Then we define the inequality indices for the avalanche dynamics and review the analytical expressions for the same. We then measure the inequality indices as obtained from the real data (FDIC data) and compare it with those obtained from the model.

Avalanches in the Fiber Bundle Model

As mentioned before, the fiber bundle consists of N elements or fibers which collectively support (through "rigid platforms" at both top and hanging end) a load $W = N\sigma$ and failure threshold (σ_{th}) of the fibers are assumed to be different for different fibers in the bundle. Initially, when a stress or load per fiber (σ) is applied, the fibers having failure threshold (σ_{th}) lower than the applied stress breaks immediately and the entire load then gets redistributed among the surviving fibers. In case of the Equal Load Sharing or ELS FBM considered here, the load is uniformly redistributed. The dynamics stops either when there is no fiber having threshold within this increased load per fiber or when all the N fibers have failed. For simplicity, we assume here the threshold distribution of the fibers to be uniform within the range 0 to 1 (normalized). If $U_t(\sigma)$ denotes the fraction of surviving fibers at time (load redistribution iteration) t, then the further broken fiber fraction U_{t+1} is given by load per fiber at that time $\sigma_t = W/NU_t$. Hence,

$$U_{t+1} = 1 - \sigma_t = 1 - \frac{\sigma}{U_t} \tag{10.1}$$

For $\sigma \leq \sigma_c$, the fixed point ($U_{t+1} = U_t \equiv U^*$) becomes,

$$U^*(\sigma) - \frac{1}{2} = (\sigma_c - \sigma)^{\frac{1}{2}}; \sigma_c = 1/4 \tag{10.2}$$

For $\sigma > \sigma_c$, the dynamics stop at $U(\sigma) = 0$. If the order parameter is defined as $O \equiv U^*(\sigma) - U^*(\sigma_c)$ then,

$$O = (\sigma_c - \sigma)^{\beta}; \beta = 1/2. \tag{10.3}$$

One can also consider the failure susceptibility χ, defined as the change of $U^*(\sigma)$ due to an infinitesimal increment of the applied stress σ

$$\chi = \left| \frac{dU^*(\sigma)}{d\sigma} \right| = \frac{1}{2}(\sigma_c - \sigma)^{-\gamma}; \gamma = \frac{1}{2} \tag{10.4}$$

Employing Josephson's identity in the Rushbrooke equality, we get $2\beta + \gamma = d\nu = 3/2$, with ν being the correlation length exponent for the ELS FBM and d denoting its effective dimension.

Inequality of Avalanches: Lorenz Function

Up to a given time i.e., step of load increment in the simulation of the FBM, the series of the avalanches can be arranged in the ascending order of their sizes. Then the Lorenz function (Lorenz 1905) $L(p, t)$ can be calculated by the cumulative fraction of the avalanche mass (sum of all avalanche sizes) coming from the p fraction of the smallest avalanches up to time t. . Note that if all avalanches were of equal sizes, then the Lorenz function would be a diagonal line from the origin (0,0) to (1,1). This line is called the equality line (see Fig. 10.1). Since the avalanches are, in general, not of equal sizes, the Lorenz function in non-linear, always staying below the equality line and monotonically increasing, with the constraints that $L(0, t) = 0$ and $L(1, t) = 1$ for any t. . The area in between the equality line and the Lorenz function, therefore, is a measure of the inequality in the avalanche sizes (the shaded area in Fig. 10.1). The ratio of this area and that under the equality line ($\frac{1}{2}$ by construction) is called the Gini index g (Gini 1912).

On the other hand, the ordinate value of the crossing point of the opposite diagonal (straight line between (0,1) to (1,0)), gives the value of the Kolkata index k (Ghosh et al. 2014), which gives the fraction $1 - k$ of the total number of avalanches that collectively account for the k fraction of the total avalanche mass up to that time. It is a generalization of the Pareto's law (Pareto 1897) that says about 80% of "attempts" account for 20% of "successes". It was previously noted that in the case of breakdown in the FBM, at the terminal point $t = t_f$, k approaches a value close to 0.62 ± 0.03, irrespective of disorder strengths and system sizes.

It was also noted elsewhere that for a broad class of systems, predominately of socio-economic nature, the early time variations of $g(t)$ and $k(t)$ follow a linear relation $k(t) = 1/2 + 0.37g(t)$. This relation is empirical and seen in data. It turns out that such linearity is also observed in the simulations for the FBM.

Fig. 10.1 The schematic diagram shows the Lorenz function (in red) and the equality line (black diagonal). The shaded area, as mentioned in the text, is therefore a measure of inequality of the events concerned. That area (normalized by maximum inequality) is called the Gini index (g). On the other hand, the intersection of the opposite diagonal and the Lorenz function gives the Kolkata index (k), denoting 1-k fraction of the largest events accounting for k fraction of the total damage

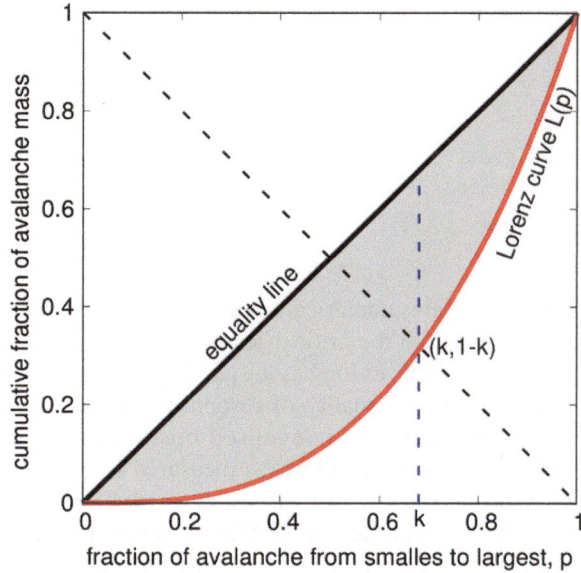

Calculating the Gini and Kolkata Indices for Fiber Bundle Model

Given its mean-field nature, it is possible to calculate the avalanche size distribution and the critical threshold (load at which the system collapses) for the fiber bundle model. Therefore, by extension, it is also possible to calculate the Gini and the Kolkata indices as follows:

We start from the definition of susceptibility, as mentioned above: $\chi = |dU^*(\sigma)/d\sigma| \propto (\sigma_c - \sigma)^{-1/2}$. Physically, this implies that a small change in the load, results in a "large" response in terms of breaking of fibers, particularly when $\sigma \to \sigma_c$. Naturally, the "responses" i.e., the avalanches are highly unequal and can thus be quantified using the indices mentioned above.

In doing so, one can write the Lorenz function, for all avalanches until the catastrophic breakdown at $\sigma = \sigma_c$, as

$$L_f(p) = \frac{\int_0^{p\sigma_c}(\sigma_c - \sigma)^{-1/2}d\sigma}{\int_0^{\sigma_c}(\sigma_c - \sigma)^{-1/2}d\sigma} = 1 - \sqrt{1-p} \qquad (10.5)$$

From the above Lorenz function, the Gini index at the point of catastrophic failure can be calculated as

$$g_f = 1 - 2\int_0^1 L_f(p)dp = 1/3 \qquad (10.6)$$

The Kolkata index then is

$$1 - k_f = 1 - \sqrt{1 - k_f} \tag{10.7}$$

which gives

$$k_f = \frac{\sqrt{5} - 1}{2} \approx 0.618 \tag{10.8}$$

This is what is numerically seen as well. Note that the values of these indices are not dependent on the critical point. This means that these are universal quantities, which will be seen as long as the particular power-law divergence is seen. Therefore, monitoring the inequality of responses can act as a good indicator of imminent failure. Indeed, in sand-piles or self-organized fiber bundle models (see e.g., Manna et al. 2022) the inequalities (as measured by the indices g and k) of the toppling or avalanche sizes universally show $g = k \simeq 0.86$ just preceding the arrival or following the departure of the SOC dynamical states.

It is this universal character of the inequality indices that we wish to utilise in failure statistics of interconnected network such as banking and the systemic risks, as observed from historic data and fiber bundle like model.

Avalanches in Bank Collapse

The banking system can be viewed as an interconnected network, where an overall financial stress can cause failures of individual banks and, if continued, can lead up to a systemic risk of catastrophic failure (such as the housing bubble of 2008–09).

From the data available with the Federal Deposit Insurance Corporation (FDIC) in the US, the bank failure sizes can be seen to be highly unequal. Indeed, in several cases, there have been events of growing sizes, leading up to major or "catastrophic" failures. In what follows, we will analyse the data for bank failures in terms of the inequality of their sizes and model the dynamics using simple fiber bundle like structure.

Inequality Indices for Bank Collapse Data

The FDIC website (Federal Deposit Insurance Corporations (FDIC)) lists bank failure (in assets size of deposits S) data from 1934 to March, 2023. The sizes of the failures, measured in terms of the deposits, vary widely (from several tens of thousands to over a billion dollars). It is, therefore, natural to measure the inequality in those failure data.

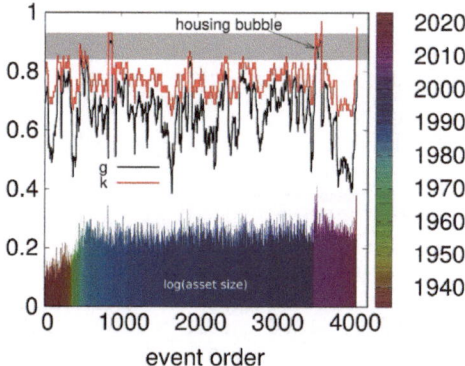

Fig. 10.2 The inequality of the bank failure sizes (shown in log of sizes S) are measured over a moving window of 50 events. The colours indicate the year of a particular failure. It is seen that the inequality indices g and k come very close to each other during the early 80s, then around 2009 and then very recently! The crossing of g and k is highlighted at the time of the housing bubble. The grey shaded region indicates the values for which the crossing might happen and therefore can be indicative of imminent large failure

In Fig. 10.2, the inequality of the bank failures was measured over a moving window of 50 events. This ensures that only the events of the recent past influence the values of g and k. It is seen that g and k cross a couple of times and most prominently near the housing bubble of 2008–09. Also, it is noted that the current failures are again leading g and k values close to each other.

In Fig. 10.3, the yearly average of the inequality indices was shown with the (normalised) cumulative sizes of failure each year and the (normalised) cumulative number of failures each year. It is seen that there have been three periods (until very recently), where major banking failures happened in the US—in 1930s, 1980s and the housing bubble of 2008–09. In all those cases, $\langle g \rangle$ and $\langle k \rangle$ came very close to each other.

A Model for Bank Failure

Informed by the intermittent nature of the bank failure data, we mention a minimal model inspired by the fiber bundle model mentioned before.

The model consists of N nodes (fibers), each having a failure threshold, which denotes the load carrying capacity. If the load exceeds the threshold, the node is broken, and the load is shared equally by the surviving nodes.

It is realistic to assume that the failure thresholds have a lower cut-off. For our purposes, we take the lower cut-off to be 0.5 and the thresholds are uniformly distributed between (0.5, 1.5). The system is then loaded until the minimum threshold i.e., 0.5. The dynamics of the model then continues as follows: At a stable config-uration, one node is randomly selected, and its threshold is set to zero. This means

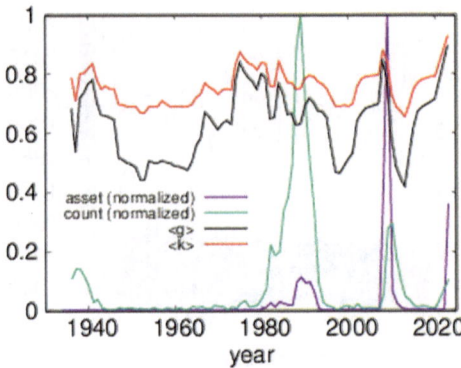

Fig. 10.3 The yearly average of g and k values noted in Fig. 10.2, denoted by $\langle g \rangle$ and $\langle k \rangle$, , are shown along with the yearly cumulative failure sizes (normalised by the maximum size) and the number of failures (also normalized by the maximum number). This shows, until very recently, there were three major "waves" of bank failures in the US, in 30 s, 80 s and the housing bubble of 2008–09. In all such cases, $\langle g \rangle$ and $\langle k \rangle$ came very close to each other

Fig. 10.4 The inequality indices g and k are studied for the SOC fiber bundle model discussed in Sect. 3.1.1. The failure avalanches (S) in assets are also indicated in the log scale. The crossing of g and k within a range of value 0.82 ± 0.05 occurs near the major cascading failures in the bank network

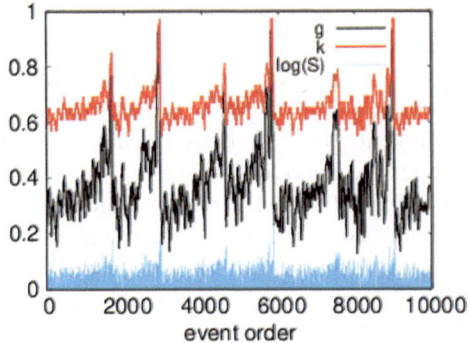

that the node collapses and the load is redistributed equally among all the surviving nodes. This may lead to further failure and an avalanche can start. The avalanche will continue until all nodes have loads lower than their respective thresholds. When such a stable configuration is reached, one more node is randomly selected, and its threshold is set to zero. Note that the system now is already "stresses" following the previous avalanche. So, in general an avalanche of higher size is expected. This process continues, until a macroscopic fraction of the nodes is eliminated (we set the threshold at 0.2 or 20% of nodes here). Following this, the broken nodes are restored with randomly chosen thresholds in the same range as before. In this way, the dynamics can continue as long as needed, with avalanches with different sizes. A similar form of the model was studied for the self-organized state of power grids.

Summary and Conclusions

Systemic failures in banking networks have been modelled here employing the Fiber Bundle Models or FBMs (Pierce 1926) for which the failure dynamics have been studied extensively in the recent physics literature in the context of fracture (in inhomogeneous materials) and earthquakes (see e.g., Pradhan et al. 2010; Kawamura et al. 2012; Biswas et al. 2015; Hansen et al. 2015) and traffic jams (see e.g., Chakrabarti 2006; Zheng et al. 2008; Batool et al. 2022). This mapping here helps us, compared to some earlier studies (see e.g., Lorenza et al. 2009), to provide some precise analysis of systemic or collective failure dynamics of the networks in the so called mean field (or long-range load reallocation) limit, as discussed in section "Avalanches in the Fiber Bundle Model". We also extend here mapping to the self-organizing critical (SOC) limit, and study the universal SOC behaviour of bank failures in section "Avalanches in Bank Collapse".

As discussed earlier, in a road network if the traffic load in one road goes beyond its capacity that road gets jammed, and the diverted traffic load gets redistributed to its link roads. This extra load share may induce jams in one or more of these link roads which, in turn, may induce further jams and may result in cascading failure of the entire network. An exactly solvable FBM model, in the mean field limit (Chakrabarti 2006), has been given in section "Avalanches in the Fiber Bundle Model". As the load σ gets increased, the steady state fraction $U^*(\sigma)$ of the intact (on service) banks (fibers/roads) decreases following a critical or power law behaviour (see Eq. (10.3)) for the uniform load level or stress on the links σ in the range $0 \leq \sigma < \sigma_c$, where $\sigma_c = 1/4$ (normalised), and a discontinuous or catastrophic failure collapse (from the steady state bank or link fraction $U^*(\sigma_c) = 1/2$ to $U^*(\sigma_c) = 0$) occurs at the critical load level $\sigma_c = 1/4$. Next, we discuss in section "Avalanches in Bank Collapse", how the cascading bank collapse in the US (data for the period 1930–2020 from Federal Deposit Insurance Corporations (FDIC)) can be comprehended using such fiber bundle models. In particular, in the SOC limit (Bak 1996; Manna et al. 2022) of such FBM models, where the broken fibers are replaced by intact fibers having breaking thresholds from the same distribution (see Sect. 3.1.1), one can search further precursors (see Figs. 10.2 and 10.3). Here the systemic bank collapses are detected by searching the event points (or times) where the Gini (g) and Kolkata (k) indices (defined in Fig. 10.1) in the failure avalanche statistics come very close ($g = k$ in the range 0.82 ± 0.05). It may be noted here again that $k = 0.8$ corresponds to Pareto's 80–20 law (Ghosh et al. 2014; Pareto 1897) of inequality. It is also interesting to note that inequality (measured in terms of Gini index) in accumulated gain or wealth through a Minority Game, reaches the highest unequal point, with g slightly above 0.8, at the point of maximum cooperation (minimum dispersion) (Ho et al. 2004).

Acknowledgements BKC is grateful to the Indian National Science Academy for their Senior Scientist Research Grant. Numerical analyses were performed using HPCC Surya at SRM University-AP. The authors are grateful to Parongama Sen for discussions and critical reading of the manuscript.

References

Abergel, F., Chakrabarti, B.K., Chakraborti, A., Ghosh, A. (eds.): Econophysics of Systemic Risk and Network Dynamics, New Economic Windows Series. Springer (2013)

Bak, P.: How Nature Works: The Science of Self-Organized Criticality. Springer-Verlag, New York (1996)

Batool, A., Danku, Z., Pal, G., Kun, F.: Temporal evolution of failure avalanches of the fiber bundle model on complex networks. Chaos Interdiscip. J. Nonlinear Sci. **32,** 063121 (2022)

Biswas, S., Ray, P., Chakrabarti, B.K.: Statistical Physics of Fracture, Breakdown, and Earthquake: Effects of Disorder and Heterogeneity. Wiley-VCH (2015)

Chakrabarti, B.K.: A fiber bundle model of traffic jams. Physica A: Stat. Mech. Appl. **372,** 162–166 (2006). https://doi.org/10.1016/j.physa.2006.05.003

Chakrabarti, B.K.: Story of the development in statistical physics of fracture, breakdown and earthquake: a personal account. Rep. Adv. Phys. Sci. **1,** 1750013 (2017). https://doi.org/10.1142/S24 2494241750013X

Diksha, K.S., Chakrabarti, B.K., Biswas, S.: Inequality of avalanche sizes in models of fracture. Phys. Rev. E **108,** 014103 (2023)

Federal Deposit Insurance Corporations (FDIC), https://www.fdic.gov/bank/statistical/index.html

Ghosh, A., Chattopadhyay, N., Chakrabarti, B.K.: Inequality in societies, academic institutions and science journals: Gini and k-indices. Physica A **410,** 30–34 (2014)

Gini, C.W.: Variabilitá e Mutabilitá: Contributo allo Studio delle Distribuzioni e delle Relazioni Statistiche. Cristiano Cuppini: Bologna, Italy (1912)

Hansen, A., Hemmer, P.C., Pradhan, S.: The Fiber Bundle Model: Modeling Failure in Materials. Wiley-VCH, Singapore (2015)

Ho, K.H., Chow, F.K., Chau, H.F.: Wealth inequality in the minority game. Phys. Rev. E **70,** 066110 (2004)

Kawamura, H., Hatano, T., Kato, N., Biswas, S., Chakrabarti, B.K.: Statistical physics of fracture, friction, and earthquakes. Rev. Mod. Phys. **84,** 839 (2012)

Lorenz, M.O.: Methods of measuring the concentration of wealth. Publ. Am. Stat. Assoc. **9,** 209–219 (1905)

Lorenza, J., Battiston, S., Schweitzer, F.: Systemic risk in a unifying framework for cascading processes on networks. Eur. Phys. J. B **71,** 441–460 (2009). https://doi.org/10.1140/epjb/e2009-00347-4

Manna, S.S., Biswas, S., Chakrabarti, B.K.: Near universal values of social inequality indices in self-organized critical models. Physica A **596,** 127121 (2022)

Pareto, V.: Cours d'Economie Politique, Lausanne, Rouge (1897)

Pierce, F.T.: The weakest link. J. Textile Inst. **17,** 355 (1926). https://doi.org/10.1080/19447027. 1926.10599953

Pradhan, S., Hansen, A., Chakrabarti, B.K.: Failure processes in elastic fiber bundles. Rev. Mod. Phys. **82,** 499 (2010)

Zheng, J.F., Zhao, X.M., Fu, B.B.: Extended fiber bundle model for traffic jams on scale free networks. Int. J. Phys. C **19,** 1727–1735 (2008)

Chapter 11
Measuring Systemic Risk: A Review of the Main Approaches

Francesca Pampurini and Anna Grazia Quaranta

Abstract Scholars, Regulatory and Supervisory Authorities have always been engaged in the search for efficient approaches to measuring systemic risk. Such procedures are extremely useful, first and foremost, in understanding and managing the stability and resilience of a financial-economic system as a whole, in forecasting possible crisis situations, and in implementing effective macro-prudential policies in response to the turbulence that can be generated by systemic risk in the financial system. Actually, over time, different approaches to measuring systemic risk have been defined. Undoubtedly, these methods are difficult to compare and often result in assessment parameters that are difficult to cointegrate. This chapter describes and analyses the main approaches for measuring systemic risk currently used in the literature. In more detail, it analyses the Probability Distribution Measures, the Network Analysis Measures, the Illiquidity Measures, the Contingent Claims and Default Measures and, last but not least, the Macro-economic Measures.

Keywords Systemic Risk Measures · Probability Distribution Measures · Network Analysis Measures · Illiquidity Measures · Contingent Claims and Default Measures · Macro-economic Measures

Introduction

As is well known, the new millennium has been characterised by multiple financial crises; this has led to an exponential growth in interest in its measurement and monitoring in order to prevent situations of financial instability.

In fact, systemic risk is not characterised by a remote possibility of occurrence, since its detection is rather frequent. If there was a need for confirmation of what

F. Pampurini
Catholic University of the Sacred Hearth of Milan, Milan, Italy
e-mail: francesca.pampurini@unicatt.it

A. G. Quaranta (✉)
University of Macerata, Macerata, Italy
e-mail: annagrazia.quaranta@unimc.it

© The Author(s) 2025   191
V. Pacelli (ed.), *Systemic Risk and Complex Networks in Modern Financial Systems*,
New Economic Windows, https://doi.org/10.1007/978-3-031-64916-5_11

has just been asserted, it is symptomatic of the fact that, in the wake of the notorious and close failures that affected a number of American banks in March 2023, it has once again become the focus of everyone's interest. Indeed, regulatory and supervisory authorities and leading economists have always been engaged in the search for efficient approaches to measuring the level of systemic risk.

In this chapter, an analysis of the main approaches currently used to measure systemic risk will be conducted. Undoubtedly, these are extremely useful in quantifying financial instability as well as in ensuring crisis prevention and the implementation of macro-prudential policies to tackle this type of risk.

In more detail, in this part of the volume, after a preliminary classification of the main approaches for quantifying systemic risk, within each of them, the measures currently most widely used in the literature and in empirical analyses will be described. Thus, the following approaches will be analysed: (i) the Probability Distribution Measures—namely the Delta Conditional Value at Risk (Delta CoVaR), the Marginal Expected Shortfall (MES), the Systemic Expected Shortfall (SES) and the Systemic Risk Measure (SRisk); (ii) the Network Analysis Measures; (iii) the Illiquidity Measures; (iv) the Contingent Claims and Default Measures (CCA) and, finally, (v) the Macro-Economic Measures.

A Preliminary Overview of the Main Approaches to Quantify Systemic Risk

The problem of assessing and measuring systemic risk is one of the fundamental topics of research in the economic and financial field. In order to measure this kind of risk, many authors proposed different approaches based on mathematical and econometric models that differ widely and are therefore difficult to compare. For this reason, both the results produced by the different procedures and the evaluation parameters cannot be directly compared.

The numerous measurement metrics have different characteristics and, depending on these, can be grouped into the five types represented in Fig. 11.1.

In the following, the five approaches will be described in detail and, for each, the main measures proposed in the literature will be illustrated.

Probability Distribution Measures

Probability Distribution Measures (also known as *Tail Measures*) quantify an individual intermediary's contribution to the risk of the whole system starting from measures of expected marginal losses. In fact, they are obtained on the

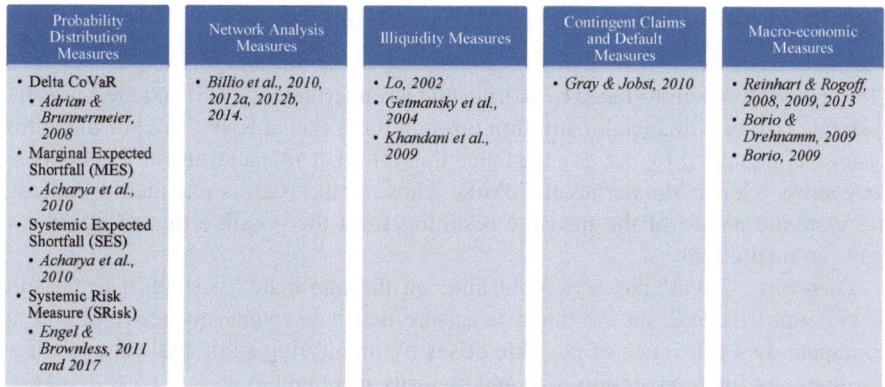

Probability Distribution Measures	Network Analysis Measures	Illiquidity Measures	Contingent Claims and Default Measures	Macro-economic Measures
• Delta CoVaR • *Adrian & Brunnermeier, 2008* • Marginal Expected Shortfall (MES) • *Acharya et al., 2010* • Systemic Expected Shortfall (SES) • *Acharya et al., 2010* • Systemic Risk Measure (SRisk) • *Engel & Brownless, 2011 and 2017*	• *Billio et al., 2010, 2012a, 2012b, 2014.*	• *Lo, 2002* • *Getmansky et al., 2004* • *Khandani et al., 2009*	• *Gray & Jobst, 2010*	• *Reinhart & Rogoff, 2008, 2009, 2013* • *Borio & Drehnamm, 2009* • *Borio, 2009*

Fig. 11.1 The main approaches to measure systemic risk

basis of the joint distribution of losses of a homogeneous set of financial institutions, with particular weighting given to those pertaining to systemically important intermediaries.

In more detail, an important role in the construction of all measures falling into this category is played by the co-dependence in the tails of the distributions of risk variables (such as equity returns) of two or more financial institutions. It is for this reason that another term used for this type of measure is *Cross Sectional Measures* to indicate the cross-sectional link between the health of the different intermediaries. Included in this type of approach are the measures proposed by Adrian and Brunnermeier (Adrian and Brunnermeier 2008, 2011, 2016) and Brownless and Engle (2011).

The construction of these measures is rather challenging due to the difficulties related both to measuring inter-tail dependence in a context of scarcity of data linked to the particular lack of extreme values of yields resulting from the occurrence of systemic crises and to the circumstance that large aggregate shocks in economies are generally treated as a single conglomerate, consequently not making it easy to understand the structure of co-dependencies. In this last regard, it is emphasised that, on the other hand, understanding the transmission mechanisms of a shock is crucial for the definition of successful economic policy responses.

In what follows, the four main measures that belong to the Probability Distribution Measures family will be analysed in detail, according to the chronological order in which they appeared in the literature: the Delta Conditional Value-at-Risk (ΔCoVaR), the Marginal Expected Shortfall (MES), the Systemic Expected Shortfall (SES) and the Systemic Risk Measure (SRISK).

Delta Conditional Value at Risk ($\Delta CoVaR$)

This approach was introduced by Adrian and Brunnermeier (2008) and measures the systemic risk of a financial institution through the Value at Risk (VaR) of the entire system, constrained by the circumstance that all other financial institutions experience stress. Hence, the acronym CoVaR, whose prefix (Co) is intended to indicate the systemic nature of the measure resulting from the possible mutual contagion between institutions.

Therefore, CoVaR has a twofold aim: on the one hand, to suggest a measure of systemic risk and, on the other, to ensure that it is countercyclical and able to anticipate the occurrence of possible crises by monitoring some indicators such as the degree of interconnection, size and maturity mismatch.

Due to the definition just given for CoVaR, of fundamental importance is the difference between the CoVaR of a financial institution in a particularly stressed situation and the CoVaR of the same intermediary under normal conditions. This difference is called ΔCoVaR and indicates the marginal contribution of the single institution to the overall systemic risk (Adrian and Brunnermeier 2011, 2016). It follows that the approach proposed by the aforementioned authors does not focus on the risk borne by individual institutions, but rather on the contribution that each financial intermediary makes to the overall systemic risk.

Several advantages can be attributed to the use of CoVaR as a measure of systemic risk. First of all, two financial institutions with the same level of Value at Risk could show different values of ΔCoVaR that would make it possible to detect which of the two is more dangerous to the system and, consequently, which one should be monitored more closely in order to contain systemic risk.

A further benefit lies in the possibility of measuring the extent of contagion affecting two financial institutions, and thus the strength with which a negative shock that initially manifests itself in only one of them affects the activity of the other (ΔCoVaR$^{i|j}$). The latter meaning of the risk measure is not necessarily symmetrical since, if for one intermediary the increase in risk derives from a worsening of the other's conditions, the reverse does not necessarily also occur.

Adrian and Brunnermeier rightly pointed out that systemic risk arises mainly from so-called SIFIs, i.e. *Systemically Important Financial Intermediaries* characterised by a high degree of interconnection with other players in the system and their large size.

Denoting by (R^i) the market return on the financial assets of intermediary i and starting from the well-known definition of its Value at Risk with a level of significance q

$$\Pr\left(R^i \leq VaR_q^i\right) = q, \tag{11.1}$$

CoVaR$_q^{j|i}$ is defined by the value of the quantile q of the following conditional probability distribution

$$\Pr\left\{ R^j \leq \text{CoVaR}_q^{j|i} \middle| R^i = \text{VaR}_q^i \right\} = q. \tag{11.2}$$

Starting from (11.2), it is possible to obtain the risk that financial institution i brings to j as

$$\Delta\text{CoVaR}_q^{j|i} = \text{CoVaR}_q^{j|i} - \text{CoVaR}_{50\%}^{j|i} \tag{11.3}$$

Adrian and Brunnermeier also demonstrated how the risk measure they proposed can assume low values in times characterised by particularly positive economic conditions and higher values from when the difficulties of an intermediary begin to spread to the rest of the system and thus the marginal contribution of each individual financial institution to the overall risk also gradually increases.

This dynamic is in fact characterised by what are defined as spill-over effects; in particular, the latter can be direct, insofar as they derive, for example, from a contractual link, or indirect, generated by the exchange of assets that gradually deteriorate as they are ascribed to intermediaries under stress.

CoVaR can be estimated in various ways even if, in reality, quantile regression[1] is generally used for this purpose as it is considered the substantially more efficient procedure.

Marginal Expected Shortfall (MES)

Acharya, Pedersen, Philippon and Richardson (Acharya et al. 2010) proposed a measure of systemic risk called *Marginal Expected Shortfall* (MES) because they considered it able to provide a result that could also be used as an input in an economic model.

This measure is still widely used to the extent that it regularly feeds an online database (https://vlab.stern.nyu.edu/welcome/srisk) aimed at monitoring the systemic risk of major SIFIs.

In a nutshell, MES measures, over a given time horizon and with assigned probability, the expected loss of a financial institution calculated on all yield values below a threshold equal to VaR.

In other words, the authors started from the well-known definition of *Expected Shortfall*

$$ES_q^i = -E\left[R|R \leq -\text{VaR}_q^i \right] \tag{11.4}$$

---

[1] Quantile regression is a kind of regression analysis aimed at estimating the conditional median or a specific quantile of the dependent variable. This method provides a better representation of how the distribution is constrained by an explanatory variable.

Subsequently, the total return R was decomposed so that the different k (k = 1... m) trading desk activities performed by the same financial institution could be attributed their specific contribution to it and thus the different values R_k. Hence,

$$R = \sum_{k=1}^{m} r_k w_k \qquad (11.5)$$

where w_k represents the contribution of the individual trading desk to overall performance.

As a direct consequence it is possible to obtain a modified expression of the previous (11.4), i.e.

$$ES_q^i = -\sum_k w_k E\left[r_k | R \le -VaR_q^i\right] \qquad (11.6)$$

MES, as a marginal measure, is therefore derived by computing the first derivative of (11.6) with respect to w_k and thus

$$MES_q^k = \frac{\delta ES_q^i}{\delta w_k} = -E\left[r_k | R \le -VaR_q^i\right] \qquad (11.7)$$

through which it is possible to quantify the sensitivity of each financial institution's total risk to the contribution made by each of its individual trading desks.

Systemic Expected Shortfall (SES)

Systemic Expected Shortfall (SES) is one of the most widely accepted approaches for measuring systemic risk in the literature, first proposed by Acharya, Pedersen, Philippon and Richardson (Acharya et al. 2010).

As one can easily imagine from the name of this systemic risk measure, it is inevitably linked to the concept of Expected Shortfall previously mentioned and described in (4).

Using this indicator, the contribution to the systemic risk of each intermediary can be measured through the quantification of its (possible) state of undercapitalisation in a context of generalised difficulties for the financial system and, therefore, when the system, as a whole, is undercapitalised.

In more detail, to obtain the SES, it is needed to start from the consideration that when the aggregate capital of an intermediary falls below a given threshold—quantified via a fraction z of its total assets (A^i)—a financial crisis is triggered. In order to tackle the negative effects arising from the latter (in essence, fire sales and credit crunch), it is therefore essential that the entire system has an adequate amount of aggregate capital.

In the aforementioned contribution, it is shown how, in fact, SES increases with the degree of leverage of a financial institution and with the expected loss calculated in the tail of the reference system's loss distribution. The authors also observed that, if a tax were imposed on financial institutions based on their SES, they would inevitably have an incentive to internalise the negative externalities arising from their possible crisis, thus avoiding heavy spillovers to the economy.

For the definition of SES, it is necessary to consider two time horizons ($t = 0$ et $= 1$) and a finite number i ($i = 1 \ldots N$) of intermediaries. To take into account the different size of each of them, the measurement of SES implies that it is calculated by weighing the current equity level of each of them, and thus

$$\frac{\text{SES}^i}{K_0^i} = \frac{zA^i - K_0^i}{K_0^i} - E\left[\frac{K_1^i - K_0^i}{K_0^i}|K_1 < zA\right] \tag{11.8}$$

where A^i is the amount of the assets of the i-th intermediary and K_0^i is its net worth (*current equity*) at time zero.

The first part of the formula represents the *ex ante leverage ratio*, i.e. the level of undercapitalisation before the crisis. If $\frac{zA^i - K_0^i}{K_0^i} \geq 0$, the intermediary immediately starts from a state of undercapitalisation. The second term in the Eq. (11.8) consists of two parts. The first ($\frac{K_1^i - K_0^i}{K_0^i}$) measures the expected return on the intermediary's shares at time 1. However, it is constrained by the possibility of the manifestation of a systemic crisis, given that inequality $K_1 < zA$ identifies precisely the fact that the system is globally undercapitalised at the time $t = 1$. It follows that, the contribution to systemic risk of each intermediary is measured by the gap between the two previous terms, thus indicating whether, and by how much, the latter will be undercapitalised during the crisis.

Although it is based on a series of formal assumptions that are difficult to observe, the strength of this methodology lies in the simplicity with which the information necessary to measure a system's losses can be found. In addition, its authors demonstrated how the SES can be operationally easily quantified due to its relationship with the MES; one has indeed

$$\frac{\text{SES}^i}{K_0^i} = \frac{zA^i - K_0^i}{K_0^i} + \lambda \, \text{MES}^i + \Delta^i. \tag{11.9}$$

This equation also makes it possible to highlight the three fundamental components of an intermediary's SES: (i) the degree of pre-crisis undercapitalization $\frac{zA^i - K_0^i}{K_0^i}$, (ii) the MES measured using pre-crisis data and then multiplied by a factor λ to take into account the (worst) performance during the systemic crisis and (iii) the correction term Δ^i introduced to measure the costs of financial distress related to credit risk.

With regard to the value of the MES^i, the authors proposed to use

$$\text{MES}_{5\%}^i = E\left[\frac{K_1^i - K_0^i}{K_0^i} | I_{5\%}\right] \tag{11.10}$$

considering it able to adequately measure tail occurrences that can predict a systemic event. Hence, the proposal to obtain SES from the following expression

$$\frac{\text{SES}^i}{K_0^i} = \frac{zA^i - K_0^i}{K_0^i} + \lambda\,\text{MES}_{5\%}^i + \Delta^i. \tag{11.11}$$

Systemic Risk Measure (SRISK)

An indicator proposed Christian Brownlees and Robert Engle (2017) known as the *Systemic Risk Measure* (SRISK) can also be used to measure a financial institution's contribution to systemic risk.

In a nutshell, this measure quantifies the capital shortfall a financial institution incurs if it is hit by a shock that negatively impacts the entire market. Thus, in other words, the value assumed by SRISK is able to inform about the capital deficit that the single intermediary expects in case of a systemic event. This capital shortfall (which in the following will be referred to as *Long Run Marginal Expected Shortfall—LRMES*) is expressed as a function of the size of the financial institution and its level of leverage and risk.

The proximity between the SRISK measure and the stress tests regularly performed on financial institutions is evident.

Although the use of publicly available information may lead one to think that SRISK is easily measurable, there are in fact considerable difficulties associated with the quantification of its LRMES component, which in turn is used to quantify the *excessive credit growth* of the financial sector, understood as the latter not having sufficient capital to cover losses in a recession.

As a direct consequence, aggregating the capital deficits of all financial institutions provides a measure of overall excess credit growth at the system level. Hence, the importance of analysing the individual levels of undercapitalisation of financial institutions for two reasons: firstly, for the purpose of identifying the intermediaries most vulnerable to adverse external shocks and, secondly, in order to react promptly at the first signs of the beginning of a generalised financial crisis.

For all of the above, for the purposes of calculating the SRISK measure, it is essential to start precisely with the measurement of the capital shortfall (CCS, from *Conditional Capital Shortfall*).

Given therefore N financial institutions, the capital shortfall of the ith unit at time t is defined by also taking into account the reserves that, for regulatory/prudential and management reasons, it must hold in excess of its equity capital, and thus

$$CCS_t^i = \gamma A_t^i - V_t^i = \gamma(D_t^i + V_t^i) - V_t^i \tag{11.12}$$

where γ is the fraction of prudential capital, A_t^i is the value of the pseudo-assets (also called *quasi-assets*) of intermediary i at time t, V_t^i is the market value of capital and D_t^i is the book value of debt.

In the aforementioned paper, Brownless and Engle defined the capital deficit of a financial institution in the case of a systemic event as the contraction of the market below a threshold C over a time horizon s. This position derives from sharing the view of Acharya et al. (2010) that the capital deficit characterising a financial institution generates negative externalities if it occurs when the entire system is in a period of stress corresponding to a particularly extreme scenario.

Given this, the SRISK measure of intermediary i at time t is obtained as the expected value of the capital shortfall conditioned on the occurrence of a systemic event, and thus as

$$SRISK_t^i = E_t\left[CCS_{t+s}^i | R_{t+1:t+s}^M\right] = \gamma E_t\left[D_{t+s}^i | R_{t+1:t+s}^M < C\right] - (1-\gamma)E_t\left[V_{t+s}^i | R_{t+1:t+s}^M < C\right] \tag{11.13}$$

where $R_{t+1:t+s}^M$ and $R_{t+1:t+s}^M < C$ indicate, respectively, the multi-period market return between period $(t + 1)$ and $(t + s)$ and the condition underlying the occurrence of the systemic event.

Since, when a systemic event occurs, a debt can no longer be renegotiated, the expected value identifying the SRISK measure changes as follows

$$E_t\left[D_{t+s}^i | R_{t+1:t+s}^M < C\right] = D_t^i \tag{11.14}$$

from which, replacing this expression in (11.13) above, it is possible to obtain

$$SRISK_t^i = \gamma D_t^i - (1-\gamma)V_t^i(1 - LRMES_t^i) = V_t^i[\gamma LVG_t^i + (1-\gamma)LRMES_t^i - 1] \tag{11.15}$$

where $LVG_t^i = \frac{(D_t^i + V_t^i)}{V_t^i}$ is the pseudo-leverage ratio (also known as the *leverage ratio*) and $LRMES_t^i = -E_t[R_{t+1:t+s}^i | R_{t+1:t+s}^M < C]$ Long Run Marginal Expected Shortfall of the ith financial institution.

It is easy to see that the $LRMES_t^i$ is simply the MES of financial institution i calculated over a long-term time horizon and thus the expected value of the multi-period return on equity of that institution constrained by the occurrence of the systemic event.[2]

---

[2] It should be noted that the computation of the SRISK measure cannot be separated from the specification of a distributional model of the returns of financial institutions as this is necessary to obtain the LRMES estimates. There are actually several approaches for this purpose; Brownlees and Engle, for example, used a DCC-GARCH model (from *Dynamic Conditional Correlation GARCH Model*—Engle 2002).

From (11.15) it is possible to see how the SRISK measure increases as the size of the intermediary, its level of debt, and its sensitivity to market downturns increase. Evidently, from the above formula it is also possible to infer the prediction of the level of capital deficit that a financial entity would experience in case of a systemic event.[3]

From the SRISK measure values of each financial intermediary, it is then possible to obtain the overall systemic risk measure by adding them up (net of measures that take a negative value), thus obtaining

$$\text{SRISK}_t = \sum_{i=1}^{N} \left(\text{SRISK}_t^i\right)_+ \tag{11.16}$$

where $(x)_+ = \text{Max}(x, 0)$.

The reason why capital deficits that take on a negative value—and thus represent actual capital surpluses ascribable to reputable and/or, at the very least, prudent financial institutions—are excluded from the calculation is that, in the course of a crisis, it is very unlikely that such capital surpluses can be channeled to other, weaker intermediaries. Therefore, such funds will certainly not be available to support distressed intermediaries and therefore it goes without saying that they should be excluded from the calculation of the overall SRISK measure.

Generally, the value of the SRISK_t calculated by means of the previous (11.16) is also attributed the ability of adequately quantifying the value of the total amount of capital that a government would have to employ in order to protect its country's financial system should a systemic event actually happen.

It is often preferred to use a financial institution's SRISK measure in percentage terms so as to be able to highlight more immediately the incidence of systemic risk attributable to each intermediary with respect to the system's overall exposure.

This measure is trivially obtained by dividing the SRISK measure of an intermediary with the value referring to the entire financial system and thus by the following expression

$$\text{SRISK\%}_t^i = \frac{\text{SRISK}_t^i}{\sum_{i=1}^{N}\left(\text{SRISK}_t^i\right)_+} = \frac{\text{SRISK}_t^i}{\text{SRISK}_t}. \tag{11.17}$$

---

[3] Three calculation models of the SRISK measure are available at https://vlab.stern.nyu.edu/wel come/srisk. They basically differ in the way the LRMES is defined and calculated. In more detail, the first model is the *Global Marginal Expected Shortfall* (GMES) and employs the global ETF ACWI (*All-Country World Index*) to measure LRMES, then assuming a stress for it that is currently 40%. The MESSIM approach, on the other hand, refers to the S&P 500 Index for the quantification of the LRMES; a shock is then applied to it, which is currently 40% or more. This measure, however, is actually only available for the US. Finally, the DMES model (simply known as MES in the US) measures LRMES through a domestic stock index to which a stress value of currently 40% is then applied. It should be noted that all these values are updated on the aforementioned site on a weekly basis.

Network Models

The basic assumption of the Network Models used to measure systemic risk is that the financial system can be adequately represented through the set of connections established between the financial institutions operating within it; it is these connections that are directly responsible for the propagation of market shocks.

In other words, the systemic risk measures derived from Network Models—also known as *connectivity measures*—aim to quantify the degree of connection between the different intermediaries operating in the financial system, and this is because it is believed that the probability with which a systemic crisis can be manifested is directly proportional to the degree of correlation existing between the different players in the system.

Network Models were introduced by Billio et al. (2010) and later improved upon in further contributions (Billio et al. 2011, 2012a, b; Billio and Pellizzon 2014).

In order to describe the aforementioned network of relationships between financial institutions, Billio et al. proposed in their study to employ principal component analysis (Jolliffe 2002) and Granger causality tests (Granger 1969), as these approaches can also provide useful information about the degree of information asymmetry that characterises the system at a given point in time and which the authors, through empirical analysis, found to be higher immediately prior to the onset of a systemic crisis.

In greater detail, the insightful contribution to the identification and subsequent quantification of systemic risk provided by principal component analysis lies in the circumstance that it is a procedure able to adequately detect the correlation existing between the returns on assets of a set of financial institutions. The interpretation of the results offered by this approach can be traced back to the fact that, if the system is highly interconnected, then even a single intermediary in difficulty can represent a real danger to the stability of the entire system.

In other words, principal component analysis is used to calculate an intermediary's exposure to systemic risk under the assumption that it is highly integrated with other financial institutions.

As far as Granger causality tests are concerned, the authors made use of them because the measures they obtain make it possible to quantify the degree of correlation between the entities that make up a system unconditionally, i.e. independently of the occurrence of some events. Granger causality also manages both to combine historical data with estimates of future values and to provide specific information about the relationship that connects two entities, thus clearly highlighting the sources of propagation of a systemic event, the direction of propagation as well as any factors directly responsible.

Illiquidity Measures

The scientific literature suggested, among others, a way of measuring systemic risk that is based on the analysis of the illiquidity of a given financial institution or, more generally, of the system.

This method proposes, first, to study how a financial crisis can alter and impact the structure of assets and liabilities of financial institutions, triggering a liquidity crisis, and, second, to observe the dynamics with which these episodes propagate throughout the system. In other words, the *Illiquidity Measures* aim to estimate how exposed financial institutions are to liquidity risk since, as we know from the Global Financial Crisis, the distress of each single institution represents an important source of systemic risk.

Lo (2002) and Getmansky et al. (2004) suggested measuring the degree of illiquidity risk exposure of each subject by computing the autocorrelation coefficients ρ_k of the monthly returns, i.e.

$$\rho_k \equiv \frac{\text{Cov}\left[R_t - R_{t-k}\right]}{\text{Var}[R_t]} \tag{11.18}$$

where ρ_k is the kth auto-correlator of R_t that measures the autocorrelation level between the monthly returns R of the months t and $t - k$, of course using variance and covariance.

The use of autocorrelation coefficients as liquidity risk exposure indicators derives from the martingale model (Revuz and Yor 1999), which was one of the first pricing models for financial instruments introduced in the literature. In a nutshell, this model is based on the assumption that financial assets returns are serially uncorrelated because, in efficient markets, price changes are random and unpredictable.

However, theory revealed that markets are not perfectly efficient because there are different kinds of frictions such as, for example, those attributable to the existence of transaction costs, borrowing constraints, information gathering and processing costs, as well as short-term institutional constraints and other business practices. Therefore, it is these imperfections that explain the existence of asset returns serial correlation.

Consequently, the degree of serial correlation in a given asset returns can be seen as a proxy for the magnitude of frictions and illiquidity as one of the most common forms of such frictions.

The use of serial correlation as an illiquidity risk proxy is also justified for an additional reason. A common practice among hedge fund managers, who, as is well known, generally hold portfolios composed of highly illiquid and thinly traded securities, is to value such portfolios monthly by assigning discretionary prices to the securities that comprise them, different from the actual market value at that time. In particular, in periods characterised by very positive returns, this smoothing practice will lead to instruments being valued at prices below market values, and vice versa in the case of periods characterised by negative returns. This is possible because illiquid securities are not priced on a daily basis and therefore the manager has a wide discretion. The main consequence of smoothing practices is thus to achieve a reduction in the fund's return volatility and, as a result, a higher Sharpe ratio. This, on the other hand, leads to serial correlation as a side effect. Of course, in the case of actively traded portfolios, the manager will be forced to value them at mark-to-market, so the discretion level will be considerably lower. On the contrary, the more illiquid a portfolio is, the greater the manager's discretionary power at the time of valuation, thus generating an increasingly evident serial correlation.

The impact of uniform returns and serial correlation is considered in more detail in other works, such as those by Lo (2002), Getmansky et al. (2004), and Khandani et al. (2009).

With particular reference to credit intermediaries, Lorenz et al. (2009) show that when they operate with excessive leverage and are characterised by many interconnections with other similar intermediaries, the fragility of an entity can easily trigger a systemic shock. In this scenario, a bank run could be generated, forcing intermediaries to massively liquidate their assets; the more illiquid these assets are, the greater the risk of generating a spiral of losses that will trigger a systemic crisis.

Serial correlation can thus be considered a proxy for illiquidity and can thus serve as an indirect measure of exposure to this risk category.

Hence the motivations of the legislator, who introduced a series of rules aimed at limiting each financial intermediary's exposure to liquidity risk in order to prevent systemic problems.

However, it must be remembered that an effective systemic risk measurement system cannot limit itself to monitoring these individual cases, since liquidity risk can also be caused by adverse situations triggered by other types of risk.

Systemic Contingent Claims and Default Measures

Systemic Contingent Claims Analysis (SCCA) is one of the most popular approaches used to measure systemic risk. The authorship of this approach can be attributed to Gray and Jobst (2010) who proposed it as an extension of the *Contingent Claims Analysis* (CCA) theory. The latter, in contrast, is based on the well-known model introduced by Merton (1973), which likened the position of a company's shareholders to that of someone who holds a call option on that company's shares and, likewise, compared the position of the same company's creditors to that of someone who sold a put option on the same underlying asset (Gray and Jobst 2010).

The Contingent Claims Analysis starts from the consideration that, at time t, the market value of the assets $A(t)$ is equal to the sum of the market value of the equity $E(t)$ and the market value of the debts $D(t)$ assumed by the same company and therefore

$$A(t) = E(t) + D(t) \tag{11.19}$$

A default occurs when the assets' value is less than that of the financing sources; this limit constitutes the default threshold for the CCA. When this happens, the value of the capital is reduced to zero and the shareholders will have the convenience of declaring insolvency and leaving the company in the hands of the creditors.

As stated earlier, in this model the debt can be likened to a put option sold by the creditors to the shareholders that allows the latter to dispose of the shares before default. The value of equity, on the other hand, refers to that of a call option written on assets whose strike price is equal to the default threshold. The value $C(t)$ of this call option can be obtained as

$$C(t) = A(t)\phi(d_1) - Be^{-r(T-t)}\phi(d_2) \tag{11.20}$$

$$d_1 = \left[\ln\left(\frac{A(t)}{B}\right) + (r + \frac{\sigma_A^2}{2})(T-t)\right](\sigma_A\sqrt{T-t})^{-1} \quad d_2 = d_1 - \sigma_A\sqrt{T-t} \tag{11.21}$$

where:

- $\phi(\bullet)$ is the cumulative distribution function of a standard normal one;
- B is the debt face value at maturity T;
- r is the risk-free return;
- σ_A is the assets' value volatility.

Formulae (11.22)–(11.24) are necessary to define the value of the debt and thus the default threshold, which is obtained by subtracting from the debt market value the expected loss $E[L(t)]$ resulting from a default, i.e.

$$D(t) = Be^{-r(T-t)} - E[L(t)] \tag{11.22}$$

where:

$$E[L(t)] = Be^{-r(T-t)}\phi(-d_2) - A(t)\phi(d_1). \tag{11.23}$$

According to the CCA, the put option price—which can be derived directly from the market price of a company's shares and its balance sheet values—can be combined with data from the *Credit Default Swaps* (CDS) market in order to capture the potential losses that a financial institution would have to bear. In other words, if the CDS price reflects the value of the expected loss associated with the part of the debt not guaranteed by the government (should that guarantee be activated), the price of the put option $P_{CDS}(t)$ can be written as follows

$$P_{CDS}(t) = \left[1 - \exp\left(-\left(\frac{S_{CDS}(t)}{10000}\right)\left(\frac{B}{D(t)} - 1\right)(T-t)\right)\right]Be^{-r(T-t)} \tag{11.24}$$

where $S_{CDS}(t)$ is the CDS spread below the level that would signal the default risk implied by the put option.

Combining (11.24) with the expected loss (11.23) it is possible to obtain the total (potential) loss resulting from the default of a financial institution benefiting from the eventual presence of a government guarantee $\alpha(t)$ expressed as follows:

$$\alpha(t) = 1 - P_{CDS}(t)/E[L(t)]. \tag{11.25}$$

Starting from (11.25) above, with simple algebraic steps, it is possible to derive both the fraction of default risk covered by the government guarantee scheme $\alpha(t)E[L(t)]$, and the component $(1 - \alpha(t))E[L(t)]$ that represents the risk assumed directly by the financial institution and thus the uncovered portion of debt reflected in the CDS spreads. The estimation of the α-value depends on a wide range of variables that influence the probability of receiving or not receiving government support in times of extreme market stress.

As anticipated, the SCCA is nothing more than an extension of the CCA since, unlike the latter—which only quantifies the contribution of each individual financial institution to the systemic risk of the sector conditioned by possible state guarantees in crisis situations—through the summation of these contributions it quantifies the risk exposure at the systemic level.

In more detail, Gray and Jobst (2010), starting from the results previously illustrated for the CCA, proposed the following measure (SRM) to assess the extent of systemic risk:

$$\text{SRM} = \sum_{i=1}^{N} \alpha^{i}(t)E\big[L^{i}(t)\big] \tag{11.26}$$

which thus summarises the total value of the risk exposure of all $(i = 1 \dots N)$ financial institutions.

Macroeconomic Measures

The last category of systemic risk metrics to be described concerns the well-known *Macroeconomic Measures*, so called because they are derived from the results of macroeconomic models.

The latter have historically played a major role in the systemic risk literature for a long time. Hence, the birth of several dedicated working and research groups, including, for example, the *Macro Financial Modelling Group*.

In fact, financial market instability has always been of particular interest to institutions, even when the systemic events that occurred could in fact only be read as a physiological response to ongoing economic changes.

In a nutshell, the metrics belonging to the Macroeconomic Measures family aim to analyse the dynamics with which a crisis can actually impact the macroeconomic aggregates described by the main economic and monetary policy models. Given that a financial crisis can easily involve many aspects of the economy of one or more countries, and that there is an obvious connection between the players in the financial system and the variables described in the best-known macroeconomic models, the main monetary aggregates can then be considered as parameters able to quantify the intensity of systemic risk.

As already emphasised, the rationale behind Macroeconomics Measures is in fact based on the intrinsic link between financial institutions and the performance of the business cycle and the real economy: since systemic events are always accompanied by structural changes in the main monetary aggregates, a measure of systemic riskiness can therefore be obtained by observing the latter, their historical performance and any existing imbalances (Reinhart and Rogoff 2008, 2009, 2013a, b).

One of the most famous models proposed in this context is that of Reinhart and Rogoff (2008), which pays particular attention to several price indices (stocks, real estate, etc.), the GDP growth rate and the level of public debt. The results obtained from the application of this model revealed a common pattern of contagion dynamics in recessionary periods, under which a number of economic and/or monetary policy interventions aimed at buffering and resolving recessionary events were proposed.

The policy approach of macro-prudential regulation is different; it proposes itself as a key player in macroeconomic risk monitoring. In particular, Borio and Drehnamm (2009a, b) and Borio (2009), focused on a policy perspective by proposing risk monitoring as a key aspect in macro-prudential regulation. Therefore, a macro-prudential framework characterised by top-down supervision was defined.

Borio's work (2009) shows that there are different risk measurement methodologies that are more or less appropriate depending on whether the issue taken as a reference is of a temporal nature, and thus relates to pro-cyclical phenomena, or manifests a transversal nature, and thus is able to capture the interconnections between intermediaries and their common exposures. In particular, in the case of a time dimension, the most appropriate measure to adopt would be early warning systems and predictive indicators of the cycle, while in the presence of a cross-sectional dimension, the most appropriate measures should be those that assess the contribution of each individual financial subject to the systemic risk, and thus those able to quantify the individual marginal contribution to the latter.

The real challenge for scholars, however, would seem to be that of being able to strike a balance between the general application of a model and its peculiarity, or, in other words, that of having the ability to mediate between the generality of a model and its specificity, actually managing the trade-off inherent in these measures and thus helping to create more variety in the approaches used.

References

Acharya, V., Pedersen L.H., Philippon, T., Richardson, M.: Measuring the systemic risk. AFA 2011. Denver Meetings Paper. Working Papers series (2010)

Adrian, T., Brunnermeier, M.K.: CoVaR: A Method for Macroprudential Regulation. Federal Reserve Bank of New York Staff Report, p. 348 (2008)

Adrian, T., Brunnermeier M.K.: CoVaR. Princeton University Press. Princeton department of Economics. SSRN (2011)

Adrian, T., Brunnermeier, M.K.: CoVaR. Am. Econ. Rev. **106**(7), 1705 (2016)

Billio, M., Getmansky, M., Lo, A.W., Pellizzon, L.: Econometric measures of connectedness and systemic risk in the finance and insurance sectors. J. Financ. Econ. **104**(3), 535–559 (2012a)

Billio, M., Pellizzon, L.: Misure econometriche di connettività e rischio sistemico nei settori finanziario e assicurativo europei, SYRTO Working Paper Series, Working Paper n. 10 (2014)

Billio, M., Getmansky, M., Lo, A.W., Pellizzon, L.: Econometrics measures of systemic risk in the finance and insurance sectors. NBER. Working Paper 16223 (2010)

Billio, M., Getmansky, M., Pelizzon, L.: Dynamic risk exposures in hedge funds. Comput. Stat. Data Anal. (2012b)

Borio, C., Drehmann, M.: Assessing the risk of banking crises—revisited. BIS Quarterly Review (2009a)

Borio, C., Drehnann, M.: Towards an operational framework for financial stability: fuzzy measurement and its consequences. Bank of International Settlements Working Paper (2009b)

Borio, C.: The macroprudential approach to regulation and supervision. working paper, VoxEU.org (2009)

Brownlees, C.T., Engle, R.F.: Volatility correlation and tails for systemic risk measurement. Working Paper, NYU Stern School of Business (2011)

Brownlees, C., Engle, R.F.: SRISK: a conditional capital shortfall measure of systemic risk. Rev. Financ. Stud. **30**(1), 48–79 (2017)

Brunnermeier, M., Crocket, A., Goodhart, C., Persaud, A.D, Shin, H.: The Fundamental Principles of Financial Regulation, Geneva Reports on the World Economy 11, Preliminary Conference Draft (2009)

Engle, R.: Dynamic conditional correlation: a simple class of multivariate generalized autoregressive conditional heteroskedasticity models. J. Bus. Econ. Stat. **20**(3), 339–350 (2002)

Getmansky, M., Lo, A.W., Makarov, I.: An econometric model of serial correlation and illiquidity in hedge fund returns. MIT Sloan School of Management, 50 Memorial Drive, E52–432, Cambridge (2004)

Granger, C.W.J.: Investigating causal relations by econometric models and cross-spectral methods. Econometrica **37**, 424–438 (1969)

Gray, D., Jobst, A.: Systemic CCA—A Model Approach to Systemic Risk. International Monetary Fund. Paper presented at the conference sponsored by the Deutsche Bundesbank and Technische Universitaet Dresden (2010)

Jolliffe, I.T.: Principal Component Analysis. Springer Series in Statistics. Springer-Verlag, New York (2002). https://doi.org/10.1007/b98835

Khandani, A., Lo, A.W., Merton, R.: Systemic Risk and the Refinancing Ratchet Effect, NBER Working Paper, 15362 (2009)

Lo, A.W.: The statistics of Sharpe ratios. Financ. Anal. J. **58**(4) (2002)

Lorenz, J., Battiston, S., Schweitzer, F.: Systemic risk in a unifying framework for cascading processes on networks. Eur. Phys. J. B **71**(4), 441–460 (2009)

Reinhart, M.C., Rogoff, M.K.: Financial and Sovereign Debt Crises: Some Lessons Learned and Those Forgotten. International Monetary Fund, pp. 13–266 (2013b)

Reinhart, C.M., Rogoff, K.S.: Is the 2007 U.S. subprime crisis so different? An international historical comparison. Am. Econ. Rev. **98**(2), 339–344 (2008)

Reinhart, C.M., Rogoff, K.S.: The aftermath of financial crises. Am. Econ. Rev. **99**(2), 466–472 (2009)

Reinhart, C.M., Rogoff, K.S.: Banking crises: an equal opportunity menace. J. Bank. Finance **37**(11), 4557–4573 (2013a)

Revuz, D., Yor, M.: Continuous Martingales and Brownian Motion, 3ª Springer, Heidelberg (1999)

Part II
Empirical Insights

Chapter 12
Systemic Risk and the Insurance Sector: A Network Perspective

Stefania Sylos Labini, Elisabetta D'Apolito, and Iryna Nyenno

Abstract Unlike the banking sector, whose impact on systemic risk has been amply proven, it remains unclear whether the insurance business constitutes a source of systemic risk. The aim of this chapter is to contribute to the debate and attempt to answer this question by examining the role of insurance companies in the financial system, the existing interconnections between insurance operators, and the vulnerability factors, i.e. the elements that increase the exposure of institutions to systemic risks. In recent years, attention to systemic risk has been focused on the insurance sector and the financial activity of insurance companies, which depends heavily on the performance of the financial markets as well as the phenomenon of bancassurance in the search for new investment methods, which inevitably increases the transmission channels of systemic risk. From this perspective, the insurance sector is significant in the network of financial relations and in the global economic system where a crisis can certainly cause systemic effects and repercussions due to the highly interconnected financial activities in place. The International Association of Insurance Supervisors (IAIS) in the Global Insurance Market Report (GIMAR) of 2022 highlights how systemic risk in the global insurance sector is still moderate overall, albeit with an upward trend in insurers' scores due to increased illiquid exposures and assets, over-the-counter derivatives, short-term loans and intra-financial assets. The role of regulators and supervisory authorities is crucial in this context. Therefore, the chapter analyses both the main and most recent European regulatory and supervisory interventions in the field of systemic risk management in insurance and the possible macro- and micro-systemic supervisory tools envisaged for the insurance sector. Based on the data available on the Acharya Volatility Lab (V-Lab) website and the Bureau van Dijk Orbis database, the insurance sector is explored in terms

S. S. Labini · E. D'Apolito (✉)
University of Foggia, Foggia, Italy
e-mail: elisabetta.dapolito@unifg.it

S. S. Labini
e-mail: stefania.syloslabini@unifg.it

Present Address:
I. Nyenno
KU Leuven, Leuven, Belgium
e-mail: iryna.nyenno@kuleuven.be

© The Author(s) 2025
V. Pacelli (ed.), *Systemic Risk and Complex Networks in Modern Financial Systems*,
New Economic Windows, https://doi.org/10.1007/978-3-031-64916-5_12

of systemic risk management and in relation to the global economic and financial system.

Keywords Systemic Risk · Insurance sector · Bancassurance

Systemic Risk: Concepts for Insurance Sector and Transmission Channels

In recent years, attention to systemic risk has been focused on the insurance sector in consideration of the following: (i) international insurance groups focused on non-traditional businesses, which are often mutually connected by complex reinsurance and retrocession relationships, vectors of systemic risk; (ii) investments of assets which increase the degree of interconnection with banks (for example by investing directly in their capital or in securitised credit products) (iii) the close interconnection with the banking system originating from the bancassurance and accentuated through the development of IT innovations and the spread of financial conglomerates (iv) the increasingly widespread offer of certain product categories (in the life sector), where the purely financial component takes on decisive importance.

Such phenomena inevitably increase the transmission channels of systemic risk. The literature on systemic risk has mainly analysed the different measurement methodologies, neglecting the transmission channels and causes.[1]

Following the public bailout of AIG in September 2008 due to exposed credit default swaps at the English branch, which affected the entire insurance group, awareness has been raised about the systemic importance of crisis risks. It was clear that an insurance company can introduce risk into the entire financial market in what has been defined as a new architecture of prudential regulation.[2] The rationale for such regulation is the awareness that the increasing number of life products are significant financial components which, in turn, means the financial activity of insurance companies is highly dependent on how markets function, as illustrated by the AIG

---

[1] Engle et al. (2015).

[2] Wan (2016). About non-bank sector, the author states that "The global financial crisis of 2008 exposed the weaknesses of a heavily interconnected financial system and revealed systemic risk in unexpected areas. While large banks and securities firms were known to pose risks to the financial system, it was largely unforeseen that nonbank financial institutions such as AIG could be equally dangerous. As a result of the crisis, a new wave of prudential regulation spearheaded by the Dodd-Frank Wall Street Reform and Consumer Protection Act (Dodd-Frank) attempted to bring about sweeping reforms to the financial sector, with a strong focus on financial stability. Through Dodd-Frank, the [US] Congress vastly expanded regulatory oversight, and has in many ways drastically changed the regulatory landscape for the financial industry. One issue surrounding post–Dodd-Frank prudential regulation that has garnered a great deal of attention has been systemic risk regulation in nonbank financial sectors, particularly the asset management industry".

crisis.[3] At the same time, insurers are part of the banking system's chain of relationships, and the phenomenon of bancassurance, seeking new investment methods,
inevitably leads to increased transmission channels of systemic risk.

The International Monetary Fund's (2016) statement indicates that life insurance
companies are increasing contributors to systemic risk as a result of the sector's
constantly rising exposure to overall risks, higher sensitivity to interest rates and
strong links with banks and asset managers. The insurance industry also clearly
plays a significant role in the financial relations network where a crisis could
have an overarching effect with global repercussions, due to the high levels of
interdependencies.

Given the degree of impact the insurance business has on the financial system, the
role of insurance companies should not be restricted to mere providers of life and non-
life insurance services. The 2022 Global Insurance Market Report (GIMAR) identified the interconnected subjects, parties and even business owners of the insurers,
based on their transactions within the financial system. Data was collected from the
aggregate insurance market from the IAIS members of 27 jurisdictions, comprising
more than 90% of global gross written premiums. Looking at the regulatory aspects
of the global insurance market, the network was analyzed to identify and observe
the relations of all the links in the financial value chain that participate in the insurance business, in order to prevent the factors of systemic risk which also include
speculations and unfair use of capital.

It is advisable that regulators be more attentive to mandating ownership transparency in partnerships with private equity. This would prevent gaps in the value
chain creation of financial markets when the insurance business, rather than being
used for its intended purposes, becomes a type of pseudo-insurance as is the case
with captive insurance companies. For instance, in lesser reliable companies, it is
advisable to introduce a networking partnership reporting map to clearly identify the
parent company and its links to affiliated structures.

Looking at a company's solvency and asset allocation measures revealed that
they are more frequently caused by a specific situation rather than being the result
of market position and global trends. According to the IAIS, the main factor driving
growth was the rising interest rates, which led most insurers towards increased
profitability and solvency.

The overall cash positions remained stable under the condition of tightened monetary policy. Total assets as reported in the SWM[4] rose by 4.9% to $44 trillion at
year-end 2021, whereas total liabilities increased by 4.3% to $38 trillion. In the
Asia and Oceania region, liabilities (+6.5%) increased more than assets (+4.3%) in
2021, which explains the decrease in the excess of assets over liabilities. Liabilities
at year-end 2021 were mostly composed of gross technical provisions for life insurance (55%), gross technical provisions for unit-linked insurance (14%) and gross
technical provisions for non-life insurance (11%). The overall amount of borrowing

---

[3] Paulson and Rosen (2016).

[4] The acronym SWM indicates aggregate data from sector-wide monitoring from supervisors across
the globe.

Table 12.1 Changes in the key global insurance market developments, covering assets, liabilities, solvency, profitability, and liquidity

#	Changes in %, world	Y19/20	Y20/21
1	Total assets	5.7	4.9
2	Total liabilities	7.4	4.3
3	Credit quality of assets	0.5	0.6
4	Solvency	−2.8	4.0
5	Excess of assets over liabilities	−1.4	0.4
6	Return on assets	−0.24	0.08
7	Revenue on assets	−5.9	5.5
8	Cash on assets	0.32	1.37

Source Own elaboration on Global Insurance Market Report (GIMAR) 2022. International Association of Insurance Supervisors (IAIS) (2022), pp. 6–14

remained limited at 2%, showing no change compared to the previous year. In terms of the geographic distribution of gross written premiums, according to the SWM data, at year-end 2021 most were underwritten in the United States (US) (38.1%), followed by China (11.7%), Germany (5.4%) and Japan (5.3%).

Solvency ratios improved in all regions again in 2021, after a slight decrease in 2020. In many jurisdictions, a slightly higher aggregate solvency level was seen compared to 2019. No jurisdiction in the SWM reported any major concerns with their local aggregate solvency requirements in 2021. On the global level, the ratio of revenues over assets increased at year-end 2021. Supervisors indicated that the main impact on life insurers can be attributed to the rise in interest rates. Changes in the global insurance market developments are presented in Table 12.1.

As briefly mentioned, the risk assessment by the prudential supervisors revealed that the core impact was due to the rise in interest rates, which resulted in higher profitability for both life insurance products and capital resources. Non-life insurers were under the pressure of higher costs, higher reimbursements, and the necessity to revaluate insurance reserves. Interest rate changes have a more modest impact on non-life insurers since their liabilities are less affected by interest rates. The Russian invasion of Ukraine has had a relatively limited impact on claims for selected insurance products, such as aviation, trade, and credit insurance. However, it is clear that this situation could change rapidly and is thus unpredictable.

The most influential aspect of macroprudential supervision was related to systemic risks in the financial market. The COVID-19 pandemic and the supply-related food and energy shocks from the war in the Ukraine led to increased monitoring requirements, for solvency and liquidity, in different jurisdictions. The monitoring tools such as risk management, ALM, stress testing, and liquidity contingency planning became important aspects of the supervision data. Supervisors collected additional information through member questionnaires, dialogues, and/or surveys. The monitoring was carried out to determine how increased interest rates and inflation could be incorporated into the pricing, actuarial, and business models of the insurers. All of these elements required the engagement of senior prudential supervisors. Observation revealed that the business models of most insurers had recently changed in the direction of being based more on capital-light business-linked products and away from the range of products with long-term interest rate guarantees.

A continuing trend for the insurance industry is the involvement of private equity (PE) companies, including asset management firms. There are different ways to attract these companies, beginning with investment partnerships that progress to reinsurance, project and asset management, and finally to mergers and acquisitions. We mention PE companies as they exemplify a strategy shift that insurance companies have made through the above-mentioned forms of cooperation. These cooperative measures require close supervision as they may have a broader impact on the financial market and greater structural shifts. Consequently, financial systemic risk should be considered in the scope of this PE-related approach in the insurance market. This concentration of the insurers together with private businesses often brings insurers higher profits, generates new internal stable cash flows as well as management and consulting for asset allocation activity, and creates new income that may lead to merger and acquisition transactions.

Systemic Risk Management of the Insurance Market

The possible concentration of systemic risk was also taken into account by the Global Monitoring Exercise (GME) as well as at the level of the individual insurer. For the Insurer Pool, the aggregate systemic risk score has been increasing over the past five years. This increase is primarily being driven by the interconnectedness and asset liquidation categories, which account for most of the total systemic risk score and have risen by 21% and 44%, respectively (Fig. 12.1).

A cross-sectoral analysis was performed to compare the systemic footprint of insurers with banks using a systemic risk scoring methodology based on indicators that are common to the Global Systemically Important Bank methodology. The

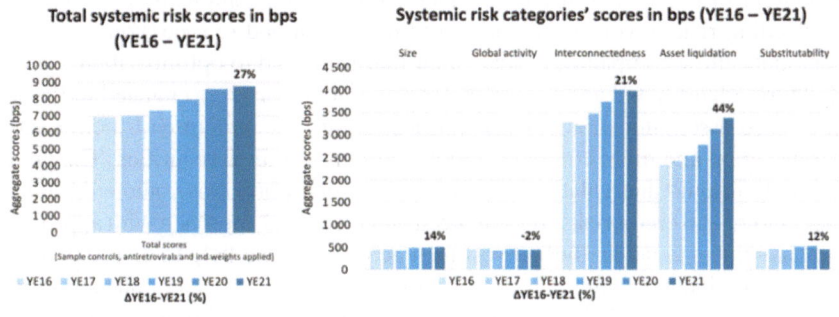

Fig. 12.1 Systemic risk scores by category (in bps). *Source* IAIS (2022)

results showed that by keeping the pool of banks and insurers stable over time, the total cross-sectoral scores for banks are significantly higher than for insurers. Although the supervisors recognized the data are still limited in the US, for 2021 the PE-owned life insurers reported an 800 billion USD increase, a significant increase from the previous year. The interest in PE firms in the UK and Asian markets is promising. The motivation for these changes is to reduce the burden of guaranteed rates and render the strategy for the capital-intensive and retirement segments of the market more sustainable and flexible. These innovative structures are also a source of additional income given the investment expertise and the variety of approaches used to optimise the profit returns on the insurance business in addition to the capacity to generate yields from the complex illiquid assets. The insurers involved in PE are mainly focused on corporate bonds supported by the opportunities of asset-backed securities and collateralised loan obligations linked to the internal asset origination platforms.

Reinsurance with the involvement of a third party may serve as a risk mitigation tool that would provide an increased solvency ratio and accelerate future income generation. For the reinsurer, the transferred insurance could be considered a means of acquiring certain blocks of the insurance business, which are sometimes similar to M&A transactions. The consideration of affiliated and non-affiliated reinsurance should be done separately. Affiliated reinsurance may lead to risk concentration and should thus be supervised more closely by authorities.

Outsourcing as a form of economy of scale to decrease expenses is another incentive for engaging in PE. Better and more attractive single or complementary products may also offer a competitive advantage in the market. PE firms consider insurance asset management desirable because these assets are stable and more resilient to market fluctuations. At the same time, these interactions involve greater complexities and require monitoring by the macroprudential bodies of the different jurisdictions. Higher returns on investments may lead to reduced liquidity and risk a lack of transparency. The adverse conditions of an economic downturn, geopolitical instability and energy uncertainty are factors that push insurance companies to seek new regionally diversified business models, and PE partnerships readily support this kind of opportunity. Such partnerships can be achieved by involving cross-border insurers and reinsurers, which increases the complexity because of the higher ceding risks involved in concentrating cession and retrocession. New business models allow companies to scale their platforms and to optimise their working capital through affiliated and non-affiliated companies. Insurers engaged in PE have an increased asset allocation towards infrastructure. The structure of their assets is less transparent and more complex compared with structures securities. The greater exposure to illiquid and volatile assets in the insurers' portfolios could contribute to higher market and credit risks. If a balanced and measured approach with appropriate risk management is taken, these investment activities could help to diversify insurer portfolios and increase internal rates of return. However, a greater concentration of such assets on insurers' balance sheets also increases potential liquidity spirals (e.g. margin calls that lead to forced sales in an illiquid market), especially in a situation of economic uncertainty where fungibility and transferability of assets decrease. The PE pressure may have an impact on the insurance market in the form of higher risk-taking as a way to become more competitive and attract more investment resources. The more complicated structure of an investment and insurance (reinsurance) portfolio makes monitoring and supervision more difficult. As well as the complex outsourcing arrangements and the opaque power structures that can also distort the decision-making processes of the insurer's board or management. All these issues should be reviewed to avoid the systemic risk caused by the conflict of interest between maintaining insurance market stability and PE aims of increased earnings. This is only possible under a cross-border supervisory relationship with written regulator-to-regulator enquiries, bilateral exchange of information from different jurisdictions, filling any information gaps and, last but not least, understanding the main motivation for this PE-influenced transformation of the insurance market.

The supervisor measures to be taken as suggested in the GIMAR include the following:

1. Introducing higher solvency ratio requirements to defend against unexpected losses.
2. Implementing better liquidity positions based on the stress-testing scenarios.
3. Restrictions on dividend extractions in the context of long- and short-term strategies to protect the policyholder.
4. Corporate governance monitoring to avoid conflicts of interest within these complicated structures.

Our point of view is that to prevent systemic risk, it is important to inspect the creation and functions of captive insurance companies, which are often not created solely for the purpose of lowering insurance costs; the other motivation is to obtain risk distribution among the pool of the affiliated insurers. These kinds of companies may also be created under the umbrella of certain risk coverage, such as the terrorism risks of the Pan American case, which were refused by traditional insurers. At the same time, there have been criticisms concerning the circular flow of investment funds. Usually insurance companies do not engage in this activity. These circular flows can occur if the premiums charged are excessive compared to existing actuarial practice. If captive insurance companies act in ways such that the policies have contractionary rules and charge unreasonable premiums, even the reimbursement itself cannot show that the insurance scheme is valid. Thus, some features indicate the PE-owned captive company may not be bona fide. Captive insurers that do not comply with the practices of insurance activity negotiation cannot go to taxation regulation for the insurance markets in different jurisdictions. This can be seen when the regulator defines an implausible risk, which is atypical for insurance practice, the insurance contract rules are unclear, there is no risk distribution, and ambiguous and illusory or risk duplication occurs related to the other insurance policies. The schedule of the insurance premium payment should be followed and the foundations of actuarial calculations should be the basis for premiums. A captive insurance company may still be a useful planning and tax reduction tool and is attractive to the PE partnership, provided it does not abuse regulations and norms.

Monitoring should be carried out if a parent company does not make direct reimbursement to the captive insurer in the holding; it is advisable to check the model if the parent company is making payments to a third-party insurer who reinsures the majority of the parent's risk with the captive; and there is a model when the captive reinsures a big share of risk with a third party.

The other rationale for PE may be that the captive is not functioning as an insurance company but instead plays a role of an accounting reserve, which is a 'segregation of retained earnings to provide for such payouts as dividends, contingencies, improvements, or retirement of preferred stock'. A wholly owned PE insurance company with all the outer trappings of insurance was merged (created) in substance as an accounting reserve, to reduce tax burden. This practice can be shared among other affiliated beneficiaries to avoid taxes. The PE-owned insurers may be bona fide in cases where the transactions initiated by the business entities are related to legal tax optimization. A parent company may explore the ability to shift self-insured risks to the captive company with policies tailored to fit the parent company's unique demand. The underwriting evaluation policy should be present to support the captive insurer's activities. Thus, risk sharing with the help of a captive company will be differentiated from simply the transferring of wealth in a tax-efficient manner.

Doubts concerning the quality of policy language, unavailability of claims processing, unclear explanation of the policy pricing, and absence of claims are indicators that the insurer is not only captive but is also illegitimate.

We have provided a regulatory and supervisory intervention map in the field of systemic risk management of the insurance market, as shown in Fig. 12.2.

As the map illustrates, the regulator is a key player in preventing systemic risk in the insurance market. The motivations of PE companies entering the insurance market and their activities should be monitored by regulators. Supportive information for regulators is the Global Insurance Market Report (GIMAR) indicators, signs, and symptoms of bad faith companies. The supervision connected with cross-border activities may not be efficient in cases where cooperation in different jurisdictions is not mutually understood. Accordingly, in-time efficient supervisory measures will result in a healthy environment with competitive insurance products, and strategies and a bona fide development of the insurance market as an important risk management link of the financial system.

In-time interventions will decrease the spread of a destructive environment across the insurance market. We consider a destructive environment to be a shadow economy and relations that involve corrupt intentions, where the insurance market fails to fulfill the risk management function and instead becomes a source of systemic risk to the financial system and the economy as a whole.

According to the Methodology of the International Association of Insurance Supervisors (IAIS) the scoring of systemic risk indicators aims to assess potential systemic risks for individual insurers that may arise from industry concentration. Looking at this issue, we can see the in-depth serious approach to assess the systemic risks in insurance is based on quantitative accounting indicators. The trend analysis of the insurers market follows the comparative principle of the results of the individual insurer versus the development of all the Insurers Pool.

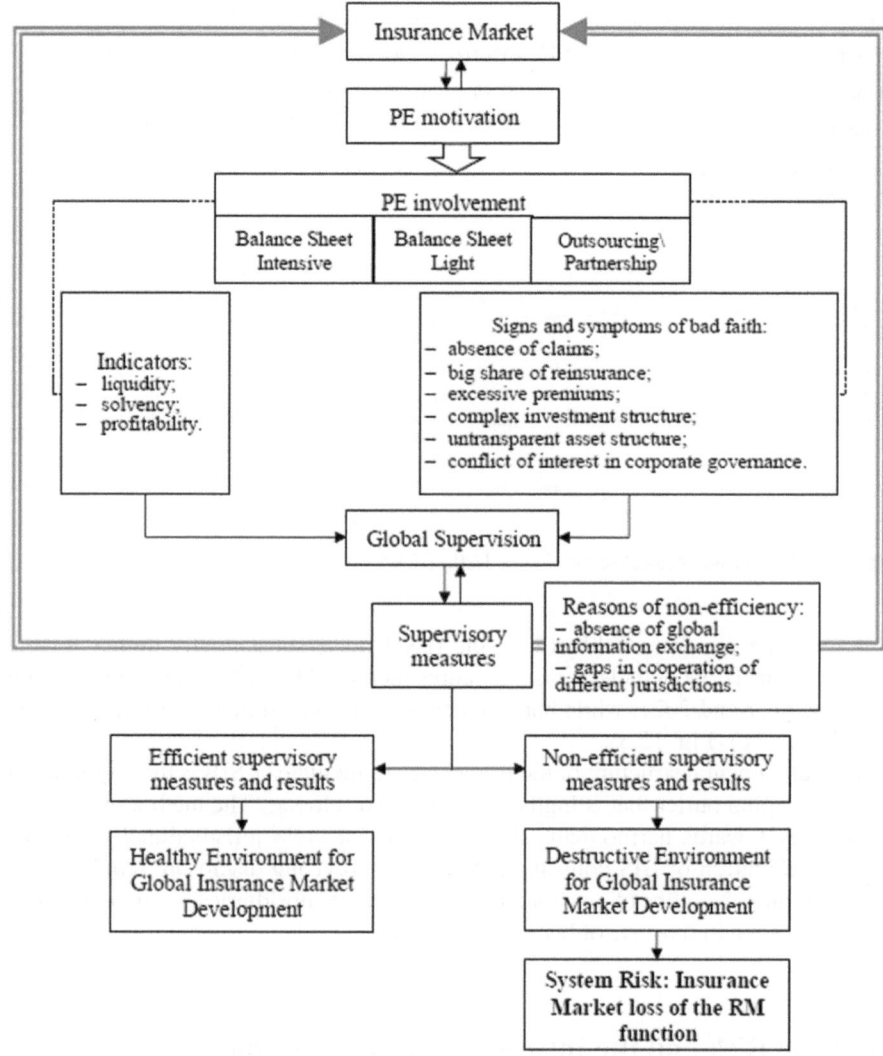

Fig. 12.2 Regulatory and supervisory intervention map in the field of systemic risk management of the insurance market. *Source* own elaboration

Basel Committee on Banking Supervision (BSBC) provided the methodology: "Global systemically important banks (G-SIB): revised assessment methodology and the higher loss absorbency requirements". It is based both on quantitative and qualitative approaches. The systemic impact is evaluated in relation to the data of the consolidated group, not individual companies (banks) differently from the IAIS risk scoring methodology.

The comparative data for 2016–2021 are presented in Table 12.2 for the insurance and banking industry.

Table 12.2 Total cross-sectoral score according to IAIS and BSBC, 2015–2021

Weighted cross-sectoral score	Y2015	Y2016	Y2017	Y2018	Y2019	Y2020	Y2021
Banks	0,77	0,78	0,76	0,77	0,76	0,75	0,75
Insurers	0,16	0,16	0,18	0,17	0,19	0,18	0,18

Source Global systemically important banks: revised assessment methodology and the higher loss absorbency requirement (bis.org), p. 42

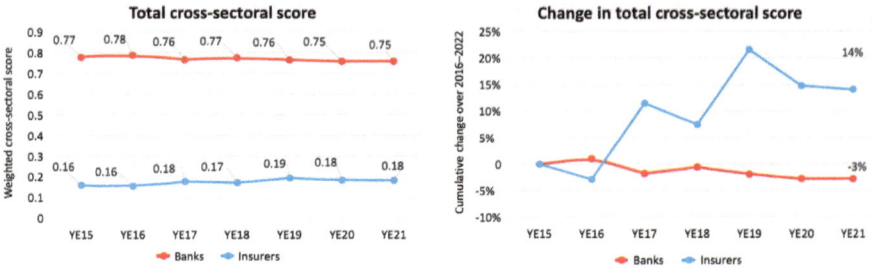

Fig. 12.3 Total cross-sectoral score. *Source* IAIS, BCBS (2022)

As we can see from Table 12.2—the risk load is substantially higher for the banking industry. Nonetheless, insurer scores increased by 12.5% between year-end 2016 and year-end 2021, while bank scores decreased marginally during that same period (−3.8%) (Fig. 12.3).

The fundamental structure of the framework to regulate G-SIB is the requirement to hold a capital buffer that is high enough to absorb losses. The methodology aims particularly towards harmonizing the definitions of cross-jurisdictional indicators with the definition of BIS consolidated statistics, revising disclosure requirements and expanding the scope of consolidation to insurance subsidiaries. Further guidance should be given in the case of bucket migration.

Systemic Risk and the Insurance Sector: An Analysis of the European Market

A wide range of issues relating to the systemic relevance of insurance is discussed in academia and among regulatory authorities and practitioners. Three main approaches can be used to classify the available literature: the market-based, the fundamental-based and the mixed approach. *Market-based models* use market data to measure externalities arising from a single institution that impact the rest of the system or vice versa and aim at measuring volatility in equity prices or CDS spreads. Provided that all relevant company information is taken into account, these measures allow for a comparison of the cross-sectional systemic importance of different industry markets with the limitation of neglecting the specific characteristics and determinants of the industry. The most commonly used systemic risk measures form part of these arrangements such as: the Conditional Value at Risk (Adrian and Brunnermeier 2016), the Marginal Expected Shortfall (MES) and Systemic Expected Shortfall (SES) (Acharya et al. 2010), the Distressed Insurance Premium (Huang et al. 2012), the Contingent Claims Analysis (Gray and Jobst 2011) and the linear and nonlinear Granger causality test (Billio et al. 2012). The *fundamental-based approach* consists in analysing accounting data, with an emphasis on the specifics of the investigated business and a combination of Theoretical and Empirical analyses. This model can identify the determinants of systemically important institutions through an analysis of their specific traits by looking at asset distribution and investment strategies as well as operational activities on the liability side. Some important contributions include the works of Cummins and Weiss (2014), Harrington (2009), Bell and Keller (2009) and The Geneva Association (2010a, b). In order to determine which specific factors are driving the systemic relevance of each trade, the *mixed approaches methods* consider the systemic importance obtained by applying market-based measures and accounting data (Weiss and Muehlnickel 2014; Bierth et al. 2015 and, focused on the European insurance market, Berdin and Sottocornola 2015a, b).

The contribution made in this chapter falls under the methodologies centred on market-based models. The most well-known measures in the literature at an international regulatory level to effectively measure systemic risk quantitatively is the Marginal Expected Shortfall (MES) or marginal expected loss proposed by Acharya et al. (2010), which was subsequently extended into the long-term analysis methodology (LRMES) that estimates the increase in systemic risk, measured by the extreme expected loss (Expected Shortfall, ES), determined by a marginal increase in the weight (capitalization of stock exchange) of the institution within the market.[5] This is followed by other important metrics such as the Delta Conditional Value at Risk (ΔCoVaR) by Adrian and Brunnermeier (2016), which expresses the value at risk (VAR) of the financial system conditioned by the default state of an institution and the Systemic Risk Measure (SRISK) by Brownlees and Engle (2017), which extends the MES metrics as it also includes other discriminating variables to the systemic risk of an institution being analysed such as the value of the assets managed, i.e. the company size and the amount of financial liability. The empirical analysis focuses on the value of the Long-Run Marginal Expected Shortfall (LRMES) proposed by Acharya et al. (2012) to estimate the marginal contribution of European insurance to global systemic risk, given the higher expected losses of capital that the latter could experience in the event of an extreme loss (expected shortfall) of the market, and specifically of the S&P 500 index, within the six months following the reference date. This is essentially the long-term MES indicator of the institutions analysed.

The analysis was conducted based on the data available on the Acharya Volatility Lab (V-Lab) website, the composition of the sample of Bureau van Dijk's Orbis database, and the ranking value in the period 2017–2022. Specifically, the empirical analysis covers 37 European insurance companies as highlighted below. Figure 12.4

---

[5] Banca d'Italia (2018). Acharya et al. (2010).

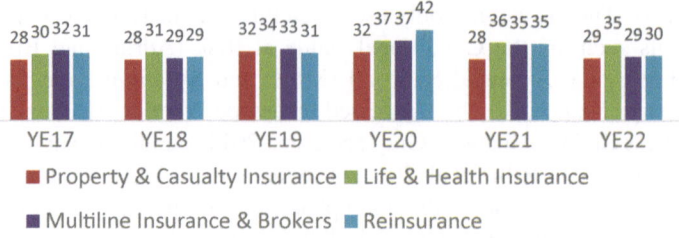

Fig. 12.4 Average LRMES by sector (YE 2017–2022). *Source* Authors' data elaboration

presents the average LRMES by sector. This measure allows us to enhance the interconnections between financial institutions and the financial system by estimating the expected marginal loss of capital in the long term. In analytical terms, Acharya et al. (2012) propose to estimate the value of LRMES using as a proxy the beta coefficient of the financial institution defined for a simulated default threshold (d) of 40% as follows: 1-exp(log(1–d)*beta).

The basic idea is that companies with a higher MES value contribute more to stock market falls and are therefore the main drivers of systemic risk. The analysis conducted made it possible to observe a high growth in the average score in 2020 especially for the reinsurance sector and on average a considerable contribution from life insurance in the period considered.

Systemic Risk and Insurance Sector: Policies Approach and Financial Connections

Systemic risk is highly complex and thus difficult to quantify. It has long been the subject of extensive interventions by regulators and supervisors to identify possible hedging methods and prudential structures.[6]

A new set of policies aimed at avoiding cascade effects and contributing to Financial Stability was required during the 2007–2008 economic crisis. A majority of the initiatives undertaken in the aftermath of this crisis have been directed at banks, which played a central role in the financial crisis. Although there is a common understanding in the banking sector that is plays an important role in the system, this issue remains under discussion in the insurance industry.

The European Systemic Risk Board (ESRB), established by the regulation of the European Parliament and Council No 1092 of 2010, is responsible for supervising the EU financial system as well as preventing and mitigating systemic risks, that is, the risk of disruption in the financial system with negative consequences for the real economy and its functioning as a whole.[7] The ESRB's macroprudential mandate

---

[6] Giesecke and Baeho (2011).

[7] IMF, BIS, FSB (2009).

covers banks, insurers, asset managers, financial market infrastructures and other financial institutions and markets. In its annual report for 2022, published between 1 April 2022 and 31 March 2023, the ESRB reviewed and updated its assessment, to take new systemic risks into consideration from the increased political and economic uncertainty, especially due to Russia's war in the Ukraine. Specifically, the risk assessment carried out by the ESRB includes: (i) a general alert regarding vulnerabilities in the EU financial system, (ii) a recommendation based on the difficult macroeconomic context for banks, insurance companies and pension systems, (iii) a report which defines the tools and elements necessary to promote cyber resilience, and (iv) the lack of liquidity in the financial markets. The overall risk assessment carried out by the Committee also includes possible threats arising from systemic cyber incidents, climate change and transition risks and disruptions to critical financial infrastructures. To improve the referenced macroprudential framework, the ESRB contributed to the review of the prudential framework of the insurance sector carried out by the European Insurance and Occupational Pensions Authority (EIOPA)[8] and highlighted how the regulatory framework for insurers, Solvency II, should be strengthened, with a particular regard to liquidity management tools.

As a contribution to this debate, EIOPA published a Discussion Paper entitled "Systemic Risk and Macroprudential Policy in Insurance" on 29 March 2019, which provided a comprehensive approach to the following questions, which is subsequent to its policy development:

1. Does insurance create or increase systemic risk?
2. If yes, what instruments are currently in place within the framework of Solvency II and how do they contribute to mitigating these sources of systemic risk?
3. Is there a need for additional tools and, where appropriate, which ones could be encouraged?

To address these questions, the EIOPA published documents, namely the "Systemic risk and macroprudential policy in insurance" of 2017, "Solvency II tools with macroprudential impact" of (2018a), and "Other potential macroprudential tools and measures to enhance the current framework" of (2018b).

To address the first question, the EIOPA developed a conceptual approach to illustrate the dynamics that allow insurance to create or amplify systemic risk (EIOPA 2017). The EIOPA's approach is based on the observation that a "triggering event" will initially have an impact at entity level with regard to one or more insurers through their risk profile (Fig. 12.5). Potential individual or collective inconveniences can then generate systemic implications, which will be more or less significant depending on the existence of different "systemic risk factors" incorporated into insurance companies. According to the EIOPA, systemic events can be generated through two main sources. The first, a 'direct effect', originates from the failure of a systemically important insurer or from the collective failure of multiple insurers, which

---

[8] European Systemic Risk Board (2022), Annual Report; EIOPA (2022), Financial Stability Report. Article 22 of EIOPA Regulation defines 'systemic risk' by reference to Article 2(c) Regulation (EU) No 1092/2010.

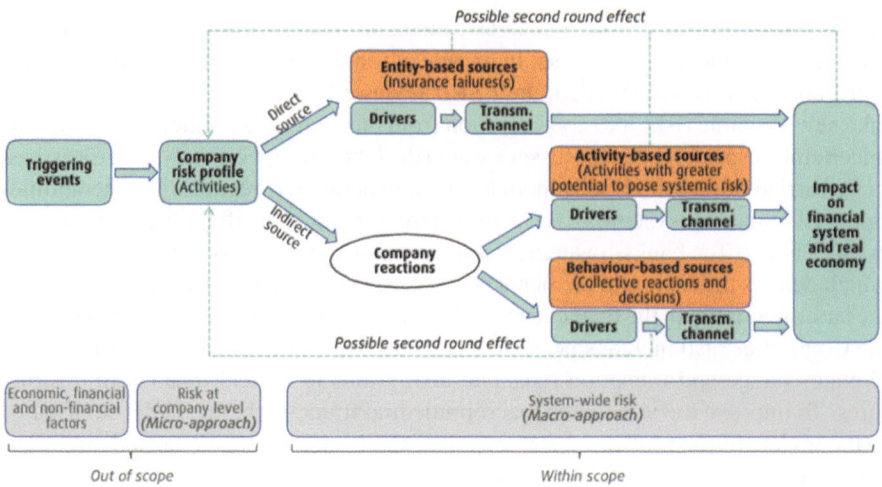

Fig. 12.5 The approach to systemic risk in insurance companies. *Source* EIOPA (2017)

generates a cascade effect. This systemic source is deemed to occur on an 'entity basis'. The second source, an 'indirect effect', is a risk in which possible external-ities are strengthened by the presence of potentially systemic activities (activity-based sources) or by widespread common reactions of insurers to exogenous shocks (behaviour-based source). Based on these sources of systemic events, attention is placed on the planning and management of intrinsically systemic activities by insur-ance companies. Potential externalities generated through direct and indirect sources are then transferred to the rest of the financial system, and the real economy as a whole, through specific transmission channels that could induce changes in the risk profile of insurers, possibly generating potential second impact effects.

The main elements of EIOPA's approach are highlighted below. A *Triggering event* is an exogenous event that can have an impact on one or more insurance companies and trigger the process of creating systemic risk. Examples include macroeconomic factors (e.g. rising unemployment), financial factors (e.g. changing yields) or non-financial factors (e.g. demographic changes or cyber-attacks). A *Company risk profile* includes the risks a company is exposed to and the potential vulnerabilities this poses to financial markets. *Systemic risk drivers* are elements that can generate negative repercussions due to the presence of company-specific stress factors that have a systemic effect, i.e. they can transform a company-specific stress into a system-level stress. *Transmission channels* represent the process through which sources of systemic risk can influence financial stability and/or the real economy. The EIOPA distinguishes five main transmission channels: (a) exposure channel; (b) asset liqui-dation channel; (c) failure to provide insurance products; (d) a channel similar to that of banks; and (e) expectations and information asymmetries. *Sources of systemic risk* affect systemic risk factors and transmission channels. These are direct or indirect externalities for which insurance constitutes a systemic threat to the financial system as a whole. These direct and indirect externalities lead to three categories of potential sources of systemic risks which are not mutually exclusive: entity-based related sources, activity-based sources and behaviour-based sources.

The EIOPA (2018a) analysed and classified the potential macroprudential tools and measures to manage systemic risk and submitted a preliminary evaluation of those tools and measures that are currently in place under Solvency II. The EIOPA preliminary analysis shows that the instruments which have a direct macroprudential impact under Solvency II provide a significant contribution to limiting procyclicality. However, several sources of systemic risk do not seem to be sufficiently addressed by the existing tools. Therefore, there is room for further tools and measures to be considered, to improve the existing framework.

In its third paper, the EIOPA (2018b) identifies other potential macroprudential tools and measures to enhance the current framework. The analysis focuses on capital instruments and specific liquidity requirements. Regarding additional capital for systemic risk, EIOPA believes that an additional capital buffer would enable resistance to shocks and thus prevent the deterioration and insolvency that could lead to the bankruptcy of insurance companies. EIOPA also advocates for adopting a sequential approach to liquidity risk in the following three phases: (1) improving reporting; (2) monitoring the liquidity risk; and (3) introducing liquidity requirements. The objective is to develop a complete and significant set of indicators to monitor and evaluate liquidity risk at both the micro and macro level. This measure was designed to improve reporting and monitoring by competent authorities. One of the important proposed indicators is the liquid assets ratio, which compares the amount of liquid assets on the balance sheet with total assets (excluding assets held for unit-linked contracts). The number of liquid assets is determined by assigning each asset item a weight that reflects the liquidity characteristics of that balance sheet item. This weight varies from 0 (highly illiquid item) to 1 (highly liquid item).

Ultimately, the EIOPA considers how the microprudential approach could be integrated by assigning certain roles and responsibilities to the competent authority responsible for macroprudential policy, which could entail the following activities: (1) aggregation of information; (2) information analysis; and (3) provision of certain information or parameters to supervisory authorities for the management of particular macroprudential risks.

Issues relating to systemic risk management are also addressed within the Financial Stability Board (FSB). In 2011, the FSB published an integrated set of policies to tackle systemic and societal hazard risks relating to Systemically Important Financial Institutions (SIFIs). In 2013, the FSB, in consultation with the International Association of Insurance Supervisors (IAIS) and the national authorities, established an initial list of Global Systemically Important Insurers (G-SIIs) using an assessment methodology developed by the IAIS, as well as policy measures to be applied. In 2022, the FSB stopped its annual identification of G-SII, approving the Holistic Framework (HF) curated by the IAIS in November 2019 to assess and mitigate systemic risk in the insurance sector. The HF recognises that systemic risks, in addition to arising from the distress or abrupt failure of an individual insurer, may also emerge from collective exposures and activities by insurers at a sector-wide level.

This framework includes enhanced supervision measures, annual trend monitoring of the global insurance market and potential accumulations of systemic risk as well as the assessment the complete and consistent implementation of supervisory measures in all jurisdictions. In 2025, the FSB reserves the right to review the evaluation process and Holistic Framework-based systemic risk mitigation, evaluating whether to reinstate and updates the identification process of G-SIIs.

Brief Concluding Remarks

This chapter offers some food for thought on the important systemic dynamics found in the insurance sector. Reflections on the systemic risks and regulatory developments in the insurance sector are provided to build a risk-oriented discipline. This chapter attempts to understand the systemic relevance of insurance and its evolution in regulatory activities that are effectively geared toward preventing a systemic crisis, starting with the historical background as well as background on the crises experienced by some large insurers.

The analysis reveals that significant changes have been made to the rules governing insurance companies, which were designed to prevent sector-wide viruses from spreading throughout the financial system. It is important that controllers and regulators avoid understating all the activities in the insurance sector that can lead to systemic risk for finance markets.

It will only be possible to accurately identify the channels of 'contagion', understand the weaknesses and establish organisational and management structures that enable insurance companies to maintain resiliency within the system as a whole by promoting prevention measures to monitor all activities and risk, to link factors that brought them together. To that end, it is essential to coordinate the supervisory instruments of prudential and micro-prudential supervision by applying harmonised rules at the international level.

Insurance companies are no longer a marginal player in the new regulatory context described and analysed at the international level. They should certainly be considered one of the systematically relevant groups whose proper assessment of assumed risks is critical to maintaining stability and overall economic development.

In addition, what emerges is how the system is set up and experimenting with new instruments and techniques to monitoring sector stability and ensure that insurance companies are able to monitor risks and address them in such a way as to protect all stakeholders' interests, starting with with the systemic role insurance companies play in financial markets.

These observations are expected to serve as a starting point for an in-depth study, both theoretically and empirically, of systemic risk management in the insurance sector under a legal context characterised by greater clarity and uniformity. An in-depth analysis is also considered interesting in the light of natural disasters, which can be a source of systemic risk for financial institutions and financial markets. EIOPA Chairperson Petra Hielkema highlights how "insurance plays a major role in protecting businesses and people against climate-related catastrophe losses by swiftly providing the necessary funds for reconstruction. In order to efficiently protect our society, we need to address the concern of the increasing insurance protection gap by proposing and finding appropriate solutions." ECB Vice-President Luis de Guindos says, "We need to increase the uptake of climate catastrophe insurance to limit the growing impact of natural disasters on the economy and the financial system. However, to reduce losses in the first place, we must ensure that a smooth and speedy green transition is complemented by effective measures to adapt to climate change." Mutualising and transferring collateral and property losses to reinsurance companies, which are better equipped to manage their climate exposures, may increase the resilience of banks to such shocks.

References

Acharya, V.V., Pedersen, L.H., Philippon, T., Richardson, M.P.: Measuring systemic risk. FRB of Cleveland Working Paper (10-02) 2010

Acharya, V.V., Engle, R.F., Richardson, M.: Capital shortfall: a new approach to ranking and regulating systemic risks. In: American Economic Review: Papers and Proceedings, vol. 102 (2012)

Adrian, T., Brunnermeier, M.K.: Co Va R. Am. Econ. Rev. **106**(7), 1705–1741 (2016)

Banca d'Italia: A survey of systemic risk indicators. Questioni di Economia e Finanza (2018)

Bell, M., Keller, B.: Insurance and stability: the reform of insurance regulation. Zurich Financial Services Group Working Paper (2009)

Berdin, E., Sottocornola, M.: Assessing Systemic Risk of the European Insurance Industry. European Insurance and Occupational Pensions Authority. Financial Stability Report (2015a)

Berdin, E., Sottocornola, M.: Insurance Activities and Systemic Risk. ICIR/SAFE Working Paper n.121 (2015b)

Bierth, C., Irresberger, F., Wei, G.N.: Systemic risk of insurers around the globe. J. Bank. Finance **55**, 232–245 (2015)

Billio, M., Getmansky, M., Lo, A.W., Pelizzon, L.: Econometric measures of connectedness and systemic risk in the finance and insurance sectors. J. Financ. Econ. **104**(3), 535–559 (2012)

Brownlees, C.T., Engle, R.: SRISK: a conditional capital shortfall measure of systemic risk. Rev. Financ. Stud. **30**(1), 48–79 (2017)

Cummins, J.D., Weiss, M.A.: Systemic risk and the US insurance sector. J. Risk Insur. **81**(3), 489–528 (2014)

EIOPA: Systemic risk and macroprudential policy in insurance (2017)

EIOPA: Solvency II tools with macroprudential impact (2018a)

EIOPA: Other potential macroprudential tools and measures to enhance the current framework (2018b)

EIOPA: Financial Stability Report (2022)

Engle, R., Jondeau, E., Rockinger, M.: Systemic risk in Europe. Rev. Finance **19**(1) (2015)

European Systemic Risk Board: Annual Report (2022)

Giesecke, K., Baeho, K.: Systemic risk: what defaults are telling us. Manage. Sci. **17**(8) (2011)

Gray, D.F., Jobst, A.A.: Modelling systemic financial sector and sovereign risk. Sveriges Riksbank Econ. Rev. (2011)

Harrington, S.E.: The financial crisis. Systemic risk and the future of insurance regulation. J. Risk Insur. **76**(4), 785–819 (2009)

Huang, X., Zhou, H., Zhu, H.: Systemic risk contributions. J. Financ. Serv. Res. **42**(1–2), 55–83 (2012)

IAIS: Global Insurance Market Report (GIMAR) (2022)

IMF: The insurance sector—trends and systemic risk implications. In: Global Financial Stability Report: Potent Policies for a Successful Normalization (. April)

IMF, BIS, FSB: Guidance to Assess the Systemic Importance of Financial Institutions. Markets and Instruments: Initial Considerations Report to the G-20 Finance Ministers and Central Bank Governors (2009)

Paulson, A., Rosen, R.: The life insurance industry and systemic risk: a bond market perspective. Annu. Rev. Financ. Econ. **8**, 155–174 (2016)

The Geneva Association: Systemic risk in insurance: an analysis of insurance and financial stability. Special Report of the Geneva Association Systemic Risk Working Group. The International Association for the Study of Insurance Economics, Geneva (2010a)

The Geneva Association: Key Financial Stability Issues in Insurance: An Account of the Geneva Association Ongoing Dialogue on Systemic Risk with Regulators and Policy-Makers. The International Association for the Study of Insurance Economics, Geneva (2010b)

Wan, J.S.: Systemically important asset managers: perspectives on Dodd-Frank's systemic designation mechanism. Columbia Law Rev. **116**(3) (2016)

Weiss, G.N., Muehlnickel, J.: Why do some insurers become systemically relevant? J. Financ. Stab. **13**, 95–117 (2014)

Chapter 13
Damping Systemic Risk. The Role of Cooperative Banks

Vincenzo Pacelli, Francesca Pampurini, and Anna Grazia Quaranta

Abstract This chapter focuses on the countercyclical potentialities linked to the bank's relational business model. We prove that cooperative banks are less involved in the mechanisms underlying systemic risk, thus verifying that their presence somehow contributes to mitigating this risk's spread in a financial system. Referring to a group of Italian banks, an innovative methodological approach is proposed, with relative empirical implementation, based on some variables—available for all banks—considered in the literature as proxies of the systemic risk propagation speed and capacity and the banks' health managerial status.

Keywords Systemic risk · Systemic risk propagation · Propagation speed and capacity · Cooperative Banks

Introduction

This chapter focuses on the countercyclical potentialities linked to the relational business model, which enhances the competitive advantage typical of smaller local cooperative banks related to geographical proximity and, therefore, to the availability of qualitative and privileged information on customers. An important stream of the literature is inclined to believe that cooperative banks can play a fundamental role in mitigating the mechanisms of systemic risk propagation, as opposed to larger banks characterised by national or international exposure (among others, EACB 2010, 2016; Demma 2015; Barone et al. 2016; Berton et al. 2017; Pacelli et al. 2020).

V. Pacelli (✉)
University of Bari Aldo Moro, Bari, Italy
e-mail: vincenzo.pacelli@uniba.it

F. Pampurini
Catholic University of the Sacred Hearth of Milan, Milan, Italy
e-mail: francesca.pampurini@unicatt.it

A. G. Quaranta
University of Macerata, Macerata, Italy
e-mail: annagrazia.quaranta@unimc.it

© The Author(s) 2025   229
V. Pacelli (ed.), *Systemic Risk and Complex Networks in Modern Financial Systems*,
New Economic Windows, https://doi.org/10.1007/978-3-031-64916-5_13

Alongside studies that acknowledge the ability of cooperative banks to contribute positively to the achievement of greater stability of the entire banking system in which they operate, there is no lack of contributions in the literature that highlight precisely the contrary effects; from these, it emerges how the presence of cooperative banks tends, instead, to aggravate the conditions of fragility of the financial system (among others, Brunner et al. 2004; Goodhart 2004; Fonteyne 2007).

This study, therefore, is part of an extremely topical debate that is sometimes controversial due to the presence of contrasting results regarding the effective contribution of mutual banks to the stability of the banking system. To the best of our knowledge, there are no contributions in the literature that use a methodology similar to the one used in this chapter to analyse the role played by these banks in the systemic risk transmission process. Therefore, in this chapter, an innovative approach is adopted in an attempt to provide empirical evidence to the debate on the countercyclical role of local cooperative banks.

The main purpose is to understand whether these banks are actually less involved than others in the mechanisms underlying the propagation and accentuation of systemic risk, thus verifying whether their presence somehow manages to mitigate the overall magnitude of this risk in a financial system.

Given the difficulties in identifying a single definition of systemic risk, it should be noted that in this study, the term refers to the risk that the crisis, the failure or the mere perception by the market of the risk of insolvency of one or more major players in an economic system—essentially, large companies, financial intermediaries or governments—may lead to generalised phenomena of crisis, insolvency or chain failures of other operators in the same economic system. Therefore, systemic risk will be considered as the risk associated with the manifestation of an event capable of causing, through mechanisms of contagion and propagation of the crisis, structural effects on an entire economic system.

On the basis of this definition, in order to pursue the aim of the research, an innovative methodological approach is proposed, with relative empirical application, aimed at operating the clustering of a group of banks that adopt different business models, starting from the values assumed by some variables considered as proxies both of the speed and capacity of propagation of the systemic risk and of the state of managerial health of the banks analysed.

As is shown later in the discussion, the variables used for the clustering of the banks analysed will be ten and deduced from the most authoritative literature that has dealt with systemic risk over the years.

The methodology proposed in this chapter is, therefore, substantially different from what is currently found in the literature in relation to the methods of measuring systemic risk.

In fact, the study focuses on variables, however, chosen on the basis of the criteria most widely used in the literature, which have the advantage of being available for all categories of banking intermediaries (and therefore not only for the systemic ones); these variables are also able to provide information on risk propagation dynamics rather than exclusively on the valuation of the risk itself.

Finally, in order to guarantee homogeneity in the empirical analysis, only the Italian banking system is analysed, since it is characterised by a massive presence of cooperative credit banks and, more generally, of small local banks that, as is known, are particularly focused on their territory needs.

The chapter is structured as follows: in section "Literature Review" a Literature Review. Section "Data and Methodology" presents the description of the dataset used and the methodology employed. Section "Empirical Results" presents the results obtained and their discussion, while section "Conclusions" concludes.

Literature Review

For several years, a wide strand of literature highlighted the fundamental role played by local banks with a mutualistic nature in promoting local development as well as the growth of the national economy (Boscia et al. 2010; EACB 2010; Bülbül et al. 2013; Fiordelisi and Mare 2013; Chiaramonte et al. 2015; Demma 2015; Clark et al. 2018; Pacelli et al. 2019). These studies highlight how the historical success of cooperative banks is not derived exclusively from their specific business model, but also from their peculiar and distinctive governance model. These characteristics have arguably allowed local banks to bear the financial crisis's negative effects and to play the countercyclical role that the predominant literature acknowledges to them and enabled the financing of local economies in the crisis characterised by credit restriction (EACB 2010). So, these banks were able to strengthen their roots and their local commitment based on trust, reputation and mutualistic values and, therefore, to enhance their competitive information advantage. This fact allows them to benefit from quantitative and qualitative information on customers and local operators thanks to sedimented relationships that have grown up over time.

Considering, therefore, the countercyclical potential linked to the relational business model, a large part of the literature agrees that local banks can play a fundamental role in mitigating the mechanisms of systemic risk propagation thanks to their countercyclical potential deriving from their specific relational business model.

The strong attention paid by researchers and international supervisory authorities to banks is justified because they represent the main vehicle for the propagation of a systemic crisis due to their role as financial intermediaries in an economic system (Iyer et al. 2013). In fact, it is well known that two channels for the propagation of a systemic crisis operate through banks. The first takes the form of a domino effect that comes from the direct relationships that characterise the interbank market or the banks' sovereign exposures. The second is an information channel, as banks are a key information provider for the financial markets.

For several years, the literature has focused on the study of the systemic impact of large international banks, analysing the interconnections between the banking system and the other microcosms that populate the economic system, both from a microprudential and macroeconomic perspective (Acharya 2011; Hanson et al. 2011; Claessens et al. 2014). These studies highlight the strong systemic impact that large banks exert on markets due to their interconnections, both in terms of value and frequency, with other economic players (Beirne and Fratzscher 2013; De Bruyckere et al. 2013; Buch and Goldberg 2016; Constâncio 2017).

A large part of these studies focuses on the construction of a quantitative model able to provide a measurement of the level of systemic risk both with reference to the whole economic system and to the contribution of each systemic bank and this is based on a series of economic-financial variables. All these works, based on advanced mathematical-statistical models, require many inputs based on market variables, therefore they can only be used in the case of listed banks.

The present work focuses, instead, on variables available for all categories of banking intermediaries (not only listed ones) that are able to provide information on the dynamics of risk propagation rather than on the evaluation of the risk level itself. These variables have been chosen on the basis of the criteria most widely used in the literature; in particular, a study by the International Monetary Fund (Blancher et al. 2013) proposes to use the financial statements of intermediaries, especially balance sheets, from which it is possible to deduce a series of information (the so-called Financial Soundness Indicators) useful for analysing the health of banks and their interaction dynamics with the system. The variables suggested by this study and by several other empirical contributions concern capital adequacy and risk coverage as measured by the Tier 1 ratio (Hoque et al. 2015), the weight of non-performing loans on total credit exposures, profitability, liquidity, the degree of interconnectedness with the system, measured through the value of loans and debts to other intermediaries (Acharya 2011; Blundell-Wignall 2012; Glasserman and Young 2015) and the weight of sovereign risk measured through the value of public securities held in the portfolio (Blundell-Wignall 2012). Another variable that is particularly popular in the systemic risk literature, is the z-score, which is an indicator of distance-to-default, (Acharya 2011; Blundell-Wignall 2012; Blancher et al. 2013; Hoque et al. 2015), i.e. how close (or far) the intermediary under scrutiny is from a financial situation that may portend imminent failure, has also been used.

As well illustrated by the International Monetary Fund (Blancher et al. 2013), systemic risk tends to arise through sequential events that start from one or more economic/financial shocks and then propagate with a chain effect.

The initial shocks that can generate a propagation mechanism and, therefore, lead to the onset of a systemic crisis are different and are classified by the literature as follows:

- a crisis of one or more financial intermediaries (Nelson and Katzenstein 2014) or of a government (Beirne and Fratzscher 2013), or the mere perception of the insolvency risk of these players;
- fall in the price of specific real or financial assets, including—in particular—residential real estate properties (Cerutti et al. 2017);
- liquidity crisis in financial markets followed by deleveraging, which fuels the fall of financial and real assets prices. This phenomenon triggers the deflationary spiral that, through the depreciation of collaterals, feeds the vicious circle of credit rationing (Reinhart and Rogoff 2013).

The causes that, according to the literature, can foster the propagation of an initial shock to an entire economic system are:

- high interconnection between the main players of the economic and financial system, in particular, high exposure of banks to sovereign debt and to interbank markets (Blundell-Wignall 2012; Hoque et al. 2015);
- savers confidence crisis and, in the most severe cases panic, leading to a domino effect characterised by generalised sales, fall in prices, credit rationing, bankruptcy and bank runs (Calvo 2012);
- strong information asymmetries in financial markets due to the increasing complexity of financial engineering, the information scarcity and the limited financial culture (Flannery et al. 2013);
- high level of indebtedness and, therefore, high dependence of borrowers on creditors, which enhances system vulnerability in times of crisis.

Above all, it is important to underline that starting from 2008, after the burst of the financial crisis, studies on systemic risk have overall increased significantly, together with the growing attention of international supervisory authorities and governments, mainly focused on strengthening the capital solidity of financial intermediaries to ensure the stability of an economic system (Acharya and Richardson 2009; Bengtsson 2013; Lane 2012; Brunnermeier 2009; Brunnermeier and Oehmke 2013).

Data and Methodology

Description of the Dataset

The dataset used for the analysis is composed of banks characterised by different business models, specifically, the set of cooperative banks, commercial banks, savings banks and investment banks active in Italy in the period 2018–2022. The data regarding the balance sheets of each intermediary come from Orbis Bank Focus (Bureau Van Dijk).

The dataset only includes those intermediaries for whom it was possible to find the values of all ten variables that are described below, as they are considered fundamental for the purposes of the study; this is because the results of the multivariate analysis, in particular those related to the grouping techniques employed, are significant only in the absence of missing data.

The number of the considered banks varies from year to year, not only due to the elimination of missing data, but also due to the Merger and Acquisition (M&A) operations that have affected the financial sector, as well as the exit of several banks from the market.

Table 13.1 shows the composition of the dataset[1] during the period analysed.

---

[1] In the discussion, reference is always made to the dataset under analysis, avoiding the definition of "sample" since, as it is known, from a statistical point of view, a "sample" is such when it is

Table 13.1 Dataset composition in terms of specialisation

Year	Commercial banks	Cooperative banks	Investment banks	Saving banks	Total
2018	61	245	13	9	328
2019	66	241	12	9	328
2020	63	234	11	9	317
2021	59	231	11	9	310
2022	58	226	9	9	302

It is clear from Table 13.1 that cooperative banks represent the largest group in the dataset analysed. Table 13.2 shows the subdivision of banks analysed in the various years that make up the period under investigation, with reference to Total Assets. The data presented in this table offers a clear vision of the structural tendencies underway in the Italian banking system, which see an increase in the volume of activity of cooperative banks together with a substantial resizing of the weight of investment banks.

Before proceeding with the construction of the indices necessary for grouping the banks belonging to the dataset, it is opportune to observe its composition with even greater attention in order to offer some micro-economic considerations on the peculiar characteristics of the units that are analysed. This focus also makes it possible to better illustrate the variables which are considered fundamental for the study of the contagion propagation dynamics and systemic risk within the banking sector considered as a whole. Table 13.3 therefore shows the main descriptive statistics,

Table 13.2 Banks distribution in terms of total assets (thousands of euros). Years 2018–2022

Year	Commercial banks	Cooperative banks	Investment banks	Saving banks	Total
2018	2,459,006,981	304,422,682	84,912,407	101,076,340	2,949,418,410
	83%	10%	3%	3%	100%
2019	2,593,934,623	417,284,070	39,007,599	104,570,389	3,154,796,681
	82%	13%	1%	3%	100%
2020	2,964,063,629	481,106,453	38,294,545	118,583,505	3,602,048,132
	82%	13%	1%	3%	100%
2021	3,114,188,200	516,328,830	32,144,141	150,397,369	3,813,058,540
	82%	14%	1%	4%	100%
2022	2,940,142,759	509,889,339	32,139,000	147,879,396	3,630,050,494
	81%	14%	1%	4%	100%

constructed following particular probabilistic procedures, while in our case, the data simply refers to the universe of all banking intermediaries for which it was possible to obtain the values of the variables that act as a proxy for the speed and capacity of propagation of systemic risk and the state of health of a bank.

Table 13.3 Descriptive statistics of some characteristic variables. Year 2022

	Mean	Standard deviation	Skewness	Variation coefficient
In thousands of euros				
Total assets (thousands)	10,843,354	69,845,278	10.85	6.44
Equity (thousands)	783,966	4,943,981	10.81	6.31
In percentage				
Equity/total assets	9.42	4.10	2.53	0.44
Securities/total assets	13.26	9.76	2.48	0.74
Loans/total assets	72.53	13.33	−2.01	0.18
Liquidity/total assets	23.43	12.43	1.97	0.53
Total liabilities/total assets	91.58	4.10	−2.53	0.04

referred to 2022, of some variables which are useful for the qualitative framing of the units under investigation.

The heterogeneity of the considered group of banks in terms of size is particularly evident: in fact, the value assumed by the variability indices referring to Total Assets and Capital is very high. This is quite normal, given that the dataset includes both local banks of smaller dimensions and large intermediaries operating at an international level. The presence of banks of various sizes and characterised by business models that are very different from one another is particularly useful for the purposes of our investigation in that it allows us to offer various considerations on the aptitude of each banking model to amplify or, on the contrary, mitigate the propagation of systemic phenomena within the financial sector.

Methodology

In order to analyse the contribution of each intermediary to the systemic risk propagation dynamics, it was necessary to group banks into homogeneous clusters starting from the values given by ten variables which, according to the literature on systemic risk, are capable of (i) providing information on the attitude of each bank to contribute, more or less quickly, to the phenomena propagation mechanisms that generate systemic risk (the first three variables) and (ii) providing a multidimensional representation of each bank's health status (the remaining seven variables).

The ten variables used for clustering the banks analysed were deduced from the most authoritative literature that has dealt with systemic risk over the years. These variables are—as already mentioned—divided according to their informative power into two groups. The first group is made up of three indicators, which quantify amounts of deposits and interbank loans and amounts of government securities held by each bank. These variables are widely used in the literature (Blancher et al. 2013;

Acharya 2011; Blundell-Wignall 2012; Glasserman and Young 2015) to assess the degree of interconnectedness of each bank with the rest of the banking sector and with the public sector. In fact, these variables, respectively, manage to determine the risk of potential contagion arising from each individual bank's greater or lesser exposure in the interbank market and each bank's greater or lesser interconnectedness with the public sector, and thus its exposure to country risk.

The second group of variables consists of seven indicators that provide a multi-dimensional representation of each bank's health status (Acharya 2011; Blundell-Wignall 2012; Blancher et al. 2013; Hoque et al. 2015). The information provided by these variables is, like the information provided by the variables in the first group, very important for this study, as it is assumed that a bank in good health can exert a braking effect against the propagation of a systemic crisis, and thus represents a stabilising factor for a financial system.

From the methodological point of view, after having divided the banks under observation into an adequate number of groups, we proceeded to analyse the composition and characteristics of each of them in terms of specialisation and business model.

The aim of this second analysis is, in fact, to verify whether the presence of cooperative banks is actually greater in those groups for which the indicators of systemic risk propagation speed and capacity and those that explain a bank's status of health assume better values.

The methodology of analysis proposed in this chapter is, therefore, substantially different from what is currently found in the literature in relation to the methods of measuring systemic risk.

In fact, this work focuses on variables, however chosen on the basis of the criteria most widely used in the literature, which have the advantage of being available for all categories of banking intermediaries (and therefore not only for those considered systemic) while providing information on the dynamics of risk propagation rather than exclusively on the evaluation of the risk itself.

The choice of analysing Italian banks derived—as stated—from reasons of analysis homogeneity as well as from the fact that the Italian banking system has a particular morphology and structure characterised by a massive presence of cooperative credit banks and, more generally, of small local banks particularly focused on their territory needs.

Among the indicators that are considered most useful for studying the ability of financial intermediaries to contribute to systemic risk propagation, there is the percentage incidence of total loans granted to public and governmental bodies on the total activity carried out by each bank. Unfortunately, none of the providers currently available is able to give this information, especially in the case of banks not listed on regulated markets which, as already pointed out, represent almost all of the dataset being studied. This is the reason why this information was excluded from the analysis.

Table 13.4 shows the main descriptive statistics, referred to 2022, of the ten variables on which the initial part of the analysis is based.

Table 13.4 Descriptive statistics of the clustering variables. Year 2022

	Mean	Standard deviation	Skewness	Variation coefficient
In percentage				
I1 = Net loans & advances to banks/total assets	8.74	6.31	1.73	0.72
I2 = Deposit from banks/total assets	15.01	9.86	2.29	0.66
I3 = Government securities/total assets	23.84	12.61	0.10	0.53
I4 = Common equity/ core tier 1 ratio	20.73	8.32	2.57	0.40
I5 = NPLs ratio	7.67	6.58	7.06	0.86
I6 = Liquidity ratio	37.26	124.52	16.29	3.34
I7 = ROAE	3.72	6.62	−2.34	1.78
I8 = ROAA	0.35	0.52	−1.37	1.50
I9 = Cost to income ratio	71.38	13.99	1.37	0.20
I10 = Z-score	127.38	162.43	4.46	1.28

As mentioned earlier, the first three indicators (I1, I2 and I3) provide useful information on the degree of interconnection of each intermediary analysed, in that they measure the absolute and relative transactions of each bank in the interbank market and the banks' exposure to the public sector (government securities and sovereign debt). The quantities used as proxies (divided by Total Assets in order to be able to compare intermediaries of different sizes) are the value of loans to the interbank system (I1), the value of deposits on the interbank market (I2) and the value of government securities held in the portfolio (I3). The Orbis Bank Focus Provider defines these indicators respectively as: Net Loans and Advances to Banks (I1), Deposits from Banks (I2) and Government Securities (I3). As already argued above and in line with what has been supported by the prevailing literature on the subject, it is believed that these variables are able to provide particularly significant information on the attitude of banks towards contributing to the propagation mechanisms of the problems that lead to systemic risk. In particular, lower values of these variables lead to the belief that the bank in question is less exposed and interconnected and, therefore, can contribute only marginally to increasing the level of systemic risk. In fact, it is well known that the more limited the active and passive relationships with the interbank market, the less likely it is that a bank can be infected by particularly critical situations involving other banks in the system and, likewise, the less likely it is that the bank itself can, in turn, be a vehicle for contagion, and therefore for the worsening of the overall level of systemic risk, due to its own specific problems. It is also well known that the smaller the value of government securities present in the portfolio of an intermediary, the less likely it is to suffer the negative effects deriving from the default, or simple downgrading, of a sovereign State and, therefore, the less likely it is to contribute to amplifying the effects of systemic risk.

The other seven indicators shown in Table 13.4 and used for the construction of the homogeneous groups of banks are the Tier1 Ratio (I4), the Non-Performing Loans Ratio (I5), the Liquidity Ratio (I6), the ROAE (I7), the ROAA (I8), the Cost to Income Ratio (I9) and finally the Z-score (I10). These variables correspond, respectively, to the following data provided by the Provider: Common Equity/Core Tier 1 Ratio, Impaired Loans/Gross Customer Loans & Advances, Liquid Assets/Deposits & Short-Term Funding, Return on Average Equity, Return on Average Assets and Cost to Income Ratio. Finally, as is well known, the Z-score is a risk variable commonly used in the literature to indicate by how many standard deviations a bank's return must fall from its average value for the value of capital to be zero. As per usual practice, it was calculated by dividing the sum of ROAA and equity by the standard deviation of ROAA itself (referring to the last three years).

As already mentioned, these seven indicators are considered useful in providing information on the health status of each bank since they represent a good proxy, respectively, for the level of capitalisation, the quality of the credit portfolio, liquidity, profitability, the level of operating efficiency and the risk of insolvency (and therefore of instability).

From a purely theoretical point of view, a healthy intermediary does not contribute to aggravating the overall level of systemic risk in the banking sector (or, at most, it could contribute to a very limited extent and certainly dependent on other causes); therefore, banks that present somewhat contained levels of indicators I5 and I9, as well as relatively high levels of indicators I4, I6, I7, I8 and I10, should be characterised by a lower probability of acting as amplifiers of the systemic risk propagation effects in the financial sector.

As noted above, the generally high values of all the relative variability indices are justified by the presence in the dataset of banks of very different sizes, some extremely large and others particularly small.

Before proceeding to the clustering of the dataset units, the classic preliminary operations with respect to the implementation of the procedures of multivariate analysis were carried out, that is, the check for lack of outliers and for any collinearity between the variables as well as the standardisation of all the values. In particular, the multi-collinearity analysis not only included the study of the values contained in the variance–covariance matrix (and therefore of correlation), but also the calculation of the tolerance index and of the VIF (Variance Inflation Factor). In the latter case, however, reference was made to a particularly cautious threshold of 5.

Different cluster analyses (Everitt et al. 2011) of both hierarchical and non-hierarchical type (K-means) were implemented for each year considered. With reference to the hierarchical methods, various combinations of clustering algorithms and distance measures were tested. The clusters obtained with the different approaches adopted were shown to be scarcely overlapping; their composition appeared dissimilar and strongly dependent on the type of procedure used. Therefore, the groups of banks thus obtained showed such differences that no valid conclusions can be drawn in a general sense.

To overcome this problem, the aggregations between banks were carried out referring to some criteria deriving from the evidence common to the various cluster analysis approaches implemented and, therefore, taking into account results that are more robust from a methodological point of view. First of all, it was deduced that the correct number of clusters to be considered is six; this evidence is based on the results from the hierarchical method dendrograms as well as from the tests relating to them. Secondly, it emerged that all the clustering methods assigned greater importance to the first three variables (I1, I2 and I3). Therefore, separately for each year, the banks were first divided into three groups, taking into account those of the first three indicators that presented a value lower than their respective median (calculated considering all the units in the dataset). The choice of the median (rather than the mean) as the threshold for discriminating between sets of units was made since the variable's value distributions were strongly skewed. Applying this criterion, the first group of banks was identified, including all the units for which at least two indicators (one of which necessarily had to be I3) had a value lower than the respective median. The second group was identified by aggregating those units for which only one of the indicators showed a value lower than the median. Finally, the third group was obtained by aggregating the remaining units, that is, all the banks for which none of the three indicators showed a value lower than the median.

Subsequently, each of the three groups previously identified was divided into two parts based on the values of the other seven indicators previously mentioned. In particular, the first subset was formed by aggregating the units for which at least four of the remaining seven indicators (I4, I5, I6, I7, I8, I9 and I10) had a value better than their median value. By difference, units for which fewer than four indicators were better than their median were grouped together in the second subset.

Empirical Results

Table 13.5 shows, separately for each group, the main descriptive statistics—referred to 2022—of the ten indicators that drove the clusterisation procedure. To further increase the robustness of the procedure adopted to divide the units into six homogeneous groups, the differences between the mean value assumed by each of the ten variables in each of the different clusters were analysed with an ANOVA test and were found to be robust and statistically significant.

As already pointed out, the empirical investigation conducted aims to answer the research question introduced in Sect. 13.1, that is, to understand whether cooperative banks are really less involved than other types of banks in the mechanisms underlying the propagation and accentuation of systemic risk and, therefore, indirectly, this study seeks to verify whether their presence can prove useful in mitigating contagion phenomena and, therefore, the spread of systemic risk in a financial system.

In order to answer this question, we distinguished the different banks in the dataset on the basis of their propensity to generate and/or spread systemic risk within the market. Subsequently, a second analysis was conducted to verify whether or not the

Table 13.5 Descriptive statistics of the ten indicators with respect to each group of banks. Year 2022

	I1	I2	I3	I4	I5	I6	I7	I8	I9	I10
Group 1										
Mean	10.87	11.89	21.91	22.99	7.92	47.67	3.68	0.38	72.96	113.09
St. Dev	6.86	7.35	12.54	7.86	8.27	182.31	5.41	0.51	12.34	107.57
Skewness	1.48	0.80	0.14	1.76	6.88	11.34	−2.31	−1.29	0.91	2.03
Var. Coeff	0.63	0.62	0.57	0.34	1.04	3.82	1.47	1.35	0.17	0.95
Group 2										
Mean	6.12	13.43	24.12	19.62	7.96	24.83	3.20	0.31	71.95	136.38
St. Dev	4.73	7.74	13.35	9.48	5.99	13.32	5.85	0.50	14.66	151.38
Skewness	1.42	0.74	−0.05	3.97	4.51	1.62	−1.85	0.25	2.20	2.20
Var. Coeff	0.77	0.58	0.55	0.48	0.75	0.54	1.83	1.63	0.20	1.11
Group 3										
Mean	7.53	20.36	26.34	18.32	7.13	30.96	4.11	0.33	68.80	141.29
St. Dev	5.67	11.96	11.73	7.34	3.52	26.53	8.35	0.55	15.25	220.62
Skewness	2.33	2.75	0.30	3.01	1.33	5.02	−2.38	−2.32	1.50	4.56
Var. Coeff	0.75	0.59	0.45	0.40	0.49	0.86	2.03	1.68	0.22	1.56
Group 1.1										
Mean	11.40	11.99	25.91	24.49	7.16	70.96	6.16	0.68	67.75	135.60
St. Dev	6.24	7.76	11.73	7.22	10.69	262.95	2.98	0.35	10.88	121.25
Skewness	0.86	1.09	0.21	1.15	6.71	7.85	0.98	1.44	−0.25	1.55
Var. Coeff	0.55	0.65	0.45	0.29	1.49	3.71	0.48	0.51	0.16	0.89
Group 1.2										
Mean	10.38	11.79	18.33	21.64	8.61	26.78	1.45	0.11	77.64	92.91
St. Dev	7.35	6.95	12.17	8.19	5.15	14.63	6.19	0.49	11.69	88.89
Skewness	1.91	0.43	0.18	2.37	2.50	2.91	−2.26	−2.37	2.08	2.77
Var. Coeff	0.71	0.59	0.66	0.38	0.60	0.55	4.26	4.37	0.15	0.96
Group 2.1										
Mean	7.36	13.15	28.03	22.82	6.24	31.24	6.88	0.67	64.50	147.27
St. Dev	4.99	8.63	14.69	12.38	3.54	14.76	2.85	0.41	9.51	146.54
Skewness	0.82	1.40	−0.34	3.30	1.97	1.42	0.87	2.61	−2.00	2.09
Var. Coeff	0.68	0.66	0.52	0.54	0.57	0.47	0.41	0.61	0.15	1.00
Group 2.2										
Mean	5.13	13.66	21.02	17.08	9.33	19.75	0.29	0.03	77.87	127.74
St. Dev	4.31	6.95	11.25	5.16	7.09	9.67	6.16	0.41	15.32	154.58
Skewness	2.23	−0.21	−0.17	2.29	4.35	1.60	−1.92	−1.34	3.18	2.40
Var. Coeff	0.84	0.51	0.54	0.30	0.76	0.49	21.36	13.18	0.20	1.21
Group 3.1										

(continued)

Table 13.5 (continued)

	I1	I2	I3	I4	I5	I6	I7	I8	I9	I10
Mean	7.59	20.07	26.24	20.10	5.98	37.49	7.10	0.58	62.48	179.02
St. Dev	6.25	13.18	12.49	8.86	3.72	34.09	5.02	0.36	11.42	276.63
Skewness	2.58	2.30	–	2.73	1.98	4.15	3.78	3.39	−1.16	4.01
Var. Coeff	0.82	0.66	0.48	0.44	0.62	0.91	0.71	0.62	0.18	1.55
Group 3.2										
Mean	7.48	20.65	26.44	16.52	8.29	24.30	1.05	0.07	75.26	102.74
St. Dev	4.56	10.56	10.89	4.81	2.84	12.54	9.93	0.61	15.95	131.43
Skewness	1.55	3.66	0.78	2.00	1.52	2.46	−2.74	−3.59	2.53	2.97
Var. Coeff	0.61	0.51	0.41	0.29	0.34	0.52	9.45	8.48	0.21	1.28

presence of cooperative banks is homogeneous within the various groups identified. If the cooperative banks were evenly distributed among the groups, this would mean that they do not differ in any way from other types of banks; therefore, it would not be possible to draw any conclusions about their ability to contribute, positively or negatively, to systemic risk propagation. The situation would be different if the cooperative banks were actually more numerous in those groups for which the indicators of systemic risk propagation speed and capacity as well as the indicators representing the bank's health status assume, respectively, more limited and better values. Indeed, in this case it would be possible to conclude that the presence of cooperative banks constitutes an important shock absorber capable of hindering (or, at least, braking) the spread of systemic risk phenomena within the banking market.

Table 13.6A–E show, therefore, for each year analysed, the presence of the various categories of banks within the nine groups previously identified.

The values represented in the first part of Table 13.6 (Group 1, Group 2 and Group 3) refer to the subdivision of the banks into three homogeneous categories on the basis of the values assumed by indicators I1, I2 and I3. It should be noted that these indicators provide information on the degree of interconnectedness of each bank, that is, the exposure of the banks analysed to the interbank market and to sovereign debt.

Moving from Group 1 to Group 3, it is possible to find banks for which the above indicators take on progressively worse values, thus indicating a more pronounced inclination to contribute significantly to the transmission of systemic problems among market participants.

From Table 13.6A–E it emerges that the probability of finding a cooperative bank within the groups decreases significantly when moving from Group 1 to Group 3. In particular, with reference to the most recent data (Table 13.6E), this probability goes from 87% for Group 1 to 80% for Group 2, reducing to 56% for Group 3. This dynamic, which shows a progressive and marked reduction, is confirmed in each of the years considered in the analysis (Fig. 13.1). This means that most of the cooperative banks in the dataset are characterised by a lower relative exposure to the

Table 13.6 (A) Banking group and sub-group composition in terms of business model. Year 2018. (B) Banking group and sub-group composition in terms of business model. Year 2019. (C) Banking group and sub-group composition in terms of business model. Year 2020. (D) Banking group and sub-group composition in terms of business model. Year 2021. (E) Banking group and sub-group composition in terms of business model. Year 2022

(A)

Commercial banks	Investment banks	Saving banks	Cooperative banks	Total
Group 1				
11	2	1	125	139
8%	1%	1%	90%	100%
Group 2				
16	4	5	59	84
19%	5%	6%	70%	100%
Group 3				
34	7	3	61	105
32%	7%	3%	58%	100%
Group 1.1				
2	2	1	90	95
2%	2%	1%	95%	100%
Group 1.2				
9	0	0	35	44
20%	0%	0%	80%	100%
Group 2.1				
11	2	1	36	50
22%	4%	2%	72%	100%
Group 2.2				
5	2	4	23	34
15%	6%	12%	68%	100%
Group 3.1				
11	3	1	23	38
29%	8%	3%	61%	100%
Group 3.2				
23	4	2	38	67
34%	6%	3%	57%	100%

(B)

Commercial banks	Investment banks	Saving banks	Cooperative banks	Total
Group 1				
13	2	1	118	134
10%	1%	1%	88%	100%
Group 2				

(continued)

Table 13.6 (continued)

(B)

Commercial banks	Investment banks	Saving banks	Cooperative banks	Total
18	3	5	68	94
19%	3%	5%	72%	100%
Group 3				
35	7	3	55	100
35%	7%	3%	55%	100%
Group 1.1				
3	1	0	66	70
4%	1%	0%	94%	100%
Group 1.2				
10	1	1	52	64
16%	2%	2%	81%	100%
Group 2.1				
11	2	1	37	51
22%	4%	2%	73%	100%
Group 2.2				
7	1	4	31	43
16%	2%	9%	72%	100%
Group 3.1				
14	2	1	30	47
30%	4%	2%	64%	100%
Group 3.2				
21	5	2	25	53
40%	9%	4%	47%	100%
(C)				
Commercial banks	Investment banks	Saving banks	Cooperative banks	Total
Group 1				
15	4	1	121	141
11%	3%	1%	86%	100%
Group 2				
12	2	3	49	66
18%	3%	5%	74%	100%
Group 3				

(continued)

Table 13.6 (continued)

(C)

Commercial banks	Investment banks	Saving banks	Cooperative banks	Total
36	5	5	64	110
33%	5%	5%	58%	100%
Group 1.1				
5	1	0	70	76
7%	1%	0%	92%	100%
Group 1.2				
10	3	1	51	65
15%	5%	2%	78%	100%
Group 2.1				
6	2	1	28	37
16%	5%	3%	76%	100%
Group 2.2				
6	0	2	21	29
21%	0%	7%	72%	100%
Group 3.1				
14	2	1	28	45
31%	4%	2%	62%	100%
Group 3.2				
22	3	4	36	65
34%	5%	6%	55%	100%

(D)

Commercial banks	Investment banks	Saving banks	Cooperative banks	Total
Group 1				
16	2	1	127	146
11%	1%	1%	87%	100%
Group 2				
11	3	4	37	55
20%	5%	7%	67%	100%
Group 3				
32	6	4	67	109
29%	6%	4%	61%	100%
Group 1.1				

(continued)

Table 13.6 (continued)

(D)

Commercial banks	Investment banks	Saving banks	Cooperative banks	Total
3	0	0	74	77
4%	0%	0%	96%	100%
Group 1.2				
13	2	1	53	69
19%	3%	1%	77%	100%
Group 2.1				
4	2	1	16	23
17%	9%	4%	70%	100%
Group 2.2				
7	1	3	21	32
22%	3%	9%	66%	100%
Group 3.1				
10	4	2	30	46
22%	9%	4%	65%	100%
Group 3.2				
22	2	2	37	63
35%	3%	3%	59%	100%

(E)

Commercial banks	Investment banks	Saving banks	Cooperative banks	Total
Group 1				
14	2	2	116	134
10%	1%	1%	87%	100%
Group 2				
9	3	1	52	65
14%	5%	2%	80%	100%
Group 3				
35	4	6	58	103
34%	4%	6%	56%	100%
Group 1.1				
5	0	1	59	65
8%	0%	2%	91%	100%
Group 1.2				

(continued)

Table 13.6 (continued)

(E)

Commercial banks	Investment banks	Saving banks	Cooperative banks	Total
9	2	1	57	69
13%	3%	1%	83%	100%
Group 2.1				
4	1	0	25	30
13%	3%	0%	83%	100%
Group 2.2				
5	2	1	27	35
14%	6%	3%	77%	100%
Group 3.1				
14	2	3	30	49
29%	4%	6%	61%	100%
Group 3.2				
21	2	3	28	54
39%	4%	6%	52%	100%

interbank market and to the public sector and, therefore, are less interconnected with the other nodes in the financial network.[2] This is in line with the main peculiarities of cooperative banks, namely their small size and their marked attention to local needs, which leads them to concentrate almost all of their funding and financing activity on customers belonging to the local community in which they operate.

The values represented in the second part of Table 13.6 (Group 1.1 and 1.2, Group 2.1 and 2.2 and, finally, Group 3.1 and 3.2) refer to the subsequent subdivision of the first three groups into two sub-groups on the basis of the values assumed by indicators from I4 to I10. These indicators provide information on each bank's health status as they refer to capitalisation level, loan portfolio, quality, liquidity level, profitability, operating efficiency degree, and insolvency and instability risk.

Moving from Group 1.1 to Group 3.2, it is possible to find banks whose situation is increasingly problematic with reference to one or more of the aforementioned indicators and, therefore, banks characterised by an ever-increasing probability of acting as systemic risk propagators in the financial sector, due to their precarious managerial conditions and, therefore, their intrinsic instability.

Specifically, Group 1.1 includes those banks that can contribute most to containing the overall entity of systemic risk in the financial sector; indeed, they are intermediaries characterised by a particularly positive health status from a managerial point of view and therefore by a low probability of transmitting problematic situations in the economic-financial system, due to their intrinsic solidity. For these banks, the

---

[2] It should be noted that I1, I2 and I3 are expressed as percentages of total assets and therefore the values referring to banks of different sizes are directly comparable.

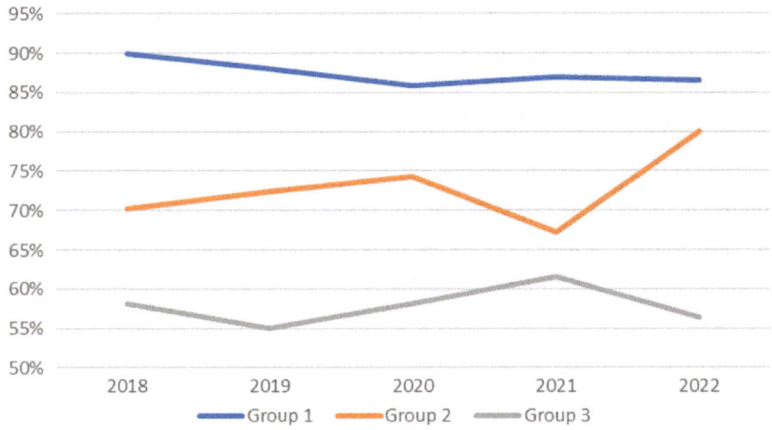

Fig. 13.1 Probability to find a cooperative bank in each group during the period 2018–2022

overall probability of contributing to the spread and generation of systemic risk is the lowest ever. In this group there is a massive presence of cooperative banks which in 2022 (numerically) represent, on average, about 91% of the total.

Group 1.2 assembles those banks which, albeit to a very slight degree, could possibly contribute to systemic risk in the financial sector since, despite their modest participation in the propagation process, they show a relatively problematic situation from a managerial point of view, making them more dangerous than the banks in Group 1.1. Consequently, although their probability of contributing to the development and spread of systemic problems is still low overall, it is nonetheless higher than that of the previous group. Even in this group, the number of cooperative banks remains relatively high at an average of 83%, which is lower than in Group 1.1.

Applying the same interpretative criteria, we observe that Group 2.1 includes those types of banks that seem to offer an average contribution to the spread of systemic phenomena. In this case, in 2022, the number of cooperative banks is similar to the case highlighted above, reaching 83%.

Group 2.2 includes those banks that are thought to contribute significantly to the systemic risk propagation in the sector, since they are characterised both by problems of a managerial nature and by a consistent propensity to act as a driving force in the diffusion of the negative effects caused, precisely, by systemic phenomena within the sector. In this Group, the percentage of cooperative banks is significantly reduced, settling at an average level of 77%.

Group 3.1 consists of those banks for which the probability of contributing to the generation and spread of systemic risk is rather high. In this group, the number of cooperative banks decreases to an average level of 61%.

Finally, Group 3.2 is made up of those intermediaries that undoubtedly play a decisive role in the dynamics of the propagation of systemic risk in the financial system. These are, in fact, banks in management disequilibrium and therefore characterised by a marked propensity to amplify the contagion dynamics and diffusion of their own difficulties as well as those of other banks. In this last group, the presence of cooperative banks is drastically reduced and reaches minimum levels (specifically 52% in 2022).

The results of our empirical analysis with reference to the Italian banking market therefore allow us to answer our research question and, in particular, confirm the countercyclical and mitigating role of cooperative banks in systemic risk. In fact, as highlighted above, moving towards groups of banks characterised by a greater probability of contributing significantly to the transmission process of systemic problems, and therefore towards groups of banks characterised by a greater aptitude to act as amplifiers of systemic risk, the presence of cooperative banks is significantly reduced.

Conclusions

The events of the last decades have highlighted how we live in a "small world", in which everything is connected to everything else and often in different ways that are variable and not easy for human rationality to understand.

It is also evident that the "Achilles' heel" of a "small world", and therefore of real networks, is represented by the vulnerability due to interconnection. An isolated shock can create chain reactions that destabilise an entire economic system, and the probability that an isolated shock will undermine an entire system is higher if the affected nodes are the most interconnected.

While not yielding to the initial temptation to believe that the systemic propagation of crisis situations is exclusively due to the difficulties caused by large institutions (since the systemic value of the various intermediaries depends not only on their size, but above all, on the degree of riskiness and correlation with others), one cannot fail to consider the fundamental anti-cyclical role played by cooperative credit banks. These banks have the intrinsic potential to interrupt the vicious circle that fuels the propagation of a systemic crisis. This intrinsic potential is due to the granularity of their relationships, their peculiar governance model, and their characteristic business model based on mutuality, long-term relationships, commitment to local development, in-depth knowledge of their customers, and greater consideration of qualitative information in the credit process.

The results of the empirical investigation, with specific reference to the Italian banking system, support the initial hypothesis at the basis of the work and allow us to answer the research question of this essay. In particular, the empirical results of our analysis confirm the countercyclical and mitigating role of cooperative banks, which are actually less involved than other categories of intermediaries in contagion phenomena deriving from the spread of systemic risk.

As the empirical analysis shows, moving towards groups of banks that are more likely to contribute significantly to systemic problem transmission, and thus towards groups of banks that are more likely to act as systemic risk amplifiers, the presence of cooperative banks is significantly reduced.

The results obtained from this study thus enrich the existing debate on the *raison d'être* of cooperative banks, which is fundamentally focused on the idea that they have withstood the various recent crises thanks to a business model that is by no means anachronistic but, on the contrary, is still capable of satisfying the needs of their customers, while also fulfilling a fundamental function of mitigating systemic risk.

The data used in this work derives exclusively from public sources, mainly financial statements, and represents the only information accessible to external researchers interested in analysing companies and the system in which they operate. Such information, however, can only partially capture actual individual banks' health status and real systemic risk propagation dynamics.

In view of this, the methodological approach presented in this chapter could prove particularly useful to the authorities and policymakers for the purposes of evaluating and monitoring systemic risk, both at a national and international level. What is more, the proposed methodology could easily be enriched with all the classified and sensitive information which was not available to this contribution, but which would certainly be useful (if not essential) for the purpose of obtaining an even more complete and up-to-date picture of the equilibrium conditions of the international banking system.

Finally, it should be pointed out that the variety and complexity (and, moreover, often lack transparency) of financial relations between the different nodes in a network tend to increase the complexity of financial systems; this circumstance produces information asymmetries, moral hazard risks and, therefore, opacity and consequences in the processes of systemic risk propagation. This complexity in the relationships between the economic agents of a financial system has been fed since the early years of the new millennium by the evolution (often uncontrolled) of financial engineering, which has made the economic-financial systems more interconnected, and therefore complex, linking operators with each other in multiple ways, and often unconsciously. In addition, network science teaches us that the mechanisms of "growth" and "preferential connection" lead the "hubs" (i.e., the largest nodes) to expand in phases of network expansion and thus encompass smaller nodes. This phenomenon, otherwise known as globalisation, leads, however, to the risk of extinction of smaller economic operators, such as local banks; their disappearance, or even their simple competitive downsizing, would lead, over time, to the loss of the extraordinary intangible and relational patrimony of these intermediaries. This situation could also lead to the financial system's impoverishment and exposure, as demonstrated in this essay, to greater systemic risks.

In addition, the events of the last fifteen years have exhaustively highlighted all the risks of a highly interconnected financial system that is disconnected from the real economy, which has grown over the years to a hypertrophic extent thanks to financial engineering and has therefore become excessively complex and exposed to human greed. This is especially true if it is left free to expand, where there is a lack of adequate controls and forms of protection and guarantee, as well as alternative models of intermediaries that are less systemic and more linked to the territory and the real economy.

Therefore, to toy with some of the paradigmatic expressions that have been in vogue for some years now, that is, since the subprime mortgage crisis in the USA in 2008, what emerges from the proposed empirical analysis should probably contribute to inducing the international supervisory authorities to shift their attention from the paradigms of "Too Big to Fail"—or "Too Central (Interconnected) to Fail"—to the notion of "Too Useful to Fail".

Indeed, in light of the results of our empirical analysis, the utility of local cooperative banks with a mutual vocation is enriched with an important connotation referring to the counter-cyclical and mitigating dimension of the contagion mechanisms deriving from systemic risk propagation. In other words, it seems evident that a model of intermediation characterised by such varied levels of utility deserves adequate and specific attention and protection from the authorities.

References

Acharya, V.V., Richardson, M.: Restoring Financial Stability: How to Repair a Failed System. John Wiley & Sons, New York (2009)

Acharya, V.V.: Systemic Risk and Macro-Prudential Regulation. Unpublished working paper (2011)

Barone, G., De Blasio, G., Mocetti, S.: The real effects of credit crunch in the Great Recession: evidence from Italian provinces. Temi di Discussione della Banca d'Italia, n. 1057 (2016)

Beirne, J., Fratzscher, M.: The pricing of sovereign risk and contagion during the European sovereign debt crisis. J. Int. Money Financ. **34**, 60–82 (2013)

Bengtsson, E.: Shadow banking and financial stability: European money market funds in the global financial crisis. J. Int. Money Financ. **32**, 579–594 (2013)

Berton, F., Mocetti, S., Presbitero, A., Ricchiardi, M.: Banks, Firms, and Jobs. IMF working paper 17/38 (2017)

Blundell-Wignall, A.: Solving the financial and sovereign debt crisis in Europe. OECD J. Financ. Market Trends **2011**(2), 201–224 (2012)

Boscia, V., Carretta, A., Schwizer, P.: Co-operative Banking in Europe: Case Studies. Palgrave Macmillan Publisher, UK (2010)

Brunner, A., Decressin, J., Hardy, D., Kudela, B.: Germany's Three-pillar Banking System—Cross-Country Perspectives in Europe, in IMF Occasional Paper 233. International Monetary Fund, Washington, DC (2004)

Brunnermeier, M.K.: Deciphering the liquidity and credit crunch 2007–2008. J. Econ. Perspect. **23**(1), 77–100 (2009)

Brunnermeier, M.K., Oehmke, M.: Bubbles, financial crises, and systemic risk. Handbook Econ. Finance **2**, 1221–1288 (2013)

Buch, C.M., Goldberg, L.: Cross-border prudential policy spillovers: How much? How important? Evidence from the international banking research network. National Bureau of Economic Research, w22874 (2016)

Bülbül D., Schmidt, R.H., Schüwe, R.U.: Savings Banks and Co-operative Banks in Europe. SAFE Policy Papers, White Paper Series 5 (2013)

Calvo, G.: Financial crises and liquidity shocks a bank-run perspective. Eur. Econ. Rev. **56**(3), 317–326 (2012)

Cerutti, E., Dagher, J., Dell'Ariccia, G.: Housing finance and real-estate booms: A cross-country perspective. J. Hous. Econ. **38**, 1–13 (2017)

Chiaramonte, L., Poli, F., Oriani, M.E.: Are cooperative banks a lever for promoting bank stability? Evidence from the recent financial crisis in OECD countries. Eur. Financ. Manag. **21**(3), 491–523 (2015)

Claessens S., Ghosh S.R., Mihet M.R.: Macro-Prudential Policies to Mitigate Financial System Vulnerabilities. International Monetary Fund, 14–155 (2014)

Clark E.A., Mare D., Radić, N.: Co-operative banks: what do we know about competition and risk preferences? J. Int. Financ. Markets Inst. Money **52**, 90–101 (2018)

Constâncio, V.: Synergies between banking union and capital markets union. In: Keynote Speech at the Joint Conference of the European Commission and ECB on European Financial Integration, Brussels, vol. 19 (2017)

De Bruyckere, V., Gerhardt, M., Schepens, G., Vander Vennet, R.: Bank/sovereign risk spillovers in the European debt crisis. J. Bank. Finance **37**(12), 4793–4809 (2013)

Demma, C.: Local banking and the financial crisis. In: Questioni di Economia e Finanza, Banca d'Italia n. 264 (2015)

EACB: European Co-Operative Banks in the Financial and Economic Turmoil, First assessments (2010)

EACB: Corporate Governance in Co-operative Banks, Key Features, 3 February (2016)

Everitt, B.S., Landau, S., Leese, M., Stahl, D.: Cluster Analysis, 5th edn. Wiley (2011)

Blancher, M.N.R., Mitra, M.S., Morsy, M.H., Otani, M.A., Severo, T., Valderrama, M.L.: Systemic Risk Monitoring ("SysMo") Toolkit—A User Guide. International Monetary Fund (2013)

Fiordelisi, F.D., Mare, S.: Probability of default and efficiency in co-operative banking. J. Int. Financ. Markets Insti. Money **26**(C), 30–45 (2013)

Flannery, M.J., Kwan, S.H., Nimalendran, M.: The 2007–2009 financial crisis and bank opaqueness. J. Financ. Intermediat. **22**(1), 55–84 (2013)

Fonteyne, W.: Cooperative Banks in Europe—Policy Issues. In: IMF Working Paper, n. 7/159 (2007)

Glasserman, P., Young, H.P.: How likely is contagion in financial networks? J. Bank. Finance **50**, 383–399 (2015)

Goodhart, C.A.E.: Some New Directions for Financial Stability. The Per Jacobsson Lecture, Bank for International Settlements, Basel (2004)

Hanson, S.G., Kashyap, A.K., Stein, J.C.: A macroprudential approach to financial regulation. J. Econ. Perspect. **25**(1), 3–28 (2011)

Hoque, H., Andriosopoulos, D., Andriosopoulos, K., Douady, R.: Bank regulation, risk and return: evidence from the credit and sovereign debt crises. J. Bank. Finance **50**, 455–474 (2015)

Iyer, R., Peydró, J.L., da-Rocha-Lopes, S., Schoar, A.: Interbank liquidity crunch and the firm credit crunch: evidence from the 2007–2009 crisis. Rev. Financ. Stud. **27**(1), 347–372 (2013)

Lane, P.R.: The European sovereign debt crisis. J. Econ. Perspect. **26**(3), 49–68 (2012)

Nelson, S.C., Katzenstein, P.J.: Uncertainty, risk, and the financial crisis of 2008. Int. Organ. **68**(2), 361–392 (2014)

Pacelli, V., Pampurini, F., Sylos Labini, S.: The peculiarity of the co-operative and mutual model: evidence from the European banking sector. J. Financ. Manage. Markets Inst. **7**(1), 1–29 (2019)

Pacelli, V., Pampurini, F., Quaranta, A.G.: Co-operative banks and financial stability. Int. J. Bus. Soc. Sci. **11**(11) (2020)

Reinhart, M.C., Rogoff, M.K.: Financial and sovereign debt crises: some lessons learned and those forgotten. Int. Monetary Fund **13**, 266–278 (2013)

Chapter 14
Shocks at Local Banks, EU GDP Growth, and Banking Sector Stability

Pasqualina Arca, Andrea Carosi, and Ornella Moro

Abstract This chapter explores whether shocks originating at local banks affect the EU's economic growth and banking system's stability. Our analysis proceeds in two steps. In the first step, we identify locally dominant banks that substantially impact their local macroeconomic environment but are not among the largest European banks. We construct a measure of idiosyncratic shocks at these dominant local banks, the so-called Local Banking Granular Residual. We show that idiosyncratic shocks to these locally dominant banks propagate nationally and explain a significant portion of aggregate EU macroeconomic fluctuations. In a second step, we relate idiosyncratic shocks at local dominant banks to EU banking system stability, as measured by an EU Bank Z-score, and find significant evidence. We show that local banks matter in the EU.

Keywords Granular residual · Idiosyncratic shocks · Banking stability · GDP growth

Introduction

Size matters in banking. Typically, banking systems are dominated by a small number of large players active in a large range of countries and market segments but also populated by small, locally oriented financial institutions. This dichotomous banking system structure is particularly prevalent in the EU, which has numerous savings

P. Arca · A. Carosi (✉) · O. Moro
University of Sassari, Sassari, Italy
e-mail: acarosi@uniss.it

P. Arca
e-mail: parca1@uniss.it

O. Moro
e-mail: omoro@uniss.it

P. Arca
CRENoS, Cagliari, Italy

© The Author(s) 2025
V. Pacelli (ed.), *Systemic Risk and Complex Networks in Modern Financial Systems*,
New Economic Windows, https://doi.org/10.1007/978-3-031-64916-5_14

and cooperative banks and only a few large and internationally active banks. In this chapter, we explore whether and how local banks affect the EU macroeconomic growth and the stability of the EU banking system. We are particularly interested in whether idiosyncratic shocks originating at local banks affect the EU GDP growth and the EU Bank Z-Score.

In our empirical analysis of the link between shocks at local banks and economic growth and banking sector stability, we merge the intuitions from Blank et al. (2009), Gabaix (2011), and Jannati et al. (2020) and construct a so-called banking granular residual of local dominant banks. Gabaix's (2011) original idea is applied to non-banks. He looks at firm-level productivity shocks and shows that the idiosyncratic volatility in the sales of the top 100 largest non-financial firms in the US—the "granular residual"—can explain a significant fraction of the volatility of US output growth. Blank et al. (2009) first apply Gabaix (2011)'s intuition to the banking system: they construct a banking granular residual for dominant banks, measuring productivity shocks at large banks and proxying for large events affecting the banking industry and look at implications of these shocks at large banks for the stability of smaller banks. On the other hand, Jannati et al. (2020) apply Gabaix (2011)'s intuition to local non-bank firms: they identify locally dominant firms that have a strong impact on their local macroeconomic environment but are not among the largest 100 U.S. firms and show that Idiosyncratic shocks to these locally dominant firms explain a significant portion of aggregate U.S. macroeconomic fluctuations. We leverage and contribute to this literature. First, we borrow the definition of granular residual for banks from Blank et al. (2009), and the empirical design meant for local firms from Jannati et al. (2020) and construct a measure of idiosyncratic shocks at dominant local banks; we called Local Banking Granular Residual. Then, we test if idiosyncratic shocks to these locally dominant banks propagate nationally and explain a significant portion of aggregate EU macroeconomic fluctuations. Finally, we relate idiosyncratic shocks at local dominant banks to the EU banking system stability and show that local dominant banks also matter.

The remainder of this research is organized as follows. Section "The Granular Residual in Banking" describes the prevailing literature dealing with granularity in banking. Section "Data and Methods" depicts the data used in the estimation and outlines our methodology. Section "Which EU Countries are Bank Granular?" identifies the phenomenon of granularity in EU27 countries. Section "Locally Dominant Banks, EU GDP Growth, and EU Bank Z Score" reports the results of the impact of local dominant banks. Section "Summary and Conclusions" concludes.

The Granular Residual in Banking

This section summarizes prevailing literature and approaches dealing with granularity in banking. An extensive review of the existing research leads us to focus on four main contributions, which we categorize into two main subfields: the first looking at the relationship between banks' idiosyncratic shocks (granularity) and real economy outcomes, the second looking at the relationship between banks' granularity and banking system stability. Our research contributes to the latter.

Relationship Between Banks' Idiosyncratic Shocks and Real Economy Outcomes

Buch and Neugebauer (2011) (BN2011) are the first to exploit the granular residual idea to analyze the real effects of financial shocks: the relation between idiosyncratic shocks to loan growth at large banks and real GDP growth is tested upon a panel of 18 Western European and 17 Eastern European countries for the pre-crisis period 1996–2006 (data are taken from BankScope—Bureau van Dijk). More in detail, BN2011 constructs the granular residual for the banking sector in each country separately and calls this variable banking granular residual (BGR). Then, the impact of shocks to loan growth at large banks (BGR) on GDP growth is tested: changes in lending by large banks is shown to have a significant short-run impact on GDP growth. To calculate BGR, BN2011 use banks' total operating income, including interest and non-interest income, as an encompassing measure of bank size; a broad measure of loans, including consumer, real estate, and investment loans, is used for shocks hitting banks. The use of loans is motivated by bank lending being relevant for transmitting monetary policy shocks to the real economy. Therefore, this BGR is, in fact, a measure of the idiosyncratic component of loan growth. BN2011 use two different ways to measure BGR.

First, BGR is proxied by changes in bank lending and is computed as follows

$$\text{BGR}_{ij,t} = \left(\sum_i^N S_{ij,t-1} \right)^{-1} \left(\sum_i^N S_{ij,t-1} \left(g_{ij,t-1} - \overline{g}_t \right) \right)$$

where, $g_{ij,t}$ is the growth rate of total loans of bank i in country j, $\overline{g}_t$ is the cross-sectional grand mean over all banks in all countries, N is the total number of banks in the sample, and $S_{ij,t-1}$ represents the bank size. Alternatively, as per Gabaix (2011), one could subtract only the mean over the K largest banks in each country.

Second, BN2011 regress the change in loan growth of bank i on the mean change in this growth rate of all banks in all countries as well as aggregate GDP growth as follows,

$$g_{ij,t} = \alpha_0 + \alpha_1 \overline{g}_t + \alpha_2 \overline{g}_{t,1} + \alpha_3 \sum_{k=0}^{2} \Delta GDP_{j,t-k} + \epsilon_{ij,t}$$

and the Banking Granular Residual is calculated as,

$$\widehat{BGR}_{ij,t} = \left(\sum_{i}^{N} S_{ij,t-1} \right)^{-1} \left(\sum_{i}^{N} S_{ij,t-1} \hat{\epsilon}_{ij,t} \right)$$

where, $\hat{\epsilon}_{ij,t}$ is the residual of the above reported regression.

At last, BN2011 regress growth in GDP per capita on the measure of shocks hitting large banks (BGR) by including a maximum of two lags to capture a typical business cycle, with the following OLS specification,

$$\Delta GDP_{j,t} = \beta_0 + \beta_1 BGR_{j,t} + \cdots + \beta_n BGR_{j,t-2} + \gamma T_t + \eta_{j,t}$$

where T_t indicates a full set of time dummies. BN2011 find evidence that large banks' lending changes have a significant short-run impact on GDP growth.

About ten years later, Bremus and Buch (2017) (BB2017) analyze whether idiosyncratic shocks affecting large banks influence the aggregate economy (GDP) and whether this link depends on the degree of financial openness. For each observation, the mean growth rate across all non-j banks in each country-year is subtracted from the total assets (or loans) growth rate of bank j. The reason for taking the average across all banks except bank j is that, for some countries, only a relatively small number of bank observations are available. Therefore, subtracting the average across all banks (including bank j) from bank j's asset (credit) growth may eliminate most of bank j's idiosyncratic variation. This holds particularly true if there is a small number of bank observations and if bank j is large. The BB2027 bank granular residual (BGR) is, therefore, as follows,

$$BGR(asset) = \sum_{j=1}^{N} Assetshock_{ji,t} \frac{Asset_{ji,t}}{Asset_{i,t}}$$

where ji,t indicates the bank j in country i at time t, and i,t indicates the aggregate bank asset in country I at time t. An equivalent granular index is also calculated for loans. BB2017 relates this BGR(asset) to macroeconomic dependent variables, such as cross-border assets and liabilities, GDP per capita growth, and domestic credit. All in all, bank-level shocks significantly impact GDP, financial openness tends to lower GDP growth, and granular effects tend to be stronger in financially closed economies.

Finally, Bremus et al. (2018) (BBRS2018) hypothesize that fluctuations in macroeconomic outcomes increase the variance of bank-specific shocks and the degree of concentration in the banking sector. BBRS2018 consider all banks (BankScope—Bureau van Dijk) with at least five consecutive years of available data to make sure that all banks in the sample are included at least for one business cycle and drop implausible observations where the loans-to-assets or the equity-to-assets ratio is greater than 1, as well as banks with negative values recorded for equity, total assets, or total net loans. To have country-level variables, they keep observations with consolidation codes C1, C2, U1, and A1 and country-year observations based on at least three banks, while only banks classified as holding companies, commercial banks, cooperative banks, and savings banks are included in the sample to represent the banking industry as a whole fully.[1] BBRS2018 calculate bank-specific credit growth by taking the difference between bank level loan growth and the mean growth rate of loans for each country and year. Precisely, the BGR index is calculated using the following,

$$BGR_{it} = \sum_{j=1}^{J} \widehat{du}_{jit} \frac{credit_{jit}}{creditet_{it}} dove \widehat{du}_{jit} = g_{jit} - \overline{g}_{it}$$

with g_{jit} is the growth rate of total credits of bank i in country j, and $\overline{g}_{it}$ is the domestic cross-sectional mean of g_{jit} excluding credit growth of bank j itself. BBRS2018 relate this BGR to macroeconomic outcome variables, such as growth in real domestic credit defined as the growth rate of log of real domestic credit in US dollars (data from the IMF International Financial Statistics) and growth rate of log real GDP per capita (WDI data). Empirically, the aggregate growth on loan growth shocks of banks (BGR) is regressed against time fixed effect and log GDP per capita and inflation as additional controls. Overall, BBRS2018 find support for the assumption that bank size follows a power-law distribution. They also demonstrate that the BGR is associated with aggregate growth in domestic credit and GDP. Hence, idiosyncratic shocks to large banks may affect macroeconomic outcomes via the concentration of banking markets.

Relationship Between Banks' Idiosyncratic Shocks and Banking System Stability

Blank et al. (2009) (BBN2009) explore whether and how the size distributions of banks affect the stability of the German banking system. Specifically, they are interested in whether idiosyncratic shocks originating at large banks affect the distress probabilities of small and mid-sized banks. They take shocks at large banks as proxies of large events that affect the banking industry. The focus is on the implications of shocks at large banks for the stability of smaller banks. They compute the granular residual for the German banking system by constructing a measure of shocks to growth in the banks' cost-to-income ratio for the largest banks. Size is measured in terms of total operating income. The critical point is to find appropriate measures of banks' outputs, which can be (i) the number of deposit accounts and loans produced, (ii) total assets or (iii) total deposits. Other measures include (i) earning assets, (ii) demand deposits or (iii) gross operating income. They need a proxy for banks' output that is available for many banks and does not suffer from potentially large measurement errors. To do so, they use total operating income (interest income plus

---

[1] See Duprey and Lé (2016) for a helpful description of the Bankscope data and its handling.

non-interest income) as output proxy. As a proxy for idiosyncratic shocks, they use the cost-to-income ratio, which measures the overheads (or cost of running banks) as a percentage of income generated before provisions. This measure can be regarded as a proxy for a bank's efficiency and is very close to the productivity measures that Gabaix (2011) uses. The banks used to calculate the shocks are the ten largest banks, ranked by total operating income. Therefore, the BBN209 banking granular residual is as follows,

$$
\text{BGR}_t = \left(\sum_i^K S_{i,t-1}\right)^{-1} \left(\sum_{i=1}^K S_{i,t-1}\left(g_{i,t} - \overline{g}_t\right)\right) = \left(\sum_i^K S_{i,t-1}\right)^{-1} \left(\sum_{i=1}^K S_{i,t-1}\epsilon_{it}\right)
$$

where, $\overline{g}_t$ is the cross-sectional mean over K banks in the country, and K represents the K largest banks, $S_{i,t}$ is the total operating income of bank i at time t, $g_{i,t}$ is the growth rate of the inverse of the cost-to-income ratio for bank i at time t and ϵ_{it} represents a shock, i.e., the deviation of the growth rate of the inverse cost-to-income ratio at time t from its mean growth rate. Empirically, BBN2009 consider all German banks from 1991 to 2005 (BankScope—Bureau van Dijk) with consolidation codes C1, C2, U1 and A1; banks have been ordered by size using operating income and information about distress events for the years 1994–2004 are confidential from Deutsche Bundesbank. They introduce the Banking Granular Residual into a stress-testing model for the German banking system.

All in all, BBN2009 show that shocks at large banks affect the probability of distress of small and medium-sized banks in Germany; positive shocks at large banks reduce smaller banks' probability of distress, while negative shocks increase this probability.

Data and Methods

Data

In line with previous research, the main data source for constructing our measure of idiosyncratic shocks to local dominant banks is BankScope—Bureau van Dijk. From BankScope, we retrieve data for all EU banks from 1994 to 2023. Some banks present both consolidated and unconsolidated accounts. To eliminate double entries, we keep only those banks with the consolidation codes C1 (consolidated and companion is not on the disc), C2 (consolidated and companion is on the disc), U1 (unconsolidated and companion is not on the disc, or if the bank does not publish consolidated accounts), and U2 (aggregated statements with no companion). Furthermore, we eliminate all entries with missing operating income, which we use to order the banks by size. We use observations on bank holding companies, commercial banks, cooperative banks, fintech banks, and savings banks to represent the banking industry as a whole. We ended with 42,622 bank-year observations.

Country-Level GDP Growth and Bank Z-Score

The primary dependent variable in our analysis is EU country-level GDP growth. We collect annual real GDP per capita from the EUROSTAT. Following Biswas et al. (2017), we calculate state GDP growth as the annual log change in the real GDP per capita. We use country GDP growth to identify granular EU countries, i.e., those countries where local dominant banks matter. Then, we relate our measure of idiosyncratic shocks at local dominant banks to a measure of EU banking system stability. To this end, we supplement our analyses using as a dependent variable the country Bank Z-score indicator from OECD, capturing the probability of default of a country's commercial banking system (capitalization and returns) with the volatility of those returns. It is estimated as (ROA + (equity/assets))/sd(ROA), where sd(ROA) is the standard deviation of ROA, calculated for country-years with no less than five bank-level observations; ROA, equity, and assets are country-level aggregate figures.

Measurement of Locally Dominant Banks

We classify locally dominant banks as those banks that are the largest top 25% in their EU country based on annual total operating income. In each country and year, locally dominant banks are banks whose total operating income is in the top quartile of last year's total operating income distribution. Although our main findings are not sensitive to the quartile cutoff, we use this cutoff to have enough locally dominant banks per country. Further, choosing locally dominant banks based on the size distribution, as opposed to fixing a specific number of banks per country, assures that we are not overestimating (or underestimating) the economic effects of banks in countries with a small (or large) number of banks. Finally, to ensure that our results are not driven by the economic effects of nationally large banks, as documented in Gabaix (2011), we drop the top 5% largest banks in each EU country-year from the sample of locally dominant banks, which we assume are EU dominant banks.

Identification of Local Bank Idiosyncratic Shocks

We follow Gabaix (2011), Blank et al. (2009), and Jannati et al. (2020) to compute idiosyncratic productivity shocks to locally dominant banks. In particular, we first measure a bank's productivity growth as the inverse cost-to-income ratio growth rate for bank i at time t. Specifically,

$$g_{j,i,t} = z_{i,j,t} - z_{i,j,t-1} \tag{14.1}$$

where, $z_{i,j,t} = \ln(\text{cost-to-income}_{i,j,t})$, and cost-to-income$_{i,j,t}$ is the cost-to-income ratio of bank i, headquartered in country j, at time t.

Next, we subtract from the bank productivity growth the average productivity growth of all banks headquartered in country j. That is, we measure idiosyncratic shocks as,

$$\xi_{i,j,t} = g_{i,j,t} - K^{-1} \sum_K g_{i,j,t} \tag{14.2}$$

where, K is the total number of banks headquartered in country j at time t. By subtracting the average productivity growth, we can isolate the cost-to-income component specific to the bank. Finally, we compute a weighted average of shocks to all locally dominant banks for each country. We denote this measure by Γ, which takes the following form for country j in year t,

$$\Gamma_{j,t} = \sum L_j \left(\frac{S_{i,j,t-1}}{S_{j,t-1}} \right) \xi_{i,j,t} \tag{14.3}$$

where, L_j is the total number of locally dominant banks in country j. Following BBN2009 and BN2011, we scale $\xi_{i,j,t}$ using the ratio of the bank's total operating income to the aggregate domestic banks' total operating income. Consistent with Gabaix (2011) terminology, we refer Γ_{jt} to as the "local bank granular residual". This local bank granular residual is our main variable of interest, and our goal is to examine the economic impact of Γ_{jt} on the country economic fluctuations.

Summary Statistics

We report the summary statistics of our main variables in Table 14.1. Specifically, Table 14.1 shows the operating income, cost-to-income ratio, annual productivity growth, granular residual, annual number of firms, annual GDP growth and bank z-score in each country throughout the sample period.

Which EU Countries Are Bank Granular?

To identify which countries are granular, we follow the methodology used by Jannatti et al. (2020). The main difference between them and our approach is that we identify granular countries by looking at local dominant banks while they look at local dominant firms. We separate European countries into two categories: (1) granular

Table 14.1 Summary statistics

Country	Variable	Mean	P25	P50	P75	Std	Country	Variable	Mean	P25	P50	P75	Std
AT	Operating income	115,874	2,667	5,980	16,867	619,567	IE	Operating income	792,133	3,254	111,813	1,029,000	1,270,465
	Cost-to-income	72.74	63.65	70.74	78.01	68.25		Cost-to-income	66.61	55.17	66.01	82.24	32.76
	Annual productivity growth	0.026	−0.047	0.008	0.063	1.130		Annual productivity growth	0.167	−0.144	−0.035	0.049	2.988
	Granular residual	−0.003	−0.003	−0.002	0.002	0.009		Granular residual	−0.052	−0.034	0.014	0.060	0.495
	Annual number of firms	342.63	272.00	359.00	384.00	63.02		Annual number of firms	22.00	10.00	28.00	29.00	9.36
	Annual GDP growth	0.04	0.03	0.03	0.05	0.03		Annual GDP growth	0.09	0.05	0.09	0.16	0.08
	Bank Z-score	29.94	24.20	30.91	34.55	5.36		Bank Z-score	11.17	6.86	11.42	15.26	4.69
BE	Operating income	1,241,499	53,297	243,819	678,994	2,566,758	IT	Operating income	442,124	11,367	28,341	92,897	2,195,547
	Cost-to-income	71.97	54.83	64.55	78.71	134.87		Cost-to-income	73.32	59.96	68.11	75.70	200.15
	Annual productivity growth	0.043	−0.098	−0.008	0.077	1.361		Annual productivity growth	−0.033	−0.075	0.005	0.084	1.968
	Granular residual	−0.007	−0.005	0.010	0.023	0.105		Granular residual	0.006	−0.007	−0.002	0.000	0.022
	Annual number of firms	43.79	34.00	51.00	52.00	11.68		Annual number of firms	308.01	316.00	346.00	363.00	94.52
	Annual GDP growth	0.04	0.03	0.03	0.05	0.03		Annual GDP growth	0.02	0.01	0.02	0.02	0.04

(continued)

Table 14.1 (continued)

Country	Variable	Mean	P25	P50	P75	Std
BG	Bank Z-score	14.12	12.45	15.45	16.53	3.87
	Operating income	126,715	18,134	61,023	206,596	139,970
	Cost-to-income	65.52	46.72	58.12	76.52	26.57
	Annual productivity growth	0.031	−0.081	0.016	0.090	0.210
	Granular residual	−0.014	−0.038	−0.004	0.010	0.033
	Annual number of firms	22.30	21.00	26.00	27.00	6.96
	Annual GDP growth	0.08	0.02	0.07	0.12	0.06
CY	Bank Z-score	8.46	8.12	8.64	8.72	0.44
	Operating income	257,004	12,376	27,871	248,134	507,686
	Cost-to-income	64.36	49.71	67.00	86.33	1,515.91
	Annual productivity growth	0.159	−0.116	−0.006	0.114	1.874
	Granular residual	−0.059	−0.094	−0.040	0.003	0.153
	Annual number of firms	30.98	32.00	35.00	36.00	9.19
LT	Bank Z-score	13.43	11.89	13.61	14.52	1.49
	Operating income	143,200	22,828	153,949	208,484	116,289
	Cost-to-income	59.54	41.99	48.36	63.23	63.81
	Annual productivity growth	0.128	−0.063	0.006	0.103	1.106
	Granular residual	−0.066	−0.025	−0.011	0.008	0.173
	Annual number of firms	6.89	6.00	8.00	9.00	2.28
	Annual GDP growth	0.08	0.04	0.08	0.13	0.07
LU	Bank Z-score	6.52	5.53	6.81	7.24	1.37
	Operating income	269,155	24,516	81,741	293,727	538,429
	Cost-to-income	−302.23	45.31	59.75	77.31	7,838.45
	Annual productivity growth	−0.401	−0.129	−0.012	0.123	10.518
	Granular residual	0.248	−0.103	−0.026	0.184	1.171
	Annual number of firms	43.66	37.00	48.00	51.00	9.96

(continued)

Table 14.1 (continued)

Country	Variable	Mean	P25	P50	P75	Std
	Annual GDP growth	0.05	0.03	0.07	0.07	0.06
	Bank Z-score	7.78	7.11	8.03	8.27	1.43
CZ	Operating income	402,026	56,019	151,736	506,169	494,313
	Cost-to-income	29.47	40.86	47.69	55.97	421.32
	Annual productivity growth	−0.074	−0.067	0.012	0.092	2.118
	Granular residual	0.029	−0.019	−0.013	0.015	0.200
	Annual number of firms	25.00	18.00	27.00	30.00	5.91
	Annual GDP growth	0.06	−0.01	0.07	0.10	0.07
	Bank Z-score	10.86	10.34	10.76	11.56	0.91
DE	Operating income	223,847	12,855	34,990	72,436	1,894,837
	Cost-to-income	73.67	65.15	71.71	79.77	106.69
	Annual productivity growth	0.041	−0.053	0.001	0.063	2.063
	Granular residual	−0.002	−0.003	−0.001	0.003	0.009
	Annual GDP Growth	0.06	0.03	0.05	0.07	0.03
	Bank Z-score	45.10	35.61	46.62	54.33	9.53
LV	Operating income	84,451	15,144	76,730	132,486	71,734
	Cost-to-income	68.76	46.98	56.55	75.34	44.70
	Annual productivity growth	0.036	−0.148	−0.008	0.111	0.386
	Granular residual	−0.011	−0.043	−0.017	0.029	0.046
	Annual number of firms	14.43	12.00	17.00	18.00	5.07
	Annual GDP Growth	0.07	0.04	0.06	0.11	0.08
	Bank Z-score	7.23	6.85	7.22	8.14	1.05
MT	Operating income	82,605	14,534	42,637	151,825	86,306
	Cost-to-income	58.33	46.72	56.59	70.48	25.67
	Annual productivity growth	−0.015	−0.122	−0.026	0.061	0.328
	Granular residual	−0.013	−0.059	−0.027	0.045	0.077

(continued)

Table 14.1 (continued)

Country	Variable	Mean	P25	P50	P75	Std
DK	Annual number of firms	1,038.46	692.00	1,207.00	1,225.00	297.45
	Annual GDP growth	0.03	0.03	0.04	0.05	0.03
	Bank Z-score	15.47	12.78	16.05	16.76	3.35
	Operating income	375,336	16,765	51,986	211,252	1,072,170
	Cost-to-income	69.83	58.87	69.10	77.59	109.02
	Annual productivity growth	−0.062	−0.092	−0.008	0.070	1.351
	Granular residual	0.021	−0.020	0.004	0.027	0.079
EE	Annual number of firms	58.13	59.00	66.00	69.00	17.87
	Annual GDP growth	0.04	0.02	0.03	0.04	0.03
	Bank Z-score	21.03	18.07	22.68	25.58	4.68
	Operating income	161,035	13,842	93,700	183,800	199,805
	Cost-to-income	60.06	41.72	49.92	71.20	28.16
	Annual productivity growth	0.030	−0.057	0.015	0.097	0.186
NL	Annual number of firms	13.30	13.00	15.00	16.00	4.05
	Annual GDP growth	0.09	0.06	0.10	0.14	0.06
	Bank Z-score	21.01	18.69	19.54	23.10	3.23
	Operating income	4,924,357	68,945	261,100	1,198,000	17,000,000
	Cost-to-income	24.84	47.70	63.66	77.93	716.33
	Annual productivity growth	−0.332	−0.116	−0.009	0.112	6.414
	Granular residual	0.097	−0.051	−0.008	0.028	0.436
PL	Annual number of firms	31.59	26.00	37.00	37.00	7.97
	Annual GDP growth	0.04	0.02	0.03	0.05	0.03
	Bank Z-score	10.36	8.94	11.03	12.21	2.40
	Operating income	237,305	1,821	4,125	111,662	545,470
	Cost-to-income	67.71	57.66	70.70	79.01	55.80
	Annual productivity growth	0.028	−0.053	0.011	0.080	0.735

(continued)

Table 14.1 (continued)

Country	Variable	Mean	P25	P50	P75	Std
	Granular residual	-0.017	-0.053	-0.013	0.017	0.072
	Annual number of firms	7.95	8.00	9.00	10.00	2.57
	Annual GDP growth	0.08	0.05	0.08	0.13	0.07
	Bank Z-score	9.79	8.42	9.11	11.64	2.14
ES	Operating income	1,485,836	7,409	48,609	468,049	5,614,053
	Cost-to-income	72.79	56.03	66.49	79.76	72.74
	Annual productivity growth	0.068	-0.102	0.004	0.098	1.832
	Granular residual	-0.020	-0.034	-0.007	0.004	0.054
	Annual number of firms	114.33	117.00	128.00	136.00	35.44
	Annual GDP growth	0.03	0.01	0.03	0.04	0.06
	Bank Z-score	18.40	16.92	18.75	20.28	1.87
FI	Operating income	214,880	2,244	6,943	41,911	966,409
	Cost-to-income	66.58	57.85	64.78	72.61	17.29
	Granular residual	-0.004	-0.009	0.001	0.011	0.031
	Annual number of firms	117.86	102.00	139.00	142.00	40.58
	Annual GDP growth	0.06	0.02	0.07	0.10	0.05
	Bank Z-score	9.09	8.11	9.40	9.75	0.96
PT	Operating income	227,747	3,735	7,304	58,393	557,420
	Cost-to-income	69.73	59.45	72.53	83.86	258.95
	Annual productivity growth	0.050	-0.102	-0.008	0.088	1.980
	Granular residual	-0.007	-0.041	-0.004	0.092	0.119
	Annual number of firms	103.26	110.00	114.00	116.00	29.28
	Annual GDP growth	0.04	0.04	0.04	0.05	0.05
	Bank Z-score	14.05	10.70	15.55	16.45	3.05
RO	Operating income	312,009	46,564	199,279	555,154	305,453
	Cost-to-income	95.31	51.08	60.00	75.02	208.32

(continued)

Table 14.1 (continued)

Country	Variable	Mean	P25	P50	P75	Std
	Annual productivity growth	0.013	−0.083	−0.012	0.077	0.228
	Granular residual	−0.004	0.000	0.001	0.003	0.039
	Annual number of firms	119.98	131.00	135.00	137.00	38.79
	Annual GDP growth	0.03	0.03	0.03	0.05	0.02
	Bank Z-score	15.46	13.97	16.28	17.20	2.60
FR	Operating income	1,472,225	86,256	292,265	522,583	4,901,260
	Cost-to-income	32.25	55.40	62.94	70.21	1,278.73
	Annual productivity growth	0.022	−0.044	0.000	0.041	4.613
	Granular residual	−0.030	−0.014	0.001	0.006	0.096
	Annual number of firms	220.84	197.00	247.00	252.00	43.03
	Annual GDP growth	0.02	0.02	0.03	0.03	0.03
	Bank Z-score	17.80	16.62	18.90	19.79	2.44
GR	Operating income	1,380,497	64,985	1,622,025	2,196,500	1,261,257
	Annual productivity growth	0.149	−0.059	0.023	0.112	1.534
	Granular residual	−0.061	−0.034	−0.022	0.013	0.147
	Annual number of firms	19.50	19.00	23.00	24.00	6.21
	Annual GDP growth	0.08	0.04	0.09	0.11	0.08
	Bank Z-score	10.28	9.76	10.84	11.07	1.05
SE	Operating income	395,952	8,868	25,952	95,324	1,155,962
	Cost-to-income	57.51	46.39	55.50	66.23	184.61
	Annual productivity growth	0.011	−0.051	0.012	0.081	0.292
	Granular residual	−0.005	−0.017	−0.010	0.009	0.019
	Annual number of firms	86.60	84.00	95.00	98.00	24.75
	Annual GDP growth	0.03	0.01	0.02	0.04	0.05
	Bank Z-score	34.77	32.08	36.47	37.74	4.54
SI	Operating income	134,966	45,467	74,237	124,962	167,218

(continued)

Table 14.1 (continued)

Country	Variable	Mean	P25	P50	P75	Std
	Cost-to-income	61.30	47.72	55.72	69.21	46.22
	Annual productivity growth	0.002	−0.103	−0.007	0.184	0.763
	Granular residual	0.067	−0.015	0.040	0.048	0.161
	Annual number of firms	11.98	11.00	12.00	14.00	3.67
	Annual GDP growth	0.01	−0.05	0.00	0.04	0.07
	Bank Z-score	5.94	5.11	6.64	7.74	2.41
HR	Operating income	174,624	10,975	82,143	276,962	210,719
	Cost-to-income	70.07	55.29	63.93	76.01	33.80
	Annual productivity growth	0.043	−0.082	0.005	0.094	0.436
	Granular residual	−0.012	−0.051	−0.021	0.022	0.061
	Annual number of firms	21.84	24.00	25.00	25.00	6.12
	Annual GDP growth	0.04	0.00	0.05	0.06	0.07
	Bank Z-score	7.51	7.39	7.64	7.88	0.63
	Cost-to-income	−72.69	55.70	62.37	71.00	2,027.96
	Annual productivity growth	−2.385	−0.076	0.003	0.092	34.187
	Granular residual	1.589	−0.018	0.002	0.061	4.966
	Annual number of firms	13.82	9.00	16.00	16.00	3.88
	Annual GDP growth	0.04	0.02	0.04	0.07	0.04
	Bank Z-score	4.11	3.50	4.52	5.11	1.53
SK	Operating income	378,848	205,674	356,532	515,125	340,751
	Cost-to-income	58.31	50.88	56.62	62.22	13.04
	Annual productivity growth	0.026	−0.053	0.011	0.067	0.235
	Granular residual	−0.001	−0.017	0.012	0.018	0.038
	Annual number of firms	12.06	8.00	14.00	14.00	3.30
	Annual GDP growth	0.05	0.02	0.05	0.07	0.05
	Bank Z-score	22.32	21.19	22.58	25.18	3.19

(continued)

Table 14.1 (continued)

Country	Variable	Mean	P25	P50	P75	Std
HU	Operating income	484,391	26,792	296,788	488,138	769,692
	Cost-to-income	76.91	59.24	70.67	84.59	48.79
	Annual productivity growth	0.035	-0.096	0.006	0.108	0.406
	Granular residual	-0.008	-0.045	0.000	0.041	0.058
	Annual number of firms	17.67	10.00	19.50	24.00	5.96
	Annual GDP growth	0.05	0.02	0.06	0.09	0.06
	Bank Z-score	8.11	7.07	8.12	9.48	1.18

Country	Variable	Mean	P25	P50	P75	Std
Total	Operating income	477,065	8,314	34,288	124,211	3,047,397
	Cost-to-income	60.26	60.21	68.98	77.91	1,124.62
	Annual productivity growth	0.004	-0.062	0.002	0.068	3.607
	Granular residual	0.009	-0.007	-0.001	0.004	0.419
	Annual number of firms	517.11	132.00	354.00	1,205.00	464.17
	Annual GDP growth	0.04	0.02	0.03	0.05	0.04
	Bank Z-score	17.91	12.42	16.09	19.98	8.25

This table presents summary statistics of the main variables used in the empirical analysis. Operating Income is the operating income of banks, Cost-To-Income is the cost-to.-income ratio, Productivity Growth is the annual log change in banks cost-to-income ratio. Granular residual is the weighted average of idiosyncratic productivity shocks for the group of locally dominant banks. Country GDP Growth is the log change country's real GDP per capita. Bank z-score is the OCED stability indicator. Bank data are from BankScope—Bureau Van Dijk. Real GDP per capita information is from EUROSTAT. The sample period is from 2004 to 2022 countries, that is, countries whose economies are affected by the productivity shocks to the locally dominant banks, and (2) non-granular countries.

Identification of Bank Granular EU Countries

We determine whether shocks to locally dominant banks explain the economic growth of their home countries by estimating country-level regressions. In particular, for each country, we regress its GDP growth on the current and lagged value of its granular residual from Eq. (14.3). That is,

$$\text{GDP Growth}_{j,t} = \alpha + \beta_1 \Gamma_{jt} + \beta_2 \Gamma_{jt-1} + \epsilon_{j,t} \tag{14.4}$$

The purpose of estimating the above regression is to obtain the estimated R^2s, which will inform us about the statistical power of the local bank granular residuals at time t and t–1, Γ_{jt} and Γ_{jt-1}, in explaining the country's GDP fluctuations. As per Jannatti et al. (2020), we consider *"granular countries"*, with an estimated R^2s above 8%. The results of the estimation of Eq. (14.4) for the granular states are reported in Table 14.2.

We find that for 16 countries out of EU27, the granular residuals of the locally dominant banks can explain a relevant portion of the country's GDP growth. These countries are Lithuania, Finland, Latvia, Italy, Sweden, Cyprus, Belgium, Czechia, Malta, Slovakia, Luxembourg, Portugal, Greece, Spain, Romania, and France. Specifically, the granular residuals explain a percentage of the GDP growth ranging from 82% for Lithuania to 9.6% for Germany.

In Fig. 14.1, we show the geographic distribution of the granular countries. Ccountries where locally dominant banks have higher economic effects (based on the estimated R^2) are shown in a darker shade. From the geographic distribution of granular countries, we see that countries for which the banking granular residual explains a significant part of national GDP growth are not necessarily the largest and most economically relevant countries in the EU27.

This figure presents the geographic distribution of EU granular countries. Specifically, productivity shocks to locally dominant banks in the identified countries explain more than 8% of the country's GDP growth. Countries where locally dominant banks have a larger economic impact are depicted in a darker shade color. Locally dominant banks per country are defined as banks that, after excluding the top 5% EU banks, are in the top quartile of the country's size distribution, where size is the prior year's bank operating income. A country's GDP growth is the log change of the country's real GDP per capita. Bank data are from BankScope—Bureau Van Dijk. Real GDP per capita information is from EUROSTAT. The sample period is from 2004 to 2022.

Table 14.2 Identification of EU granular countries

Country	Γ_τ	$\Gamma_{\tau-1}$	R^2 (%)	Adj. R^2 (%)	Average GDP growth (%)
LT	−0.170***	−0.263***	82.4	76.5	7.45
	(−6.09)	(−12.54)			
FI	0.236***	0.057*	57.0	49.8	2.97
	(6.96)	(1.94)			
LV	−0.146	−0.668**	42.1	29.3	6.93
	(−0.74)	(−2.74)			
IT	−0.923*	0.465	36.0	26.2	1.64
	(−1.92)	(1.14)			
SE	−1.412*	−0.695	32.0	21.5	3.66
	(−2.06)	(−0.96)			
CY	−0.236*	0.005	30.2	14.7	3.83
	(−2.20)	(0.04)			
BE	−0.135**	0.137***	27.5	16.3	3.52
	(−2.92)	(3.08)			
CZ	−0.099***	−0.075**	20.1	7.8	5.77
	(−3.07)	(−2.85)			
MT	0.060	0.269**	19.8	3.8	7.58
	(0.78)	(2.62)			
SK	0.133	0.505	19.6	6.2	6.40
	(0.63)	(1.27)			
LU	−0.016	0.007***	17.2	4.5	5.75
	(−0.78)	(4.27)			
PT	0.170	−0.078	17.0	3.2	2.62
	(0.91)	(−0.55)			
GR	0.205	0.073	17.0	3.1	0.43
	(1.52)	(0.58)			
ES	−0.306*	−0.016	11.2	−2.4	2.34
	(−2.07)	(−0.11)			
RO	0.239**	−0.055	10.5	−3.3	8.30
	(2.44)	(−0.44)			

(continued)

Table 14.2 (continued)

Country	Γ_τ	$\Gamma_{\tau-1}$	R^2 (%)	Adj. R^2 (%)	Average GDP growth (%)
FR	0.018	0.093	9.6	−4.3	2.43
	(0.42)	(1.24)			

This table shows EU countries in which productivity shocks to locally dominant banks explain over 8% of the country's GDP growth. Countries are ranked based on the estimated R^2 from time-series regressions with country GDP growth as the dependent variable (Eq. (14.4)). Γ_{jt} and Γ_{jt-1}, are the independent variables, and are equal to the granular residual of locally dominant banks at time t and t − 1, respectively. Locally dominant banks per country are defined as banks that, after excluding the top 5% EU banks, are in the top-quartile of the country's size distribution, where size is the prior year's bank operating income. A country GDP growth is the log change of the country's real GDP per capita. Bank data are from BankScope—Bureau Van Dijk. Real GDP per capita information is from EUROSTAT. The sample period is from 2004 to 2022. White robust t-statistics are reported in parentheses below the coefficient estimates

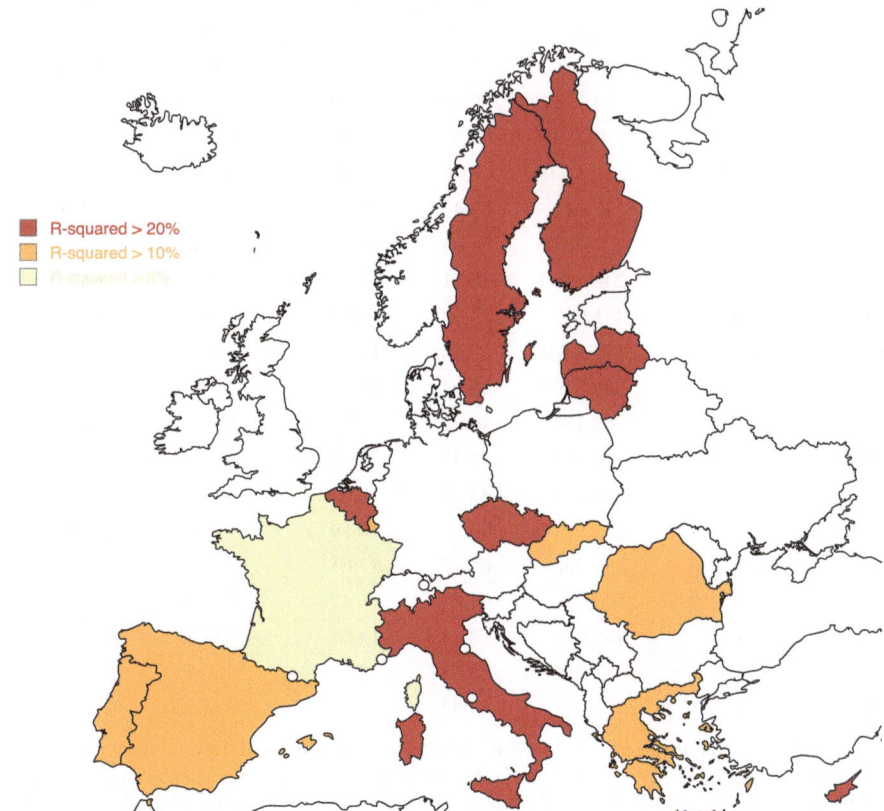

Fig. 14.1 Geographic distribution of EU granular countries

Characteristics of Locally Dominant Banks

In this section, we look closely at the sample of locally dominant banks in the granular countries. In Table 14.3, we report some relevant characteristics of the locally dominant banks and compare them with those of other domestic local non-dominant banks. Panels A and B report the average number of locally dominant and non-dominant banks, along with their operating income and cost-to-income ratio.

Table 14.3 Summary statistics of granular countries

	Panel A: locally dominant banks			Panel A: non-locally dominant banks		
(1)	(2)	(3)	(4)	(5)	(6)	(7)
Country	Number of firms	Operating income	Cost to income	Number of firms	Operating income	Cost to income
AT	68.43	73,191	69.00	254.02	5,353	73.86
CY	8.64	3,162,095	63.86	32.85	199,565	74.76
CZ	6.32	554,176	70.50	23.10	91,138	62.94
DE	5.19	1,102,303	47.55	18.97	159,931	23.82
DK	22.77	2,076,518	68.06	84.90	83,963	74.82
EE	23.91	232,019	62.47	88.86	21,716	68.10
ES	44.24	1,424,503	54.05	163.72	198,095	23.54
FI	2.63	2,886,562	50.11	9.31	989,530	64.30
GR	4.31	471,009	53.03	16.63	71,607	75.27
HU	3.67	1,138,830	63.08	13.40	202,551	81.30
IE	61.75	368,390	65.93	228.35	28,435	75.75
IT	1.74	268,434	49.07	5.37	115,532	61.85
LT	8.67	569,466	59.34	32.75	68,578	n.a
LU	2.94	167,643	46.74	10.99	60,540	74.60
LV	3.21	213,909	59.47	10.19	47,044	58.00
MT	23.32	529,298	65.75	87.61	28,911	69.31
PL	20.98	507,119	64.69	76.47	52,224	71.96
RO	3.76	740,842	52.87	14.91	174,452	108.17
SE	17.34	683,861	58.83	64.36	36,794	57.82
SK	2.60	755,976	55.82	9.49	278,946	58.97

This table presents summary statistics of firms headquartered in the granular countries. Column (1) shows the granular countries, identified in Table 14.2. Columns (2) and (5) show the an-nual average number of locally dominant and non-dominant banks per country. Columns (3) and (6) show the annual average of banks' operating income. Columns (4) and (7) show the annual average of banks' cost-to-income. Locally dominant banks per country are defined as banks that, after excluding the top 5% EU banks, are in the top-quartile of the country's size distribution, where size is the prior year's bank operating income. Operating income data are in thousands of euros. Bank data are from BankScope—Bureau Van Dijk. The sample period is from 2004 to 2022

The average number of locally dominant banks in the granular states is about one-fourth of the non-dominant ones by design. However, on average, the operating income and the cost-to-income ratio of locally dominant banks are more than ten times larger than those of non-dominant banks. For example, while there are only 22 locally dominant banks in ES (compared to 84 non-dominant banks), the average operating income of these banks is 2,076 million euros. In contrast the average operating income of the non-dominant banks is only 83 euros million. Collectively, the evidence in Table 14.3 suggests that we can identify banks that are important for the local economy.

Locally Dominant Banks, EU GDP Growth, and EU Bank Z Score

In this section, we examine whether shocks to locally dominant banks in granular countries have any aggregate effects. In particular, we examine the aggregate impact of locally dominant banks in the granular countries on the EU27 economy. In addition, we compare the predictive power of locally dominant banks with the economic power of the dominant banks at the European level. To perform this analysis, we run two sets of regressions similar to the regression of Eq. (14.4), with the dependent variable being the GDP growth of the EU27, which we use as a measure for aggregate macroeconomic fluctuations, and the OECD EU27 Bank Z-score, which we use as a measure of the soundness and stability of the EU financial system. These results are reported in Table 14.4.

In Column (1), we report the estimation results of the predictive power of the productivity shocks of the dominant banks at the European level on the EU27 GDP growth. Our results suggest that shocks of EU dominant banks (top 5% by size in EU) explain 78.2% of EU27 GDP growth (dominant at time t, coeff = 0.622*** with t-statistic = 25.74); dominant at $t - 1 = 2.158$***, t-statistic = 20.14). In Column (2), we report the same evidence with regard to locally dominant banks: results indicate that shocks of local dominant banks also matter (local dominant at time t, coeff = 0.125** with t-statistic = 2.29). Reasonably, the effect of local dominant banks on EU GDP growth appears significantly lower than that of EU dominant banks. In Column (3), we quantify the cumulative explanatory power of banks dominant at the country and European level. Specifically, we estimate a regression that includes the granular residuals of the banks locally dominant and dominant at the European level. We find the same evidence when we consider the joint effects of locally dominant and dominant at the European level.

We step forward in Columns (4) to (6), where we move to investigate the effect of EU and local dominant banks on the stability of the EU financial system. More in detail, our dependent variable is now the EU Bank Z-score, measuring the soundness of the EU27 financial system. While we expect positive coefficients for relationships between granular residuals and economic growth, we predict negative coefficients

Table 14.4 Locally dominant banks, EU GDP growth, and EU bank Z score

	EU GDP Growth			EU Bank Z-Score		
	(1)	(2)	(3)	(4)	(5)	(6)
	EU dominant banks	EU locally dominant banks	(1) + (2)	EU dominant banks	EU locally dominant banks	(4) + (5)
EU dominant banks Γ_t	0.622*** (25.74)		0.622*** (22.98)	−3.200 (−1.59)		−3.075 (−1.50)
EU dominant banks Γ_{t-1}	2.158*** (20.14)		2.165*** (20.12)	−2.200*** (−3.86)		−2.282*** (−4.03)
EU locally dom banks $\Gamma_{j,t}$		0.125** (2.29)	−0.009 (−0.24)		−0.728 (−0.73)	−0.278 (−0.28)
EU locally dom banks $\Gamma_{j,t-1}$		0.021 (0.79)	0.012** (2.00)		−0.139* (−1.76)	−0.168** (−2.33)
R^2	78.2	0.9	78.3	1.7	0.2	1.8
Adj R^2	78.1	0.2	78.0	0.9	0.5	0.4

This table presents the estimates of times-series regressions, where the EU GDP growth and the EU Bank Z-Score are the dependent variable in model (1)–(3) and (4)–(6) respectively. In column (1) and (4) the dependent variables are EU Dominant Banks Γ_t and EU Dominant Banks Γ_{t-1}, where EU Dominant Banks Γ_t is the granular residual of the dominant banks in EU. In column (2) and (5) the dependent variables are EU Locally Dom Banks $\Gamma_{j,t}$ EU Locally Dom Banks $\Gamma_{j,t-1}$, EU Locally Dom Banks $\Gamma_{j,t}$ is the granular residual of locally dominant banks in the granular countries. In column (3) and (6), the dependent variables are all the four aforementioned variables. The EU dominant banks are the largest top 5% EU banks. Locally dominant banks per country are defined as banks that, after excluding the top 5% EU banks, are in the top-quartile of the country's size distribution, where size is the prior year's bank operating income. Granular countries are identified in Table 14.2. Bank data are from BankScope—Bureau Van Dijk. The sample period is from 2004 to 2022. White robust t-statistics are reported in parentheses below the coefficient estimates

when the dependent variable is a default probability. We show that across all specifications, shocks of both EU and locally dominant banks matter in determining the implicit default probability of the EU banking system.

Summary and Conclusions

Existing research shows that adverse shocks to large financial institutions impact macroeconomic fluctuations and the soundness of the banking system. In this chapter, we change the angle and ask whether shocks originating at local but dominating banks also matter. To answer this question, we identify locally dominant banks that have a substantial impact on their local macroeconomic environment and construct a measure of idiosyncratic shocks at dominant local banks. We then show that idiosyncratic shocks at local dominant banks impact EU GDP Growth and EU banking system stability. We contribute by showing that local dominant banks also matter.

Acknowledgements The authors wish to thank the reviewers and the editor for their helpful comments. The views expressed herein are those of the authors, and they alone bear responsibility for any mistakes and inaccuracies. This work acknowledges Uniss Research Fund 2020 and D.M. 737/2021 2021-2022 financial support.

References

Biswas, S., Chakraborty, I., Hai, R.: Income inequality, tax policy, and economic growth*. Econ. J. **127**, 688–727 (2017)

Blank, S., Buch, C.M., Neugebauer, K.: Shocks at large banks and banking sector distress: the banking granular residual. J. Financ. Stab. **5**, 353–373 (2009)

Bremus, F., Buch, C.M.: Granularity in banking and growth: does financial openness matter? J. Bank. Finance **77**, 300–316 (2017)

Bremus, F., Buch, C.M., Russ, K.N., Schnitzer, M.: Big banks and macroeconomic outcomes: theory and cross-country evidence of granularity. J. Money, Credit Banking **50** (2018)

Buch, C.M., Neugebauer, K.: Bank-specific shocks and the real economy. J. Bank. Finance **35**, 2179–2187 (2011)

Duprey, T., Lé, M.: Bankscope Dataset: Getting Started (2016)

Gabaix, X.: The granular origins of aggregate fluctuations. Econometrica **79**, 733–772 (2011)

Jannati, S., Korniotis, G., Kumar, A.: Big fish in a small pond: locally dominant firms and the business cycle. J. Econ. Behav. Organ. **180**, 219–224 (2020)

Chapter 15
How Does NPLs Securitization Affect EU Banks' Systemic Risk?

Stefano Dell'Atti, Caterina Di Tommaso, Grazia Onorato, and Vincenzo Pacelli

Abstract This chapter contributes to the growing debate on the NPLs issue by addressing the challenges leading to financial stability and promoting the NPLs resolution plans for EU banks. Our main hypothesis is a U-shaped relationship between the NPLs securitizations and the systemic risk. We find that the maximum amount of NPLs securitization performed by EU banks to minimize the contribution to systemic risk shifts about Global systemic important banks (G-SIB) designation and country risk. The bank's contribution to systemic risk lies in the involvement of the bank in this instrument and not in its features. Our results contribute to the ongoing debate on the important issue of designing suitable systemic risk indicators that act as Early Warning Systems (EWS) for predicting incoming financial crises. Evaluating the bank's contribution to systemic risk is important to take into account the bank's exposure to NPL securitization.

Keywords Systemic risk · Securitization · Non-performing loans · Global Systemically Important Institutions (G-SIBs)

S. Dell'Atti · C. Di Tommaso · V. Pacelli (✉)
University of Bari Aldo Moro, Bari, Italy
e-mail: vincenzo.pacelli@uniba.it

S. Dell'Atti
e-mail: stefano.dellatti@uniba.it

C. Di Tommaso
e-mail: caterina.ditommaso@uniba.it

G. Onorato
University of Foggia, Foggia, Italy
e-mail: grazia.onorato@unifg.it

V. Pacelli (ed.), *Systemic Risk and Complex Networks in Modern Financial Systems*,
New Economic Windows, https://doi.org/10.1007/978-3-031-64916-5_15

277

Introduction

Preserving financial stability, and thereby supporting sustainable growth, requires the continued monitoring of developments in the global financial system. The recent financial turmoil has exacerbated the issue of the financial stability of the banking system. The collapse of important financial institutions has raised questions about the involvement of the banking sector in the propagation of the financial crisis. The large stock of non-performing loans (NPLs) has been and still is a pressing financial stability issue for the euro area (Fell et al. 2017). A high NPLs volume may cause micro-prudential and macro-prudential problems. From the micro-prudential point of view, the high volume of NPLs reduces the profitability and efficiency of the banks whereas, from the macro-prudential point of view, the amount of NPLs impacts economic growth by reducing the capability of banks to provide new lending. Furthermore, the banking sector's resilience to shock is harmed, which leads to an increase in systemic risk (EBA 2018). A wide range of possible resolution options to address the NPLs' problems have involved on- and off-balance sheet approaches, with the former involving the internal workout of NPLs, whereas the latter involves outright sales to investors (Grodzicki et al. 2015). Specifically, bank NPLs securitization is the process whereby distressed loans are pooled together into tradable securities, named Asset-Backed Securities (ABS), and sold to the investors. The possibility of transforming a distressed loan into tradable security allows for transferring the risks of the distressed loans to the market. For this reason, the effect of an NPLs securitization on the systemic risk could be twofold. On the one hand, NPLs securitization could enhance banks' risk appetite as they could find this tool as a good deal to sell distressed loans to the market. From this perspective, the securitization of bad loans could incentivize banks to reduce lending standards and thus, threaten financial stability and systemic risk. All this together is the basis of an unbalanced and fragile financial system. On the other hand, NPLs securitization allows banks to hold less risk and manage credit risk more effectively. Therefore, it provides a mechanism where the risks of distressed loans, concentrated in a bank, could be transferred and dispersed to the investors. NPLs securitization could operate as a means through which the risk is distributed on the market and thus it could make the financial system more stable and resilient.

This chapter aims to study the relationship between NPLs securitization and systemic risk to understand the logic behind the resolution plans of EU banks and the incentives of the supervisory authorities to develop and promote the use of this tool. The issuance of NPLs securitization might initially help the banks reduce their systemic risk by providing benefits related to liquidity, capital requirements, and NPLs resolutions. However, the nature of this relationship may change when a bank increases the use of this particular instrument. An increase in the NPLs securitization may be translated by the financial market as huge exposures of banks in distressed loans, ex-ante wrong assessment of the credit risk, and excessive risk-taking. All this together would be translated into a change in the relationship between the NPL securitization and the systemic risk. An increase in the use of NPLs securitizations beyond a certain level would imply an increase in systemic risk related to the problems explained above.

This study makes several contributions to the literature. First, we extend the very scarce literature on the impact of NPLs securitizations on systemic risk (Vuković and Domazet 2013; Pedisic 2019). The paper examines the effect of NPLs resolution plan on systemic risk and evaluates the effectiveness of the NPLs management on financial stability. NPLs securitization is widely used in the EU context because of the need to manage the stocks of NPLs in the banking system. It appears important to understand how they impact the systemic risk for financial stability issues. Second, the study tests the hypothesis of a quadratic relationship between the systemic risk and the NPLs securitizations that allows calculating of a threshold beyond or below which the use of NPLs securitization may be a detrimental tool for financial stability. To the best of our knowledge, this hypothesis has never been tested for NPLs securitizations. Only the paper of (Arif 2020) has tested the hypothesis for the covered bond and securitization market. However, we differentiate from this paper because we investigate the effect of a particular type of securitization on systemic risk. Third, our chapter contributes to the ongoing debate on the important issue of designing suitable systemic risk indicators that act as Early Warning Systems (EWS) for predicting incoming financial crises. Assessing the impact of NPLs securitization on the contribution to systemic risk provides empirical evidence on the real impact of the propagation of bank's risk on the financial system and, thus, on the effectiveness of the risk transferring process in the financial markets via securitizations. Finally, we construct an original dataset in which we include the main characteristics of the NPLs securitizations. We analyze a sample of EU banks during the period 2012–2020. We include in our sample 35 EU banks that have performed 133 NPLs securitizations.

We report different key results. First, it holds a U-shape relationship between systemic risk and NPLs securitizations implying that, though the initial positive effects of NPLs securitization on the systemic risk, the greater involvement of the banks in the NPLs resolution plans via securitizations exacerbates the bank's systemic risk and damages the financial stability. Therefore, the banks heavily involved in the NPLs securitization market experience greater exposure to a potential financial crisis and maybe a propagation mechanism of the individual financial crisis. Second, we identify the threshold below which the NPLs securitization is a good tool to transfer distressed loans to institutional investors but above this threshold, the issuance of an NPLs securitization is detrimental to bank systemic risk. On average, the securitizations performed by banks in our sample are well above the maximum amount identified by the empirical model. The only banks showing, on average, NPL securitizations lower than the estimated average are the Global Systemically Important Banks (G-SIB) in Portugal, Ireland, Italy, Greece, and Spain (PIIGS countries). These findings underline that the main problem lies in the way in which this instrument is used and not in the features of the tool itself. Shedding light on the effect of NPLs divestitures on financial stability could provide useful information on the determinants of financial contagion and, at the same time, on the involvement of the bank sector in the propagation of financial crises. This information may act as EWS and, thus, can be incorporated into systemic risk indicators to predict financial crises.

The rest of the chapter is organized as follows. In the next section, we review the relevant literature and develop our hypotheses. Section "Data and Methodology" presents our methodology and data. Our empirical results are in section "Empirical Results". Section "Robustness" verifies the robustness of our empirical analysis and section "Conclusion" concludes.

Literature Review

Despite the rapid development of securitizations as a tool to solve the NPLs problems in EU banks, the dynamics of the relationship between NPLs securitizations and systemic risk have been partially unexplored. The very scarce literature examining the effects of the NPLs resolution plans on systemic risk (Vuković and Domazet 2013; Pedisic 2019) has underlined the important role played by the NPLs reduction measures on systemic risk. Vuković and Domazet (2013), focusing on macroeconomic contagion with non-performing loans and the infection of the financial sector with non-performing loans, find that the NPLs are the main generator of systemic risk in the financial and real sectors of Serbia. Pedisic (2019) highlights that the NPLs reduction measures and the statutory framework affect the EU systemic risk.

Despite the very few studies focusing on the impact of NPLs securitization on the systemic risk in EU banks, our chapter can be related to the literature examining the use of securitization and its impact on the banks' systemic risk. In this direction, different studies have analyzed the issue of financial stability related to the use of credit derivatives, especially in the aftermath of the US financial crisis. The advent of the US financial crisis has changed the previous positive role associated with credit derivatives in supporting financial stability (Wagner and Marsh 2006; Loutskina 2011) and in managing and diversifying effectively the credit risk portfolio of banks (Cebenoyan and Strahan 2004; Duffie 2008). The US financial crisis has highlighted that securitization may undermine financial stability by weakening the bank's credit standards and increasing risk-taking (Diamond 1984; Chiesa 2008; Minton et al. 2009; Keys et al. 2010; Kara et al. 2016) and by increasing the complexity of the financial markets and reduced the monitoring role of US banks (Halili et al. 2021). In the aftermath of the US financial crisis, several studies have demonstrated the negative impact of credit derivatives on financial stability. Specifically, focusing on the Italian listed banks, Battaglia and Gallo (2013) show that the use of securitization increases the expected losses in case of extreme events. They add that the impact of securitization on systemic risk does not change with the inception of the financial crisis by concluding that there is a severe implication of securitization for financial stability both before and after the financial crisis. Focusing on US banks' contribution to systemic risk, Mayordomo et al. (2014) find that the impact of financial derivatives on systemic risk differs among the types of financial derivatives. There is a positive association between systemic risk and foreign exchange and credit derivatives and a negative association between systemic risk and interest rate derivatives. Furthermore, the NPLs and leverage ratios have a stronger impact on systemic risk than financial derivatives. Studying the impact of bank competition in the run-up to the 2007–2009 crisis on the banks' systemic risk, Altunbas et al. (2022) highlight that the use of securitization acts as a transmission mechanism channel and exacerbates the effects of market power on the systemic dimension of bank risk. Ivanov and Jiang (2020) underline the different roles of the underlying assets on systemic risk by showing that systemic risk is more sensitive to the securitization of residential mortgages. Finally, Arif (2020) explains the negative and positive association between the use of securitization and systemic risk through the theory of the "scalability view" of securitization suggesting the impact of the securitization on the systemic risk depends on the involvement of the bank in this market. On the same line, the paper of Mazzocchetti et al. (2020), by developing an agent-based model including the securitization position of banks, highlights that the involvement of a bank in securitizations weakens the financial stability of banks with relevant effects on different sectors of the economy.

Based on this literature, several assumptions can be made to build the conceptual framework of this study. Previous theories have underlined that the use of securitization made banks more resilient and, thus, reduced systemic risk (Greenspan 2005). The development of NPLs securitization has provided banks with a range of flexible instruments for selling distressed loans, transferring loan risk, and managing credit risk. The use of NPLs securitizations has helped to mitigate informational problems and acted as a mechanism to clean up the bank's balance sheet resulting in an increase in credit supply and a reduction of systemic risk. Therefore, our hypothesis is a negative association between NPLs securitization and systemic risk. Nevertheless, securitization creates an alternative funding source for banks that is less stable if compared to deposits. This may increase systemic risk because the banks are more vulnerable to changes in financial markets (Loutskina 2011; Laeven et al. 2016). Also, the view that banks reduce the credit standards and increase risk-taking may turn the relationship between systemic risk and NPLs securitization. The excessive involvement of a bank in NPLs securitizations could result in the weakening of the bank's credit standards and increasing risk-taking. Against this backdrop, the following hypotheses hold:

H1: There exists a quadratic relationship between NPL securitization and systemic risk.

H2: There is a threshold beyond or below which the issuance of NPLs securitizations reduces the systemic risk.

Our chapter is also closely related to the literature studying the determinants of systemic risk. Previous literature has used several bank-specific factors (Laeven et al. 2016; Bostandzic and Weiß 2018; Brunnermeier et al. 2020; Mazzocchetti et al. 2020) to evaluate how these can affect financial instabilities within the financial system. Brunnermeier et al. (2020) show a positive association between noninterest income and the total systemic risk of U.S. banks. Laeven et al. (2016) underline the role of bank size on systemic risk showing that, in EU and U.S. banks, the systemic risk grows with bank size and is inversely related to bank capital, and this effect exists above and beyond the effect of bank size and capital on standalone bank risk. Bostandzic and Weiß (2018) by investigating the reasons why some banks are more exposed and contribute more to systemic risk in the global financial system find that the quality of loan portfolio and the interconnectedness with the rest of the global financial system increase the contribution to the global systemic risk. Furthermore, they show that the average exposure of banks to systemic risk decreases in response to the higher capital regulations.

In light of the above literature, the threshold previously identified may be affected by different factors such as the complexity of the financial system, the country's financial condition, and the bank's network (Bakkar and Nyola 2021). These factors may increase the contribution of a bank to systemic risk. Therefore, we formulate our third hypothesis as follows:

H3: NPL, country, and bank characteristics change the threshold that minimizes the systemic risk of EU banks.

Data and Methodology

Data

To investigate the relationship between the NPLs securitizations and the systemic risk, we use a Panel of quarterly data spanning between Q1 2012–Q3 2020 for 35 EU-listed banks. This study focuses on European banks because they are the most active players in the NPLs market (EBA 2019). It has deep roots in Europe, especially in Portugal, Ireland, Italy, Greece, and Portugal (PIIGS countries).

Our analysis uses data coming from multiple sources. First of all, we collect data regarding SRISK and LRMES from V-lab maintained by the NY Stern Business School.[1] After identifying the banks with data on the V-lab website, we check Debtwire's NPL Coverage database and, one by one, the website of each bank to collect information about the NPLs securitizations.[2] Bank-level data, comprising the information from the financial statements, are obtained from Datastream. Non-listed banks are excluded from the sample. Banks with missing information about total

---

[1] https://vlab.stern.nyu.edu/welcome/srisk.

[2] Only for the Italian banks, we use the Securitization website (www.securitisation.it/index.htm) in which we can collect information about the securitizations performed by Italian banks.

assets, loans, and non-performing loan ratios are also excluded. All the variables are winsorized at a 1% level. Specifically, we replace all the data points less than the 1st percentile of each variable equal to the 1st percentile and all the data points exceeding the 99th percentile equal to the 99th percentile, thereby excluding extreme observations from the sample.

Table 15.7 in the appendix reports a detailed description of our variables whereas Table 15.1 provides the descriptive statistics of the dependent variables (Panel A) and independent variables divided by variable of interest and NPLs-specific characteristics (Panel B) and control variables (Panel C). Furthermore, Table 15.8 in the appendix provides detailed descriptive statistics of our dependent variables and variables of interest.

Table 15.1 Descriptive statistics

	Mean	Median	Std	Min	Max
Panel A: Dependent variable- Bank's level systemic risk					
SRISK (%)	25.98	19.75	26.07	0	100
LRMES	46.98	48.21	13.36	−6.1	89.13
Panel B: Variable of interest-NPLs securitizations					
Securitization of NPLs	1,560	1,000	1,680	15	11,000
Number of deals	2.55	2	1.75	1	11
Guarantee	14				
Type of loans					
CRE	29				
Consumer	21				
Corporate	22				
Legacy	2				
Mixed	19				
Mortgage	34				
Shipping	6				
Panel C: Control variables					
Banks size	19.57	19.49	1.54	16.45	22.59
Funding structure	48.9	48.33	16.83	5.92	96.6
Leverage	0.43	0.05	7.35	0	1.42
Capital adequacy	12.72	12.78	3.87	−7.3	27.9
NPL ratio	7.89	4.1	10.48	1.03	64.07
ROA	0.54	0.61	1.29	−12.4	4.99
Stock Price volatility	31.84	29.48	11.32	10.92	70.5
Sovereign CDS spread	703.4	48.5	455.9	5.95	3703.5

In line with the previous literature, as a dependent variable, we use two different measures of systemic risk: SRISK and Long-run marginal expected shortfall (LRMES) (Laeven et al. 2016; Arif 2020; Halili et al. 2021). First, we use the SRISK which measures the systemic risk contribution of a financial firm to the overall systemic risk. The systemic contribution of EU banks to the overall systemic risk ranged from 0 to 100%. On average, the EU banks show a systemic risk contribution of 25%. This indicates that banks will need around 25% of capital to cover the losses in case of a decline in the market index. The second measure of systemic risk is the LRMES which indicates the decline in equity values to be expected if there is a financial crisis. The analysis of the LRMES shows that the values ranged from − 6.10% to 83.13%. On average, the equity values of EU banks will decline by around 47% in case of a financial crisis. The country-level data of our systemic risk variables shows that the SRISK is lower for banks in PIIGS countries whereas the LMRES is more or less equal across the countries. Therefore, the systemic risk contribution of a bank in PIIGS countries to the overall systemic risk is lower than in other EU countries, whereas the decline in equity values in case of the financial crisis in banks in PIIGS countries is in line with the EU average.

Focusing on our variable of interest, Table 15.1 shows that the NPL securitization values ranged from USD 15 million to 11,000 USD million. The country-level values of NPLs securitization show that the most active banks in the NPLs market are those in PIIGS countries performing more than 33,998 USD million. The total gross book value (GBV) of NPLs securitizations performed by Italian banks is the highest (GBV of 14, 835 USD million) among EU banks immediately followed by Greek banks (GBV of 9,150 USD million). The Danish, Austrian and Norwegian banks have performed a lower amount of NPLs securitization compared to the other EU banks. The NPL-specific variables indicate that the EU banks, on average, have performed more than 2 NPLs securitizations. The maximum number of deals performed by one bank is 11 over the analysis period. Among the 133 NPL securitizations announced by EU banks, 14 are guaranteed by the government. In particular, the State-backed guarantee is from the Italian and Greek governments.[3] Different types of loans are the object of NPLs securitization. Based on a quantitative approach, in EU banks, 21.60% of the collateral are Commercial Real Estate (CRE) loans, 15.79% consumer loans, 16.54% corporate loans, 1.50% legacy loans, 14.29% mixed loans,[4] 25.56% mortgage loans and 4.51% shipping loans.

---

[3] In Italy, the public scheme that guarantees the senior tranche of NPL securitization is named *GACS- Garanzia sulla Cartolarizzazione delle Sofferenze* whereas in Greece, it is named *Hercules*.

[4] The mixed loans represent a mixture of the other loan types in an unknown proportion.

Empirical Methodology

This chapter aims to study the relationship between NPLs securitization issuance and bank stability and investigates the possible non-linearity in the target relationship. The main idea is that the relationship may vary with the level of involvement of a bank in the issuance of NPLs securitizations.

The issuance of NPL securitization can initially assist banks in mitigating systemic risk by offering advantages such as enhanced liquidity, meeting capital requirements, and facilitating NPL resolutions. However, the dynamics of this relationship may shift as a bank increases its use of this specific financial instrument. The financial market may interpret a rise in NPLs securitization as a sign of substantial bank exposure to distressed loans, potential errors in the ex-ante assessment of credit risk, and excessive risk-taking. These factors alter the correlation between NPL securitization and systemic risk. Beyond a certain threshold, an escalation in the use of NPLs securitizations implies an increase in systemic risk, attributed to the outlined issues. For these reasons, we examine a quadratic relationship between NPLs securitization and systemic risk. The extent of a bank's involvement in NPLs securitizations may influence systemic risk dynamics. EU banks derive systemic risk benefits up to a specific level of NPLs securitizations; however, surpassing this level results in drawbacks for EU banks engaged in further NPLs securitizations.

To mitigate potential endogeneity concerns,[5] we estimate our model employing the system GMM instrumental variables approach suggested by (Arellano and Bond 1991) and (Arellano and Bover 1995). We run two specification tests. The first is the Hansen test of over-identifying restrictions, which examines the validity of the instruments by analyzing the sample analog of the moment conditions used in the estimation procedure. The second test is the AR2 test (Arellano and Bond 1991) for the hypothesis of no autocorrelation in the error term where the presence of second-order autocorrelation in the differenced residuals implies that the estimates are inconsistent. The applied model is the following regression model:

$$
\begin{aligned}
SRISK_{i,t} =& \alpha + \beta * SRISK_{i,t-1} + \gamma * NPLs_{Sec\ i,t} \\
& + \delta * NPLs^2_{Sec\ i,t} + *PIIGS * NPLs_{Sec\ i,t} \\
& + \theta * G - SIBs * NPLs_{Sec\ i,t} + \vartheta * Z_{i,t} + \mu * X_{i,t-1} \\
& + \tau * SovereignCDS_{j,t} + \varepsilon
\end{aligned}
\tag{15.1}
$$

where the dependent variable, $SRISK_{i,t-1}$, is the systemic risk measure of the ith bank in period t–1, $NPLs_{Sec i,t}$ is a variable that measures the ith bank's GBV of NPLs securitization at the time t.

To measure the systemic risk, we employ various proxies, with the primary measure being SRISK, which is calculated by V-Lab at the NY Stern Business

---

[5] Endogeneity might arise, for example, from inverse causality between some of the covariates or because of omitted variable bias.

School.[6] SRISK assesses the capital required by a bank in the event of a 40% market index decline over six months. The bank's contribution to systemic risk is quantified as its systemic expected shortfall (SES), reflecting its likelihood of being undercapitalized when the entire system faces undercapitalization (Acharya et al. 2017). SRISK represents the bank's percentage of the financial sector's capital shortfall, capturing its sensitivity to a market index decline. As a secondary measure of systemic risk, we employ LRMES, an extension of the Marginal Expected Shortfall (MES) introduced by Acharya et al. (2012). While MES serves as a short-term indicator, LRMES functions as a long-term indicator by categorizing a crisis as a 40% decline in the market index over the subsequent six months (Acharya et al. 2012). For these events, the LMRES is the expected loss of equity value of the firm ith.

Our main interests in Eq. (15.1) are the coefficients on $NPLs_{Sec_{i,t}}$ and $NPLs_{Sec_{i,t}}^2$ (γ and δ). If our hypothesis is verified, we would observe $\gamma < 0$ and $\delta > 0$. In this case, the function in Eq. (15.1) has a minimum that represents the maximum amount of NPLs securitizations that a bank can perform to minimize the systemic risk. Furthermore, we insert two binary variables, the PIIGS indicator,[7] and the G-SIBs indicator,[8] that allow us to understand if the impact on systemic risk may change with the bank's country and the G-SIBs designation by the Financial Stability Board (FSB).[9] We consider the PIIGS dummy because these countries have been shown to be in an ongoing systemic crisis by the European Systemic Risk Board (ESRB)[10] and the G-SIBs dummy to understand how the size of the bank can impact the transmission channel of a systemic crisis. The coefficients ∂ and ϵ measure the additional effect on systematic risk when the NPLs securitization is performed by a bank in a PIIGS country and/or by a G-SIBs bank.

The vector $Z_{i,t}$ includes key characteristics of the NPLs transaction. Our interest is to verify whether NPL characteristics are more or less conducive to risk-taking behavior. Thus, the vector $Z_{i,t}$ includes a dummy variable taking value 1 if the NPLs securitization is guaranteed by the government (only for Italian and Greek banks) and 0 otherwise, the number of deals for each bank and the type of securitized loan.

The vector $X_{i,t-1}$ contains a set of control variables consisting of bank-specific characteristics. We include indicators of bank size, leverage, capital adequacy, profitability, funding structure, nonperforming loans, and stock price volatility. Finally, we control for the country risk ($SovereignCDS_{j,t}$).

After studying the relationship between systemic risk and the use of NPLs securitization to manage the banks' NPLs stocks, our focus lies on assessing the quantity

---

[6] https://vlab.stern.nyu.edu/welcome/srisk.

[7] The PIIGS indicator is a dummy variable taking the value of 1 if the countries are Portugal, Ireland, Italy, Greece, and Spain and 0 otherwise.

[8] The G-SIBs indicator is a dummy variable taking the value of 1 if the banks are designed as Global systemically important banks and 0 otherwise.

[9] The *Financial Stability Board* (FSB) defines the *Global Systemically Important banks* (G-SIBs) as those companies whose default could cause the blackout of the entire financial and economic system given the breadth, complexity, and strong systemic connection. See FSB, "Policy Measures to Address Systemically Important Financial Institutions", November 2011.

[10] For more details, see https://www.esrb.europa.eu/pub/financial-crises/html/index.en.html.

Table 15.2 Possible estimated vertices

Scenarios	Conditions	Estimated Vertex
I scenario	PIIGS $= 0$ G-SIB $= 0$	$NPLs\_Sec^* = -\frac{\gamma}{2*\delta}$
II scenario	PIIGS $= 1$ G-SIB $= 0$	$NPLs\_Sec^* = -\frac{\gamma+\epsilon}{2*\delta}$
III scenario	PIIGS $= 0$ G-SIB $= 1$	$NPLs\_Sec^* = -\frac{\gamma+\theta}{2*\delta}$
IV scenario	PIIGS $= 1$ G-SIB $= 1$	$NPLs\_Sec^* = -\frac{\gamma+\epsilon+\theta}{2*\delta}$

This table provides the calculation of the possible vertex of the function in Eq. (15.1). We consider all possible scenarios

of NPLs securitizations necessary for an EU bank to minimize systemic risk. In case of a U-shaped relationship between NPLs securitizations and systemic risk, we can assess the optimal level of engagement in NPLs securitizations for minimizing systemic risk. We need to determine the minimum point of the quadratic model in Eq. (15.1), computing the first derivative of SRISK as a function of NPLs securitizations and assuming that the first derivative is equal to zero. In symbols, we would have:

$$\frac{\Delta SRISK}{\Delta NPLs\_Sec} = \gamma + 2\delta * NPLs_{Sec}^* + \epsilon * PIIGS + \theta * G - SIBs = 0 \qquad (15.2)$$

The systemic risk of a bank varies with the issuance of securitization if $\gamma \neq \delta \neq 0$. The vertex ($NPLs\_Sec^*$) of the function in Eq. (15.2) changes in relation to the country in which the bank is based and the G-SIB designation. We can have four possible scenarios in relation to different conditions. We summarize the calculation of the vertex of the function in Eq. (15.2) in Table 15.2.

Empirical Results

This section presents the empirical results of various tests. We first run Eq. (15.1) by including only the variables indicating the NPL securitization position. In the second model, we add the NPL securitization characteristics and in the third model, we add the control variables. Overall, the results show the existence of a quadratic relationship between systemic risk and the use of NPLs securitizations. Therefore, we can identify in EU banks a threshold below which the use of the NPLs securitization can lower the systemic risk. However, above it the NPLs securitization increases the systemic risk and, thus, the issuance of an NPLs securitization could be detrimental to financial stability.

Table 15.3 reports the results of the system GMM model with robust standard errors. In columns (1), (2) and (3) we report the results for the SRISK whereas in columns (4), (5) and (6) we report the results for the LRMES. The coefficients of the NPLs securitization in all specifications are negative and significant, suggesting a negative relationship between the use of NPLs securitization and the systemic risk measured by short and long-term indicators. The square term of the NPLs securitization is positive and significant in all our specifications suggesting that the issuance of NPLs securitizations initially helps the bank to control its systemic risk, but this relationship is reversed when the bank increases its NPLs securitization issuance. These results endorse our first hypothesis H_1. The issuance of NPLs securitizations initially helps the banks reduce their systemic risk by providing multiple benefits related to the management of distressed loans, liquidity, funding cost and risk transfer. However, the nature of this relationship changes when banks increase their reliance on this particular instrument. The reliance on ABS may result in main effects: (i) the bank's incentive to monitor the loans decreases because the bank can use the securitization to clean up the balance sheet (Chiesa 2008); (ii) the reduction of banks' incentives to work to a more efficient procedure to internally work out NPLs; (iii) the increase of banks' risk-taking behavior (Cordella et al. 2018).

In models (2) and (5) we insert in Eq. (15.1) the NPLs characteristics variables and we find a different effect on SRISK and LRMES. First, the positive effect of a State-backed guarantee is incorporated only in the long–term implying that the involvement of the government in the management of NPLs acts as a mitigation mechanism for systemic risk in the long term (Broccardo and Mazzuca 2017; Bolognesi et al. 2020). Second, an increase in the number of NPLs securitizations leads to a systemic risk reduction, suggesting that the decision of a bank to manage the NPLs via securitizations is beneficial in terms of contribution to a systemic crisis. Furthermore, the types of impaired loans sold through securitization impact the systemic risk of EU banks differently. Specifically, the sale of consumer loans narrows the SRISK indicator more than the sale of Commercial Real Estate (CRE) loans. This difference in impact may be due to the guarantee that is associated with the CRE loans. They are a particular type of mortgage secured by a lien on a commercial property. On the contrary, the sale of residential loans narrows the LRMES indicator more than the sale of mixed loans. Finally, the results show country and size effects on the systemic risk. Specifically, the banks in PIIGS countries and designed as G-SIB banks show a higher LRMES than those in non-PIIGS countries and no G-SIB banks. Taken at face value, these results provide empirical evidence in favor of the view of (Laeven et al. 2016) that large banks in riskier countries pose excessive systemic risk. However, the excessive systemic risk is mitigated by the NPL's resolution plans. The issuance of NPLs securitization by G-SIB banks and banks in PIIGS countries would narrow the expected shortfall.

The country and size effects previously identified are absorbed by the banks and country-specific variables in columns (3) and (6). Indeed, among the control variables, the coefficients on size and sovereign CDS spreads are shown to be positive and statistically significant whereas the coefficients on PIIGS countries and G-SIB banks lose significance. This suggests that the contribution to the overall systemic risk of a big bank is greater than the contribution to the overall systemic risk of a small bank (Laeven et al. 2016) and the country risk is incorporated in the systemic

Table 15.3 System GMM estimates the impact of NPL securitization on systemic risk

	Srisk			LRMES		
	(1)	(2)	(3)	(4)	(5)	(6)
Lag of systemic risk	1.012*** (0.007)	1.015*** (0.010)	0.945*** (0.030)	0.583*** (0.041)	0.407*** (0.058)	0.119** (0.054)
NPL securitization	−0.081** (0.038)	−0.112*** (0.040)	−0.196** (0.085)	−0.586*** (0.095)	−0.316** (0.130)	−0.619*** (0.177)
NPL securitization square	0.004** (0.002)	0.005** (0.002)	0.008** (0.004)	0.026*** (0.004)	0.012** (0.006)	0.033*** (0.009)
Public guarantee		0.001 (0.011)	−0.006 (0.009)		−0.153*** (0.055)	−0.040** (0.017)
Number of deals		−0.063* (0.036)	−0.186** (0.074)		−0.262 (0.102)	−0.205* (0.120)
CRE loans		−0.483** (0.240)	−0.876** (0.382)		−1.104 (0.812)	0.221 (0.817)
Consumer loans		−0.876*** (0.313)	−1.571*** (0.542)		−0.631 (0.779)	1.357 (0.948)
Corporate loans		0.288 (0.278)	0.626 (0.392)		−1.040 (0.912)	−0.203*** (0.068)
Legacy loans		0.383 (0.786)	4.345 (3.539)		−0.468** (0.237)	−1.509 (1.817)
Mixed loans		−0.249 (0.322)	−0.701 (0.490)		0.069 (1.024)	−1.426** (0.695)
Residential loans		−0.610*** (0.218)	−0.136 (0.334)		−0.015 (0.731)	−1.816*** (0.579)
Shipping loans		0.262 (0.680)	−1.306 (1.310)		−1.818 (3.247)	−1.075 (3.687)
PIIGS countries		−1.309* (0.714)	−0.622 (0.618)		1.081*** (0.229)	−0.099 (1.380)
PIIGS countries* NPL securitization		−0.074*** (0.021)	−0.064** (0.027)		−0.056*** (0.020)	−0.265*** (0.024)
G-SIBs banks		3.557*** (1.276)	−0.591 (1.673)		0.182*** (0.037)	−0.079 (0.066)
G-SIBs banks*NPL securitization		−0.177*** (0.065)	−0.075** (0.036)		−1.023*** (0.187)	−0.491** (0.184)
Size $_{t-1}$			0.246** (0.106)			0.503*** (0.150)
Funding $_{t-1}$			−0.039*** (0.011)			−0.024 (0.021)
Leverage $_{t-1}$			0.005 (0.012)			0.041 (0.042)

(continued)

Table 15.3 (continued)

	Srisk			LRMES		
	(1)	(2)	(3)	(4)	(5)	(6)
Capital ratio$_{t-1}$			−0.023 (0.039)			−0.187*** (0.063)
NPL ratio$_{t-1}$			0.008 (0.012)			0.060** (0.028)
ROA$_{t-1}$			−0.143 (0.183)			−0.193 (0.286)
Price volatility $_{t-1}$			0.023* (0.012)			−0.014 (0.021)
Sovereign CDS spread$_{t-1}$			0.309** (0.157)			0.004** (0.002)
Intercept	0.012 (0.140)	0.353* (0.214)	−4.306* (2.579)	2.565*** (0.412)	4.041*** (0.684)	−2.475 (4.034)
Observations	626	626	626	626	626	626
AR2 test (p-value)	0.302	0.643	0.521	0.234	0.653	0.114
Hansen test (p-value)	0.543	0.875	0.876	0.832	0.843	0.622

This table reports the results of the model in Eq. (15.1). The dependent variable is bank Srisk (in %) in columns (1), (2) and (3) and LRMES (in %) in columns (4), (5) and (6). ***, **, and * indicate statistical significance at the 1%, 5% and 10% levels, respectively. The Hansen test reports p-values for the null hypothesis that the instruments used are not correlated with the error term. The Arellano and Bond (1991) test reports p-values for the null hypothesis that the errors in the first difference regression exhibit no second-order serial correlation

risk contribution of a bank to the overall systemic risk. Furthermore, a less stable funding and price volatility have a significant and positive relationship with the SRISK, suggesting that the stand-alone risk of a bank increases the systemic risk (Laeven et al. 2016) and the bank's funding structure based on retail deposits is a more stable source of funding able to lower the contribution to the overall systemic risk. Despite a less stable funding structure helping the recovery of the economy by supporting the credit expansion, it may also increase default risk and, thus, undermine financial stability (Shleifer and Vishny 2010). Finally, the coefficients on capital and NPL ratio impact the LRMES. Specifically, bank with a higher capital ratio has the financial resources to cover the fall of the market index by 40% over six months and banks with greater non-performing loans are more exposed to crises than banks with an unimpaired loans portfolio.

We estimate the vertex of the quadratic function in Eq. (15.1) based on the results in columns (3) and (6) of Table 15.3 for SRISK and LRMES, respectively. The estimated vertices, reported in Table 15.4 and compared to the average sample, underline different situations among the banks designed as G-SIB and based in PIIGS countries. Specifically, No G-SIB EU banks in both PIIGS and No PIIGS countries

Table 15.4 Estimated vertices

Scenarios	Conditions	Average sample	Estimated vertex (SRISK)	Estimated vertex (LRMES)
I scenario	PIIGS = 0 G-SIB = 0	20.51	$NPLs_{Sec^*} = -\frac{(-0.196)}{2*0.008} = 12.25$	$NPLs_{Sec^*} = -\frac{(-0.619)}{2*0.033} = 9.38$
II scenario	PIIGS = 1 G-SIB = 0	20.81	$NPLs_{Sec^*} = -\frac{(-0.196-0.064)}{2*0.008} = 16.25$	$NPLs_{Sec^*} = -\frac{(-0.619-0.265)}{2*0.033} = 13.39$
III scenario	PIIGS = 0 G-SIB = 1	19.71	$NPLs_{Sec^*} = -\frac{(-0.196-0.075)}{2*0.008} = 16.94$	$NPLs_{Sec^*} = -\frac{(-0.619-0.491)}{2*0.033} = 16.82$
IV scenario	PIIGS = 1 G-SIB = 1	20.81	$NPLs_{Sec^*} = -\frac{(-0.196-0.064-0.075)}{2*0.008} = 20.94$	$NPLs_{Sec^*} = -\frac{(-0.619-0.265-0.491)}{2*0.033} = 20.83$

This table provides the estimates of the vertex of the function in Eq. (15.1). We consider all possible scenarios

show no systemic risk benefits in conducting additional NPL securitizations. These banks derive no advantages from new issuances. This conclusion holds true for G-SIB banks in non-PIIGS countries as well. In contrast, the average scenario for G-SIB banks in PIIGS countries indicates that they have not fully utilized their capacity to cleanse their balance sheets of impaired loans. These banks have a margin within which they can accrue systemic risk benefits from the issuance of new NPL securitizations.

Robustness

To further investigate the effect of NPL securitizations on systemic risk, as in (Chiaramonte et al. 2013) and (Arif 2020), we measure the bank's risk by adopting the modified version of the Altman Z-score (Demirgüç-Kunt and Huizinga 2010). The modified version of the Altman Z-score is an accounting measure of bank solvency that reflects the distance to default, and it is measured as:

$$Z - score_{i,t} = \frac{ROA_{i,t} + CAR_{i,t}}{\delta_{ROAi,t}}$$

where ROA is the return on assets, CAR is the capital assets ratio and δ_{ROA} is the standard deviation of ROA. The modified version of Altman Z-score shows the number of standard deviations that a bank's rate of return of assets has to fall for the bank to become insolvent (Demirgüç-Kunt and Huizinga 2010). Higher z-score means that the firm is in a "safe" zone and, thus, the probability of default is low. A lower z-score suggests that the firm is in a "distress" zone implying an increase on the probability of default.

We report the results of the GMM model in Table 15.5. In column (1) we insert only the indicator of NPLs securitizations whereas, in columns (2) and (3) we add the NPLs-specific characteristics and the control variables, respectively. In all our specifications, the results confirm a quadratic relationship between the issuance of NPLs securitizations and the bank's risk. Specifically, the results of the quadratic model estimated show a positive and significant coefficient of the NPLs securitization indicator and a negative and significant coefficient of the square of the NPLs securitization indicator. These results suggest that the issuance of ABS linked to the NPLs securitizations may increase bank stability in the beginning but this relationship turns into a negative one when the ABS issuance is above a certain level in the bank. However, the maximum identified threshold of ABS is shifted for banks designed as G-SIB by the FSB. These results reinforce the earlier findings on the relationship between systemic risk and the use of NPLs securitizations.

The results in columns (2) and (3) of Table 15.5 show that the bank's risk incorporates the risk of ceded loans. The z-score of the bank improves when securitization involves consumer, corporate, and, only partially, mixed loans. By ceding riskier loans on the markets, the banks can obtain a reduction of bank risk because they reduce their exposure to credit risk by using the NPLs securitizations for capital relief purposes. Furthermore, the results provide evidence of the "too big to fail" concept. The positive and statistically significant coefficient on the G-SIB indicator suggests the existence of a size effect on the accounting measure of bank solvency. This size effect is absorbed by the size variable in the estimates reported in column (3). Finally, we find that the sovereign risk is incorporated into the bank's Z-score. Therefore, banks in risky countries experience greater insolvency.

The identification of a U-shape relationship between the issuance of NPL securitizations and the Z-score enables the determination of the optimal amount of NPL securitizations a bank can undertake to mitigate insolvency risk. The calculated vertices, as detailed in Table 15.6, are based on the findings from column (3) in Table 15.5. These estimated vertices indicate that banks in our sample have surpassed the maximum threshold of NPL securitization necessary for Z-score improvement. To derive risk-related benefits and prevent financial instability, only banks designated as G-SIB are advised to continue engaging in NPL securitizations.

Table 15.5 Robustness test: system GMM estimates the impact of NPL securitization on the bank's Z-score

	Z-score		
	(1)	(2)	(3)
Z-score$_{t-1}$	0.335***	0.323***	0.090*
	(0.056)	(0.048)	(0.052)
NPL securitization	0.532***	0.372***	0.292***
	(0.164)	(0.120)	(0.111)
NPL securitization square	−0.023***	−0.015***	−0.012**
	(0.007)	(0.006)	(0.005)
Public guarantee		0.068*	0.044
		(0.038)	(0.035)
Number of deals		−0.055	0.008
		(0.089)	(0.096)
CRE loans		0.105	−0.107
		(0.643)	(0.589)
Consumer loans		2.028***	0.927
		(0.622)	(0.595)
Corporate loans		0.869	1.462**
		(0.827)	(0.745)
Legacy loans		2.829	0.519
		(2.179)	(1.967)
Mixed loans		0.566	1.324*
		(0.887)	(0.798)
Residential loans		0.453	−0.463
		(0.672)	(0.621)
Shipping loans		−0.502	0.586
		(1.932)	(1.800)
PIIGS countries		−0.306	0.149
		(0.993)	(0.942)
PIIGS countries* NPL securitization		−0.008	−0.035
		(0.047)	(0.043)
G-SIBs banks		5.543*	0.649
		(3.216)	(3.225)
G-SIBs banks*NPL securitization		0.433***	0.488***
		(0.168)	(0.168)
Size$_{t-1}$			0.511***
			(0.109)
Funding$_{t-1}$			−0.008
			(0.013)
Leverage$_{t-1}$			0.011
			(0.015)
Capital ratio$_{t-1}$			−0.029
			(0.054)

(continued)

Table 15.5 (continued)

	Z-score		
	(1)	(2)	(3)
NPL ratio$_{t-1}$			0.001
			(0.018)
Price volatility$_{t-1}$			−0.026
			(0.016)
Sovereign CDS spread$_{t-1}$			−0.003**
			(0.001)
Intercept	0.045	0.509	11.401***
	(0.481)	(0.572)	(2.642)
Observations	446	446	446
AR2 test (p-value)	0.543	0.643	0.133
Hansen test (p-value)	0.895	0.721	0.241

This table reports the results of the model in Eq. (15.1). The dependent variable is the bank's Z-score (in %). ***, **, and * indicate statistical significance at the 1%, 5% and 10% levels, respectively. The Hansen test reports p-values for the null hypothesis that the instruments used are not correlated with the error term. The Arellano and Bond (1991) test reports p-values for the null hypothesis that the errors in the first difference regression exhibit no second-order serial correlation

Table 15.6 Robustness estimated vertices

Scenarios	Conditions	Average sample	Estimated vertex (Z-score)
I scenario	PIIGS = 0 G-SIB = 0	20.51	$NPLs_{Sec^*} = -\frac{(0.292)}{2*(-0.012)} = 12.17$
II scenario	PIIGS = 1 G-SIB = 0	20.81	$NPLs_{Sec^*} = -\frac{(0.292)}{2*(-0.012)} = 12.17$
III scenario	PIIGS = 0 G-SIB = 1	19.71	$NPLs_{Sec^*} = -\frac{(0.292+0.488)}{2*(-0.012)} = 32.50$
IV scenario	PIIGS = 1 G-SIB = 1	20.81	$NPLs_{Sec^*} = -\frac{(0.292+0.488)}{2*(-0.012)} = 32.50$

This table provides the estimates of the vertex of the function in Eq. (15.1). We consider all possible scenarios

Conclusion

The goal of this chapter is to investigate the relationship between the use of NPLs securitization and systemic risk. The main idea behind this chapter is to examine whether the use of NPL securitization increases the contribution to systemic risk and acts as a transmission mechanism channel for ongoing financial crises. Our purpose is to analyze the effect of NPL securitization on banks systemic risk of EU banks and the impact of NPLs characteristics on the systemic risk. For this purpose, we analyze a sample of EU banks over the period 2012–2020 by building on a unique dataset of banks which includes NPLs securitizations information, and bank and country-level data. We focus on banks due to the potentially systemic nature of these firms. We examine whether the effect of NPLs securitizations on a bank's systemic risk is quadratic and depends on the G-SIB designation and country risk by employing a panel of 35 banks from European countries.

Our results suggest that the issuance of NPLs securitization may provide benefits and drawbacks. Since we find evidence of a quadratic relationship between NPLs securitization and systemic risk, the huge involvement of a bank in NPLs securitizations may trigger a banking and financial crisis by acting as a transmission mechanism channel (Karim et al. 2013). The management of NPLs volume by adopting internal solutions is useful for strengthening financial stability and restructuring the banking sector. Despite the use of NPLs securitization preserves financial stability, the huge involvement in this instrument undermines financial stability. The heavy involvement of EU banks in NPLs securitization may be translated by the financial market as huge exposures of banks in distressed loans, resulting from an ex-ante wrong assessment of the credit risk and excessive risk-taking. Therefore, we identify the threshold below which the NPLs securitization is a good tool to transfer distressed loans to institutional investors but above this threshold, the issuance of an NPLs securitization is detrimental to the financial stability of EU banks. Furthermore, we find that there is a significant interaction effect of NPLs securitizations on PIIGS countries and G-SIB designation in the systemic risk. The country risk and the G-SIB designation shift the maximum amount of NPLs securitization that an EU bank can perform to minimize the systemic risk. The turning point for NPL securitizations identified by our model is lower than the average of NPL securitization in our sample in three scenarios. As a result, Regulatory authorities should adopt policies targeting EU banks that curtail the excessive use of NPL securitization to alleviate the concerns on the systemic risk. This result holds whether we consider alternative definitions of bank systemic risk, and when we control for potential differences in country characteristics and banks designation. According to the recommendations of ESRB (ESRB 2013), the key policy implication of our result is that actions aimed at reducing NPLs to sustain financial stability should enhance the control of the use of securitization instruments and encourage the use of internal workout measures to reduce the NPL volume of EU banks to avoid that NPLs securitizations become a transmission mechanism of financial crises. Furthermore, our results contribute to the ongoing debate on the important issue of designing suitable systemic risk indicators that act as EWS for predicting incoming financial crises. Given our findings, we believe that to have a complete vision of the contribution of a bank to the systemic risk, the indicators should take into account the bank's exposure to securitizations (in line with Mazzocchetti et al. 2020) but also the bank's NPL resolution plans.

Appendix

See Tables 15.7 and 15.8.

Table 15.7 Variable definition

	Description	Source
Panel A: Dependent variable-Bank's level systemic risk		
SRISK(%)	Systemic risk contribution of a financial firm to the overall systemic risk	V-Lab
LRMES	Decline in equity values to be expected if there is a financial crisis	V-Lab
Panel B: Variable of interest-NPLs securitizations		
Securitization of NPLs	NPLs securitization amount in the US $ million	Banks web site
Number of deals	Number of NPLs securitization performed by EU banks	Banks web site /Author calculation
Guarantee	Dummy variable taking value 1 if the securitization has a public guarantee, 0 otherwise	
Panel C: Control variables		
Banks size	Natural logarithm of total assets	Datastream
Funding structure	The ratio of deposits to total liabilities	
Leverage	The ratio of liabilities to the sum of liabilities and equity	
Capital adequacy	The ratio of Tier 1 capital to total risk-weighted assets	
NPL ratio	The ratio of non-performing loans to total loans	
ROA	Return on assets	
Stock price volatility	The quarterly variance of the bank's stock price	
Sovereign CDS spread	5-year sovereign CDS spreads in bps	

Table 15.8 Summary statistics on SRISK, LMRES and NPLs securitization for sample

Country	SRISK			LMRES			NPLs sec
	Mean (std. dev.)	Min	Max	Mean (std. dev.)	Min	Max	Gross book value (GBV)
Austria	49.73 (0.23)	49.57	49.89	47.13 (4.30)	44.09	50.17	815
Belgium	34.31 (15.67)	14.62	91.14	51.39 (11.42)	31.41	80.79	1,900
Denmark	84.08 (0.54)	83.7	84.46	40.97 (5.76)	36.89	45.04	420
France	29.71 (2.97)	18.65	35.04	52.49 (7.75)	32.69	76.2	1,890
Germany	37.65 (22.85)	16.38	60.79	54.88 (6.63)	47.29	64.03	1,743
Greece	24.33 (3.28)	19.06	30.27	46.06 (9.36)	36.12	74.06	9,150
Ireland	19.77 (32.43)	0.00	93.99	41.75 (11.79)	23.11	62.42	3,545
Italy	19.64 (11.11)	4.08	34.29	46.59 (5.56)	35.07	55.10	14,835
Netherlands	27.44 (30.07)	0.00	88.70	35.14 (15.70)	7.42	77.14	3,060
Norway	72.58 (5.30)	54.25	82.74	46.66 (8.92)	30.27	64.46	1,100
Portugal	57.19 (37.02)	14.94	100.00	34.70 (17.10)	13.86	59.09	1,598
Spain	22.42 (15.95)	4.79	49.98	47.54 (6.49)	36.59	59.85	4,870
UK	12.28 (6.34)	0.00	34.09	41.29 (8.67)	21.75	81.73	3,372
Total	25.98 (26.07)	0.00	100.00	46.98 (13.36)	7.42	81.73	48,298
PIIGS countries	21.38 (21.31)	0.00	100.00	47.74 (12.62)	13.86	74.06	33,998
Non-PIIGS countries	29.49 (28.69)	0.00	91.14	46.40 (13.87)	7.42	81.73	14,300

The table reports summary statistics on SRISK, LMRES and NPLs securitizations for the 36 sample banks over the period January 2012–September 2020. Mean, minimum (Min.) and maximum (Max.) of SRIRSK and LMRES are expressed in percentage. NPLs' securitization amount is expressed in USD millions

Source Datastream database, bank website and V-Lab in the NY Stern Business School

References

Acharya, V., Engle, R., Richardson, M.: Capital shortfall: a new approach to ranking and regulating systemic risks. Am. Econ. Rev. **102**, 59–64 (2012)

Acharya, V.V., Pedersen, L.H., Richardson, M.: Measuring systemic risk. Rev. Financ. Stud. **30**, 2–47 (2017). https://doi.org/10.1093/rfs/hhw088

Altunbas, Y., Marques-Ibanez, D., van Leuvensteijn, M., Zhao, T.: Market power and bank systemic risk: role of securitization and bank capital. J. Bank Financ. (2022). https://doi.org/10.1016/j.jbankfin.2022.106451

Arellano, M., Bond, S.: Some tests of specification for panel Carlo application to data: evidence and an employment equations. Rev. Econ. Stud. **58**, 277–297 (1991)

Arellano, M., Bover, O.: Another look at the instrumental variable estimation of error-components models. J. Econ. **68**, 29–51 (1995)

Arif, A.: Effects of securitization and covered bonds on bank stability. Res. Int. Bus. Finance **53**, 101196 (2020). https://doi.org/10.1016/j.ribaf.2020.101196

Bakkar, Y., Nyola, A.P.: Internationalization, foreign complexity and systemic risk: evidence from European banks. J. Financ. Stab. **55**, 100892 (2021). https://doi.org/10.1016/j.jfs.2021.100892

Battaglia, F., Gallo, A.: Securitization and systemic risk: an empirical investigation on Italian banks over the financial crisis. Int. Rev. Financ. Anal. **30**, 274–286 (2013)

Bolognesi, E., Compagno, C., Miani, S., Tasca, R.: Non-performing loans and the cost of deleveraging: the Italian experience. J. Account. Public Policy (2020). https://doi.org/10.1016/j.jaccpubpol.2020.106786

Bostandzic, D., Weiß, G.N.F.: Why do some banks contribute more to global systemic risk? J. Financ. Intermediat. **35**, 17–40 (2018). https://doi.org/10.1016/j.jfi.2018.03.003

Broccardo, E., Mazzuca, M.: New development: can ' public ' market-based solutions restore the banking system ? The case of non-per-forming loans (NPLs). Public Money Manage. **37**, 515–520 (2017). https://doi.org/10.1080/09540962.2017.1338430

Brunnermeier, M.K., Dong, G.N., Palia, D.: Banks' noninterest income and systemic risk. Rev. Corp. Finance Stud. **9**, 229–255 (2020). https://doi.org/10.1093/rcfs/cfaa006

Cebenoyan, A.S., Strahan, P.E.: Risk management, capital structure and lending at banks. J. Bank Financ. **28**, 19–43 (2004). https://doi.org/10.1016/S0378-4266(02)00391-6

Chiaramonte, L., Poli, F., Oriani, M.E.: Are cooperative banks a lever for promoting bank stability? Evidence from the recent financial crisis in OECD countries. Eur. Financ. Manag. **21**, 1–33 (2013). https://doi.org/10.1111/j.1468-036X.2013.12026.x

Chiesa, G.: Optimal credit risk transfer, monitored finance, and banks. J. Financ. Intermediat. **17**, 464–477 (2008). https://doi.org/10.1016/j.jfi.2008.07.003

Cordella, T., Dell'Ariccia, G., Marquez, R.: Government guarantees, transparency, and bank risk-taking. IMF Econ. Rev. **66** (2018)

Demirgüç-Kunt, A., Huizinga, H.: Bank activity and funding strategies: the impact on risk and returns. J. Financ. Econ. **98**, 626–650 (2010). https://doi.org/10.1016/j.jfineco.2010.06.004

Diamond, D.W.: Financial intermediation and delegated monitoring. Rev. Econ. Stud. **51**, 393–414 (1984)

Duffie, D.: Innovations in credit risk transfer: implications for financial stability (2008)

EBA: Draft Guidelines on disclosure of non-performing and for-borne exposures (2018)

EBA: EBA report on NPLs: progress made and challenges ahead (2019)

ESRB: Recommendation of the European Systemic Risk Board of 4 April 2013 on intermediate objectives and instruments of macro-prudential policy (2013)

Fell, J., Moldovan, C., O'Brien, E.: Resolving non-performing loans: a role for securitisation and other financial structures ? Financ. Stab. Rev. 158–174 (2017)

Greenspan, A.: Risk Transfer and Financial Stability. Federal Reserve Bank of Chicago Proceedings, p. 968 (2005)

Grodzicki, M., Laliotis, D., Leber, M., et al.: Resolving the legacy of non-performing exposures in euro area banks. Financ. Stab. Rev. **1**, 146–154 (2015)

Halili, A., Fenech, J., Contessi, S.: Credit derivatives and bank systemic risk: risk enhancing or reducing ? Financ. Res. Lett. (2021)

Ivanov, K., Jiang, J.: Does securitization escalate banks' sensitivity to systemic risk? J. Risk Finance **21**, 1–22 (2020). https://doi.org/10.1108/JRF-12-2018-0184

Kara, A., Ozkan, A., Altunbas, Y.: Securitisation and banking risk: what do we know so far? Rev. Behav. Finance **8**, 2–16 (2016). https://doi.org/10.1108/RBF-07-2014-0039

Karim, D., Liadze, I., Barrell, R., Davis, E.P.: Off-balance sheet exposures and banking crises in OECD countries. J. Financ. Stab. **9**, 673–681 (2013). https://doi.org/10.1016/j.jfs.2012.07.001

Keys, B.J., Mukherjee, T., Seru, A., Vi, V.: Did securitization lead to lax screening? Evidence from subprime loans. Quart. J. Econ. **125**, 307–362 (2010)

Laeven, L., Ratnovski, L., Tong, H.: Bank size, capital, and systemic risk: Some international evidence. J. Bank Financ. **69**, S25–S34 (2016). https://doi.org/10.1016/j.jbankfin.2015.06.022

Loutskina, E.: The role of securitization in bank liquidity and funding management. J. Financ. Econ. **100**, 663–684 (2011). https://doi.org/10.1016/j.jfineco.2011.02.005

Mayordomo, S., Rodriguez-Moreno, M., Peña, J.I.: Derivatives holdings and systemic risk in the U.S. banking sector. J. Bank Financ. **45**, 84–104 (2014). https://doi.org/10.1016/j.jbankfin.2014.03.037

Mazzocchetti, A., Lauretta, E., Raberto, M., et al.: Systemic financial risk indicators and securitised assets: an agent-based framework. J. Econ. Interact. Coord. **15**, 9–47 (2020). https://doi.org/10.1007/s11403-019-00268-z

Minton, B.A., Stulz, R., Williamson, R.: How much do banks use credit derivatives to hedge loans? J. Financ. Serv. Res. **35**, 1–31 (2009). https://doi.org/10.1007/s10693-008-0046-3

Pedisic, R.: The effects of non-performing loans reduction measures on systemic risk in European. J. Appl. Econ. Bus. **7**, 5–22 (2019)

Shleifer, A., Vishny, R.W.: Unstable banking. J. Financ. Econ. **97**, 306–318 (2010). https://doi.org/10.1016/j.jfineco.2009.10.007

Vuković, V., Domazet, I.: Non-performing loans and systemic risk: comparative analysis of Serbia and countries in transition CESEE. Industrija **41**, 59–73 (2013)

Wagner, W., Marsh, I.W.: Credit risk transfer and financial sector stability. J. Financ Stab **2**, 173–193 (2006). https://doi.org/10.1016/j.jfs.2005.11.001

Chapter 16
The Systemic Importance of Cyber Risk in Banks

Giuliana Birindelli and Antonia Patrizia Iannuzzi

Abstract This chapter aims to analyse cyber risk with specific regard to the banking and financial sector by highlighting the progress made in academic studies (section "The Cyber Risk in Banks: A Literature Review"), the systemic impacts of this risk (section "Cyber Risk as a Systemic Risk") as well as the point of view of supervisory authorities (section "The Point of View of Financial Regulators"). At the end of the chapter, data provided by the ORBIS database are reported in order to understand current bank (and other financial institutions) exposure to this new and sophisticated risk (section "Banks' Exposure to Cyber Risk: Some Empirical Evidence").

Keywords Cyber risk · Systemic cyber risk · Regulation · Banks

Cyber Risk: Definition and Implications

Despite the growing amount of research focusing on cyber risk, scholarly definitions of the issue are relatively limited. There are several reasons for this and one of them is the complexity. Cyber risk is an interdisciplinary problem involving different research areas, from Information Technology (IT) to information security, from business to finance, and economics (Eling and Wirfs 2019). Awiszus et al. (2023) point out that the term "cyber" risk comprises many different types of risk with different root causes and types of impact. This makes it difficult to create a single definition that encompasses all the causes and effects of this risk. Cyber risk is also dynamic in nature. The continuous digital innovations, the increased use of internet-enabled devices and the ongoing sophistication of hackers have disrupted traditional business

G. Birindelli
Full Professor of Financial Markets and Institutions, Department of Economics and Management, University of Pisa, Pisa, Italy
e-mail: giuliana.birindelli@unipi.it

A. P. Iannuzzi (✉)
Associate Professor of Financial Markets and Institutions, Department of Economics, Management and Business Law, University of Bari, Bari, Italy
e-mail: antoniapatrizia.iannuzzi@uniba.it

© The Author(s) 2025 301
V. Pacelli (ed.), *Systemic Risk and Complex Networks in Modern Financial Systems*, New Economic Windows, https://doi.org/10.1007/978-3-031-64916-5_16

models, making it difficult to define cyber risk and identify the boundaries (Sheehan et al. 2021; Curti et al. 2023). Cyber threats and cybersecurity are changing very rapidly (Strupczewski 2021). At present, these issues go beyond data breaches and privacy. Indeed, the more sophisticated cyberattacks are able to block entire countries, industries, businesses and supply chains as highlighted by recent business reports (Allianz 2021; WEF 2020). The recent annual Global Risks Report from the World Economic Forum ranks malicious cyber incidents in the top five risks in terms of both likelihood and impact severity (WEF 2020).

In the banking sector, cyber risk constitutes an operational risk event, and though it is often a minor part of loss events, the consequences are very severe and pervasive (Eling and Wirfs 2019). It is generally defined as an *"operational risk to information and technology assets that have consequences affecting the confidentiality, availability, and/or integrity of information or information systems"* (Cebula et al. 2014: 1). Human behaviour (whether criminal or not) appears to be the main source of cyber risk which by nature is highly complex and heterogeneous: indeed, the time trends, the impact of company size, business sector, and the diffusion vary by cyber risk type (Bouveret 2018; Malavasi et al. 2022). Likewise, Strupczewski emphasizes how cyber risk constitutes a distinct risk category where more research is warranted, given the rising importance for the economy and society (ESRB 2022, 2023). After collecting 20 different definitions, the Author proposes his broad definition of cyber risk to include the three key components, i.e. sources, objects, and economic impacts (Strupczewski 2021).

Prominent cyber risk events for financial institutions include data breaches and cyberattacks (Agrafiotis et al. 2018), even if, over the last few years, these events have become much more diversified, ranging from distributed denial-of-service (DDoS) attacks, extortion, and fraud to the widescale exploitation of key financial infrastructure (Kopp et al. 2017; Bouveret 2018; ESRB 2022). This intensified frequency of cyber-attacks on banking institutions is linked to the increasing use of electronic data and the increasing shift to digital processes and service offerings, following the extensive lockdowns caused by COVID-19 (Frost and Shapiro 2021). These changes have made financial services companies particularly vulnerable, becoming one of the key targets for cybercriminals (Sheehan et al. 2021).

Cyber risks represent, thus, one of the biggest threats to the banking industry, resulting in important financial and organizational implications (Aldasoro et al. 2022; Bouveret 2018). Firstly, it is a very costly risk, which can result in considerable economic losses for a bank. A data breach costs a company on average $3.92MM per breach, according to the 2020 IBM security report (Ponemon Institute 2020). Considering the entire economy, Statista (2023) estimated the global cost of cybercrime in 2022 at $8.4 trillion and expects this to go beyond $11 trillion in 2023. Therefore, the cost of cybercrime had an annual growth rate of 30% during the years 2021–23. The average cost of a data breach between 2020 and 2022 increased by 13%, where the financial industry suffered the second highest average cost after healthcare (Statista 2023). Focusing on the financial sector, the IMF estimates the global average aggregated annual losses due to cyber-attacks at 9% of banks' net income, or around $100 billion. In a worst case scenario, losses could be 2½ to 3½ times higher, reaching between $270 billion and $350 billion (Lagarde 2018).

The damage to an institution's reputation resulting from a cyber-attack, which can compromise bank business continuity, has to be added to the economic costs of business interruptions (Sinanaj and Muntermann 2013; Sheehan et al. 2021; Mangala and Soni 2023). Cyber fraud also harms customers' trust towards the entire banking network (Akinbowale et al. 2020; Creado and Ramteke 2020). Creado and Ramteke (2020) recommended banking institutions make appropriate investments in cyber security and upgrade their security measures to cope with sophisticated cyber threats. Unfortunately, the availability of data on cyber risks remains limited.

Several scholars highlight this critical problem (Cremer et al. 2022). Data on cyber events and cyber losses is scarce and, for many different reasons, is not usually granular enough (Bouveret 2018), primarily because cyber risk is an emerging and evolving risk; therefore, historical data sources are still incomplete (Biener et al. 2015).[1]

Additionally, institutions that have been hacked are reluctant to publish the incidents to avoid further damage to their reputation (Sinanaj and Muntermann 2013; Eling and Schnell 2016). Finally, there is an *"aggregate cyber risks"* issue due to shared IT architectures or complex interconnections that are hard to untangle (Awiszus et al. 2023). Given the role of banks in global financial markets, there is growing concern for the potential impact of cyber breaches, not only on a specific bank but also on the entire economy and stability of the wider financial system (Santucci 2018; Berger et al. 2022; FSB 2023; besides, see paragraph 3). Banks and other financial institutions are critical to ensuring liquidity, payments and settlements, to guarantee the money supply in the economy, and to provide loans, savings and deposits. For these reasons, the impact of cyber-attacks against financial institutions can cause severe and pervasive effects (Gulyás and Kiss 2023). Unfortunately, cyber risk and its aggregate impacts cannot be eliminated, but they can be mitigated and managed. How the banking industry can best assess and manage cyber risks is therefore a topic of increasing interest in the financial, economic, and information systems literature (Akinbowale et al. 2020; Pollmeier et al. 2023).

This chapter will analyse cyber risk within the banking and financial sector by highlighting the progress made in academic studies (section "The Cyber Risk in Banks: A Literature Review"), the systemic impacts of this risk (section "Cyber Risk As a Systemic Risk") as well as the point of view of supervisory authorities (section "The Point of View of Financial Regulators"). At the end of the chapter, data

---

[1] In turn, the lack of historical data on cyber risk also creates considerable difficulties for the risk management and insurance process. Several scholars shed light this criticality (Eling 2020; Nurse et al. 2020). A greater quantity of cyber data could further the understanding and measurement of cyber risk, and thus make it more insurable (Eling 2020; Cremer et al. 2022). This is an important challenge, common to many areas, such as research, risk management and cybersecurity (Boyer and Eling 2023). Regulators also point out the importance of this topic: in April 2021 the European Council announced that a Centre of excellence for cybersecurity will be established to pool investments in research, technology and industrial development. For more information, see: European Cybersecurity Competence Centre and Network (europa.eu).

provided by the ORBIS database are reported in order to understand current banks' (and other financial institutions) exposure to this new and sophisticated risk (section "Banks' Exposure to Cyber Risk: Some Empirical Evidence").

The Cyber Risk in Banks: A Literature Review

The analysis of cyber risk in banks represents an emerging and relatively unexplored topic. Despite financial institutions tending to maintain better data collection practices due to regulatory reporting, data on cyber incidents are still limited and thus academic research on the impacts of cyber events is still in an early stage (Aldasoro et al. 2020). Cyber incidents are becoming more complex and sophisticated and this hinders their investigation (Aldasoro et al. 2022; FSB 2023). Some important trends/ evidence, however, can be identified from the existing studies.

First, it is now widely recognized that cyber risk in the banking sector is continuously growing both in terms of frequency and economic impact magnitude (Skinner 2019; Aldasoro et al. 2022; FSB 2023). This aspect is highlighted by Aldasoro et al. (2020) who, focusing on over 70 international banks, document a prominent increase in losses due to cyber events from 2002 to 2018, with a corresponding increase in risk. The largest losses occurred in 2016, which is then followed by a decrease likely aided by increased attention and investments in cyber risk management. The economic impact of cyber incidents is substantial, even though it is a relatively minor share of operational losses. Cyber losses can account for up to a third of total operational VaR value-at-risk. More in detail, the cyber value-at-risk (CVaR) is estimated to be between 0.2 and 4.2% of banks' income (Aldasoro et al. 2020). Similarly, Bouveret (2018), assessing cyber risk for the entire banking sector through historical loss data and the Basel II AMA methodology from Basel II, estimates an average loss due to cyber-attacks for the countries in the ORX database is equal to 9% of banks net income. The VaR would range between USD 147 and 201 billion (14–19% of net income) and the expected shortfall between USD 187 and 281 billion. These estimates show significantly higher potential aggregated losses in the financial sector, than the publicly reported losses by financial institutions in these jurisdictions.

Corresponding to this increased frequency and relevance of cyber risk (Akinbowale et al. 2020), the perception of this risk by financial firms is also growing. This not only leads to better cyber risk management but also a positive impact on firm value measured by Tobin's Q. These results are documented by Gatzert and Schubert (2022), based on a cyber risk consciousness score—assembled by Authors—that is applied to annual reports of large- and mid-cap US banks and insurers from 2011 to 2018 using a text mining algorithm.

Other studies focus on the effects and consequences caused by cyber risk. Some Authors analyse the relationships between this risk and specific bank accounting measures; other research focuses on economic and reputational impacts. Among the former, there is the study by Boungou (2023) who, using data from 2144 U.S. banks over the period 2011–2019, reveals a reduction in deposits in response to cyber-attacks. In turn, banks appear to have less incentive to lend for the long term. In sum, the cyber-attacks lead banks to reduce the maturity of their loan portfolios and this impact seems to be stronger for low-deposits, small, and less capitalized banks. Focusing on the same geographical area (U.S. commercial banks, from 2004 to 2019), Jin et al. (2023) show the likelihood of a cyber-attack leads banks to increase loan loss provisions. Thus, this accounting item would have a predictive value by highlighting banks most exposed to cyber risk. This occurs because discretionary loan loss provisions represent an indicator of banks' internal control weakness.

The studies focused on the economic and non-financial impacts of cyber risk are carried out by Pollmeier et al. (2023) and Akinbowale et al. (2023). Both pieces of research adopt an exploratory approach using structured questionnaires administered, respectively, to an Australian bank and staff of 17 South African banks. Both studies point out how a cyber-attack entails high costs for a bank, leading to both significant losses in reputation and customer confidence (Pollmeier et al. 2023; Akinbowale et al. 2023) and a decrease in profitability, productivity and risk management skills (Akinbowale et al. 2023). These negative effects could be mitigated if the bank is not held directly responsible for the cyber-attack (Mikhed and Vogan 2018).

In addition to causing significant operational losses and reputational damage, cyber risk can also adversely affect the bank's stability (Uddin et al. 2020b). Based on 354 banks from 43 countries from 2008 to 2017, Uddin et al. (2020a) find a marginal increase in CyberTech spending above a certain threshold decreases the stability of banks, especially in technologically mature countries. This occurs because excessive spending on information technology implies taking on high cyber risk by banks, which is followed by a deterioration of financial stability. Thus, CyberTech spending can improve bank stability up to a certain threshold, beyond which, the impact becomes negative.

At the same time, however, the financial sector is demonstrating increasing resilience to cyber-attacks. Even with a growth in the pervasiveness of cyber events (Uddin et al. 2020b), Aldasoro et al. (2022) find that U.S. banks experience lower costs, on average, than firms belonging to other sectors. This resilience is likely due to higher investments in IT by the financial sector. The size of financial institutions is also positively linked with cyber resilience. Duffie and Younger (2019) examine 12 systemically important U.S. financial institutions to estimate their exposure to a "cyber run" including the likely propagation dynamics on the entire payment system. They find that these banks hold sufficient stocks of high-quality liquid assets to cover wholesale funding run-offs in a relatively extreme cyber event.

Considering the importance of cybersecurity for the efficiency of the capital market, another strand of studies explores cyber-risk disclosure and reporting. Hence, stakeholders are increasingly asking companies for information on how they identify, measure, and manage cyber risk (Mazumder and Hossain 2023). There is less research focused on banks, but it is increasing (Bakker and Streff 2016). Overall, these studies show improved and increased cyber risk disclosure by banks, especially in emerging economies (Mazumder and Hossain 2023) and as a result of external cyber-attacks (Mazumder and Sobhan 2021; Firoozi and Mohsni 2023); a positive linkage between board diversity (especially, female and independent directors) and cyber security reporting (Mazumder and Hossain 2023) and, finally, how the cyber security mandatory disclosure regime still appears inadequate in addressing the stake-holders' interests (Skinner 2019).[2] The regulation of cyber risk disclosure is another relevant topic. Indeed, both Aldasoro et al. (2020) and (2022) note that as the quality of financial regulation and supervision of cyber risk management improves, banks suffer minor losses in terms of both frequency and amount. Similarly, An et al. (2021) find institutional quality, such as the protection of investor rights and the function of the legal system, mitigates the negative impact of cybersecurity risks on raising capital through initial coin offerings (ICOs). Others reveal that cyber risk disclosure by banks is affected by proprietary costs (Firoozi and Mohsni 2023). While during external threats there is an overall improvement in the various subcategories (Risk, Impact, Governance, Mitigation, Incident and Other), during in regular times banks are more likely to disclose information on risk and potential impact, and less on mitigation and governance. This divergence is probably due to avoiding costs related to the disclosure of company-sensitive information that may be used by hackers. Therefore, finding the right balance is critical in cybersecurity disclosure (Firoozi and Mohsni 2023).

Ultimately, how can banks protect themselves from cyber risk? Undoubtedly it is crucial to increase investment in cyber security, which, as noted by several authors (Eling et al. 2021; Aldasoro et al. 2022; FSI 2023), leads to an increase in cyber resilience. The size of the company seems to be more controversial. While larger banks would seem to be better equipped to deal with this risk as they have more economic and financial resources (Duffie and Younger 2019), greater size could intensify the potential impacts of cyber risk and be a contagion tool. Eling and Jung (2022), using data from the world's largest database of operational risk (SAS OpRisk database), document that the contagion effects of cyber risk can cause multiple losses from a single event and nearly double the loss estimates. Therefore, cyber risk management must be aligned not only with firm size but also with the level of interconnectivity of the bank. To fully cover potential extreme cyber losses larger banks should respect stricter capital requirements.

---

[2] The Smaili et al. (2023) study also highlights that board independence and financial expertise significantly increase the disclosure of cybersecurity risks in the company's financial statements. However, this study does not analyse the banking sector, but is based on 60 largest Canadian companies listed on the Toronto Stock Exchange from 2014 to 2018. Similarly, Radu and Smaili (2022), focusing on a sample of the companies listed on the S&P/TSX 60 Index over the period 2014–2018, find the presence of women on boards has a positive and significant effect on the level of cybersecurity disclosure, but only if the board has a critical mass of at least three women. To proxy the cybersecurity disclosure, Radu and Smaili (2022) use three alternative measures: (1) a dummy variable coded 1 for firms presenting cybersecurity-related disclosure in the annual report and 0 otherwise; (2) the number of words in the cybersecurity disclosure and, finally, (3) the number of paragraphs in the cybersecurity disclosure.

Cyber Risk as a Systemic Risk

As a consequence of the 2008 financial crisis, systemic risk was intensively studied in terms of relevant implications for the stability of financial institutions and the financial system as a whole (Caruana 2010; Harum and Gunadi 2022). This concept is also important in the context of cyber risk, since agents and organizations in cyber systems are strongly interconnected, for example within IT networks or via business contacts. Whether cyber-attacks remain on the level of specific institutions or are spread to other industries and critical infrastructures is currently a much-debated issue (Bouveret 2018).

Like the financial sector, cyberspace presents a system of heavily interdependent organizations connected through network ties (Welburn and Strong 2022). Meanwhile, banks and other financial institutions appear to be increasingly exposed to systemic risk of technology as a consequence of the widespread use of IT (Uddin et al. 2020b) and their reliance on cloud technologies (Aldasoro et al. 2022). Increasingly, a single breach in a banking network can have spillover effects on other banks and financial firms (Baldwin et al. 2017; Eisenbach et al. 2022; Crosignani et al. 2020) and thus shake the entire financial system (FSB 2023). This might happen through three key channels: the potential loss of data integrity, the lack of substitutability for services from key financial institutions and, finally, the plausible loss of confidence (OFR 2017). The COVID-19 pandemic and the war in Ukraine have now strongly exacerbated these effects (Adelmann et al. 2020; Aldasoro et al. 2020; ERSB 2023). Likewise, a cyber-attack on payment clearing or settlement systems— that are essential to the financial services industry—can significantly compromise the functionality of financial markets by impeding credit and financial intermediation processes (Eisenbach et al. 2022) with prominent systemic consequences (Gulyás and Kiss 2023). *"In the context of an increasingly interconnected financial ecosystem, an attack on one or more institutions or critical infrastructures can have significant ripple effects"* (DTCC and Oliver Wyman 2018: 5). These events might even be viewed as a matter of national security. Thus, in the financial sector, cyber risks are compared to a "known unknown" tail risk, one of the major threats to financial stability. The benefits of a heavily interconnected world through cyberspace have increasingly come with a cost (Aldasoro et al. 2020). Cyber incidents have risen both in prevalence and significance in their disruptions to individuals, businesses, and governments (Welburn and Strong 2022). According to the last DTCC Systemic Risk Barometer, cyber risk is among the top five systemic risks to the broader economy (Fig. 16.1). Additionally, the respondents of the latest Bank of England' systemic risk survey cite cyber risk as highest perceived source of systemic risk to the financial system (Bank of England 2023a).

These considerations have given rise to the expression *"systemic cyber risk"* as the deliberate combination of two fields, systemic risk and cyber risk. *Systemic cyber risk* is thus an emerging form of cyber risk (Welburn and Strong 2022) deemed important by leading regulatory and macroprudential institutions (WEF 2016; ESRB 2020a, 2022). In particular, the ESRB recently stated that a cyber incident could

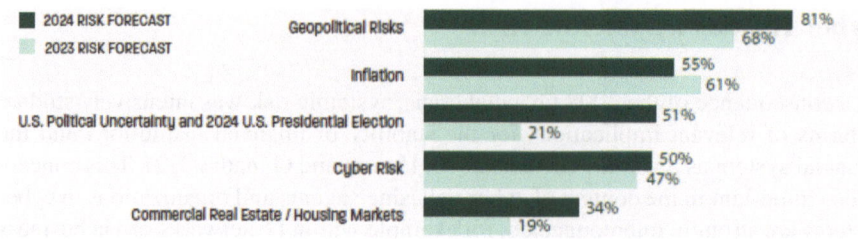

Fig. 16.1 Risks to broader economy. *Source* DTCC Systemic Risk Barometer Survey 2024

affect operational systems in the financial system and impair the provision of critical economic functions, trigger financial contagion or lead to an erosion of confidence in the financial system. If the financial system is not able to absorb these shocks, financial stability is likely at risk and a systemic cyber crisis could unfold (ESRB 2020b). Similarly, the Financial Stability Board (FSB 2023) and Financial Stability Institute (FSI 2023) assign cyber resilience a top priority for the financial services industry and a key area of attention for financial authorities (BCBS 2018). Indeed, cyber incidents pose a significant threat to the stability of the financial system and the global economy. The financial system performs several key activities that support the real economy (e.g., deposit taking, lending, payments and settlement services). Cyber incidents can disrupt the information and communication technologies that support these activities (FSB 2023; ESRB 2023).

In recent times, effort has been made to reach a common definition of the term "*systemic cyber risk*". Awiszus et al. (2023) classify cyber risks into three different types: idiosyncratic, systematic, and systemic cyber risks. While the idiosyncratic risks refer to cyber risks at the level of individual policyholders that are independent from other risks, systematic risks result from shared vulnerabilities affecting different firms at the same time, e.g., firms belonging to the same industry sector or region, or firms that utilise the same software, server, or information system. Conversely, systemic cyber risk originates from interconnected systems or strategic interactions triggering local or global contagion effects following a cyber-attack.

The World Economic Forum proposes a more structured definition. Specifically, it defines the *systemic cyber risk* as follows (WEF 2016: 5):

> Systemic cyber risk is the risk that a cyber event (attack(s) or other adverse event(s)) at an individual component of a critical infrastructure ecosystem will cause significant delay, denial, breakdown, disruption or loss, such that services are impacted not only in the originating component but consequences also cascade into related (logically and/or geographically) ecosystem components, resulting in significant adverse effects to public health or safety, economic security or national security. The adverse real economic, safety and security effects from realized systemic risk are generally seen as arising from significant disruptions to the trust in or certainty about services and/or critical data (i.e. the integrity of data), the disruption of operations and, potentially, the incapacitation or destruction of physical assets.

Summarizing this definition, it is possible to identify the following distinctive characteristics of *systemic cyber risk*. It is a risk that: (1) affects entire systems, not just individual parts or components; (2) has widespread and unexpected consequences;

and (3) can accumulate over time resulting in large aggregate effects. In the same vein, the ESRB defines systemic cyber risk as the risk of disruption to financial services that is (1) caused by an impairment of all or part of the financial system following a cyber incident, (2) has the potential to result in serious negative consequences for the real economy (ESRB 2020a).[3]

More recently, academic research has focused on identifying the potential impacts of systemic cyber incidents and how a cyber-attack may be amplified through the system. In this context, Welburn and Strong (2022) distinguish between the broader term of systemic cyber risk, encompassing probability, uncertainty and consequence, and the more specific systemic cyber failure. Eisenbach et al. (2022), using a wide network of U.S. banks, show the impairment of any of the five most active banks can result in significant spillovers to other banks, with 38% of the network affected on average. Thus, the concern that a catastrophic cyber event could cause systemic collapse in the financial industry is tangible and concrete. Zhang et al. (2023), focusing on Chinese-listed commercial banks from 2010 to 2019, find the digitalization of banks driven by the development of financial technology reduces systemic risk. In particular, bank digitization reduces liquidity risk and increases non-interest income by promoting financial innovation. In turn, these effects reduce bank systemic risk. Finally, Tian et al. (2023) offer new evidence on how systemic cyber risks modify the central banks' views on CBDC (central bank digital currency). More in detail, there is a drop in sentiment by central banks toward CBDC (central bank digital currency) following cyber-attacks as potential systemic risks to the country's financial system are uncovered.[4]

Others estimate the cumulative costs of aggregate cyber incidents. Cyber value-at-risk models, borrowed from the Value-at-risk (Var) techniques popular within financial risk management, can be used to estimate likely cyber incident-driven losses over a specific period of time (WEF and Deloitte 2015).[5]

In sum, it is clear that aggregate cyber risks represent an important and current challenge that needs to be addressed at the global level, through combined actions (Awiszus et al. 2023). Firstly, data on how cyber risk spreads should be more carefully collected. Models and assessment methods of the contagion process need to be improved. Also, the interaction effects in cyber models should be analysed not only using static frameworks but also dynamic strategic interaction. Multilayered

---

[3] Additionally, the ESRB specifies that *"while a serious cyber incident may cause system-wide disruption, this need not lead to a systemic event. However, a systemic event could occur when the system no longer has the capacity to absorb the shock and recover so that it can continue to provide key economic functions"* (ESRB 2020b: 53).

[4] Central Banks are also prominent targets of cyber-attacks. For example, on February 2016, hackers broke into Bangladesh Bank and hacked its credentials to send payment messages over the SWIFT network. They stole $81 million. In the same year (December 2016), Russia's central bank reported that hackers stole about $31 million during the year from its correspondent banks. These incidents showed the finance world that systemic cyber risks had been severely underestimated (Gulyás and Kiss 2023).

[5] In particular, using the cyber value at-risk approach, a study by Deloitte estimated an annual expected loss of €10 billion to the Dutch economy, or 1.5% of GDP (Deloitte 2016).

networks used to understand the degree of interdependence of financial operations from digital technology is another important research field to be explored. Finally, a specific approach for pricing *systemic cyber risk* should be developed.

In addition, "cyber stress tests" should be used as a number of regulators are considering implementing. For example, the Office of Financial Research (OFR) in 2017 suggested the use of "cyber stress tests" as a regulatory tool. In the same year, the Federal Financial Institutions Examination Council (FFIEC) included the use of stress tests in its cyber maturity model. More recently, the Bank of England (BoE) conducted an exploratory test through its "cyber stress test" to verify the financial institutions' capability to operationally absorb a cyber incident within a defined timeframe and to continue services without material economic impact. In March 2021, the Financial Policy Committee (FPC) of BoE identified the impact tolerance and connected it to the ability of the financial system to make payments on the date they are due (Bank of England 2023b). Based on this experience, in December 2021, the Prudential Regulation Authority (PRA) announced its plans to invite a number of systemic, as well as smaller firms, to participate in a voluntary cyber stress test which would focus on a severe but plausible data integrity scenario on a retail payment system.

Cyber stress tests are a relatively new tool that needs to be still refined. Many other challenges and approaches exist (Welburn and Strong 2022). Addressing them will help make the cyber landscape more resilient and secure in the future, although the dynamic nature of cyber risk could impede a more definitive protection and solution (Dupont 2019; Awiszus et al. 2023).

The Point of View of Financial Regulators

Coinciding with the growing academic focus on systemic cyber risk, financial regulators have also started providing valuable input to enable a better understanding of the dynamics of this risk, its impacts and financial stability implications (Anand et al. 2022). The primary contributions come from the ESRB (European Systemic Risk Board) and the FSB (Financial Stability Board). In October 2017, the European Systemic Cyber Group (ESCG)—a dedicated group established by ESRB—developed a conceptual model for systemic cyber risk aimed at identifying system-wide vulnerabilities of cyber incidents, describing the macro-financial implications and suggesting system-wide actions that could act as systemic mitigants of cyber risks (ESRB 2020b). The systemic cyber risk model consists of four different phases:

Context, aimed at identifying and describing the circumstances in which a cyber incident arises.

Shock, aimed at describing the technical and business impacts experienced when the cyber incident originates. While the technical impacts are linked to the immediate effects of disruption on the assets affected (i.e., loss of confidentiality, integrity and availability of information and/or information systems), there are different types of business impacts (*Financial, Reputational, Legal and Regulatory, Business objectives, Operational, Environmental, Human*, see Table 16.1).

Amplification, aimed at describing the systemic amplifiers and contagion channels which exacerbate the shock by increasing the extent of the cyber impacts. The characteristics of the financial system that make it susceptible to systemic disruption arising from a cyber incident are a high degree of interdependence, lack of transparency, reliance on data and reliance on confidence.

Systemic event, identifies the point at which a cyber shock leads to a systemic event. This event occurs when the system no longer can absorb the shock and recover in order to provide key economic functions.

In 2022, the ESRB identified two complementary paths to mitigate systemic cyber risk and ensure financial stability in the event of a systemic cyber crisis (ESRB 2022). The first action consists in developing a macro-prudential strategy that, in turn, should comprise:

(a) amendment of the IOs of the ESRB policy framework to include cyber-specific aspects able to prevent a cyber incident from evolving into a systemic cyber crisis;
(b) development of a complementary set of indicators to monitor the cyber vulnerabilities in addition to ESRB's cyber-inclusive IOs (FSB 2021);

Table 16.1 Technical and business impacts of cyber incidents

Impacts of cyber incidents	Technical impacts	– They are disruptive events related to the loss of each asset property (availability, integrity, confidentiality, authenticity, accountability, non-repudiation, reliability); – They often result in business impacts for related entities and the services they provide
	Business impacts	– *Financial*: financial losses due to fines, penalties, lost profits or diminished market share; – *Reputational*: negative opinion or brand damage; – *Legal and regulatory*: litigation liability and withdrawal of licence of trade; – *Contractual*: breach of contracts or obligations between organisations; – *Business objectives*: failure to deliver on objectives or take advantage of opportunities; – *Operational*: discontinued or reduced service levels, workflow disruptions, or supply chain disruptions; – *Environmental:* harmful effects on the biophysical environment; – *Human*: loss of life, injury, impact to community, short and long-term emotional impact To measure business impacts, institutions can adopt two complementary approaches: qualitative (judgement-based) and quantitative (metric-based)

Source Authors' elaboration on ESRB (2020b)

(c) development, calibration and activation of systemic cyber risk-mitigants able to increase (or to ensure a sufficient level) financial resilience in a systemic cyber crisis (Dupont 2019).

Regarding point (b), the ESRB agrees with the implementation of a "*systemic cyber resilience scenario stress testing*" since it offers financial authorities a tool to quantify the financial impact of a cyber incident by documenting its potential amplification into a systemic event (Adelmann et al. 2020). However, to do so, macro-prudential authorities need to define, in advance, the acceptable level of disruption to operational systems providing critical economic functions.

Secondly, the ESRB recognizes the need to establish a pan-European systemic cyber incident coordination framework (EU-SCICF) aimed at increasing the awareness of systemic cyber risk by European financial authorities and bridging any coordination and communication gaps between financial authorities themselves, with other sector authorities and with other key actors at international level (ESRB 2022). Indeed, the successful management of a systemic cyber crisis depends heavily on each financial authority interacting with other financial and cyber authorities. In sum, in the intentions of the ESRB, the new EU-SCICF should overcome the risk to financial stability stemming from a coordination failure during the response to a cyber incident.

More recently, the ESRB, within the context of a substantially heightened cyber threat environment across Europe, developed the concept of a systemic impact tolerance objective (SITO) which can assist in identifying and measuring the impacts of cyber incidents on the financial system, and evaluate when they are likely to breach tolerance levels and cause significant disruption. Specifically, the SITO is the point at which a cyber incident moves from the amplification phase to the systemic event phase. They could assist authorities in assessing when a cyber incident poses a risk to financial stability by representing a kind of "intervention ladder" to adopt timely response and recovery measures (ESRB 2023).

Finally, recognizing that timely and accurate information on cyber incidents is crucial for effective management of the cyber risk and its potential systemic effects, at the request of G20, the FSB formulated 16 recommendations to achieving greater convergence in cyber incident reporting (CIR) (FSB 2023). These recommendations are grouped into four categories, and encourage financial authorities to set clear reporting objectives on cyber incidents by specifying how they can be achieved efficiently, to identify common data requirements on cyber risk, to develop standardized formats for the exchange of incident reporting information, to define toolkits and guidelines to promote effective communication practices in cyber incident reports, to collaborate with Financial Infrastructures (FIs) to proactively share cyber events and vulnerability within the financial sector and finally, to implement secure forms of incident reporting handling to ensure the protection of sensitive information at all times (Table 16.2). This is the first attempt, by an international regulator, to establish common rules to harmonize cyber risk disclosure, rendering it easier to manage and mitigate, especially during systemic events. A previous attempt to expand cyber risk disclosure was made by the Security and Exchange Commission (SEC). In 2011 this

Authority released guidance for listed firms (Li et al. 2018) that was revised in 2018 to provide additional details on how and when firms should disclose information on cyber risk to investors (Li et al. 2018).

Overall, there is a recognizable joint effort by the macroprudential authorities to monitor and identify systemic risks stemming from cyberattacks (Lagarde 2018). The European Central Bank also recently provided its valuable contribution by pointing out some specific features of systemic cyber risk which need further reflection. First, cyberattacks pose systemic risk not only when they affect a critical entity, but also when they affect non-critical but strongly interconnected entities. Second, threats originating from cyberattacks also have a time dimension. Such threats seem to increase in periods of heightened political and economic uncertainty. Finally, rather than being random and idiosyncratic, systematic patterns in cyberattacks can be linked to both economic and political cycles. In sum, *"large data gaps, a fast-changing cyber landscape and the complexity of systemic cyber risks as well as growing interlinkages between technologies and the financial system make it challenging to design policies tailored to mitigate risks associated with cyberattacks. As such, policymakers should work to improve monitoring and analytical frameworks, expand the macroprudential toolkit and foster collaboration and information-sharing at both operational and policy levels to increase and safeguard resilience of the financial system and mitigate the systemic impact of cyberattacks"* (Fell et al. 2022: 128).

Banks' Exposure to Cyber Risk: Some Empirical Evidence

After analysing cyber risk and its systemic impacts from a theoretical point of view, this last section aims to provide an overview of the worldwide banking system's exposure to cyber risk. Academic studies show cyber-risk exposure predicts cyber-attacks, affects stock returns and profits, and is priced in the equity option market (Jamilov et al. 2023). To this end, the recent Cyber Risk Rating (CRR) produced by Bitsight company and provided by ORBIS database is used; of which Figs. 16.2, 16.3 and 16.4 offer a synthetic elaboration. More in detail, the CRR is a measure of an organization's security performance and, accordingly, its level of cybersecurity risk with a value range from 250 and 900. Banks with CRR > 740 fall into the "Advanced range" and pose lower cybersecurity risk than companies in the "Basic" (CRR > 250) and "Intermediate ranges" (CRR > 640). Thus, the higher the CRR, the stronger the banks' overall security posture and the lower the cyber risk exposure. This depends on having strong security practices and few compromised systems (e.g. malware) and security incidents (e.g. breaches) (Table 16.3).

Figure 16.2 compares banks belonging to seven different geographical areas showing that all banks fall into the intermediate range. It emerges that, in recent years, banks have been able to adequately equip themselves against cyber risk by lowering the level of exposure to this risk. However, banks located in the United States, Oceania (mainly Australia) and Western Europe perform better than banks

Table 16.2 FSB Recommendations for achieving greater convergence in cyber incident reporting

Recommendations	Brief description
Design of CIR approach	
Establish and maintain objectives for CIR	Financial authorities should have clear objectives for incident reporting, and periodically assess and demonstrate how these objectives can be achieved efficiently
Explore greater convergence of CIR frameworks	Financial authorities should explore ways to align their CIR regimes with other relevant authorities to minimise potential fragmentation and improve interoperability
Adopt common data requirements and reporting formats	Financial authorities should identify common data requirements, and, where appropriate, develop or adopt standardised formats for the exchange of incident reporting information
Implement phased and incremental reporting requirements	Financial authorities should implement incremental reporting requirements, balancing the authority's need for timely reporting with the affected institution's primary objective of bringing the incident under control
Select appropriate incident reporting triggers	Financial authorities should explore the benefits and implications of a range of reporting trigger options as part of the design of their CIR regime
Calibrate initial reporting windows	Financial authorities should consider potential outcomes associated with window design or calibration used for initial reporting
Provide sufficient details to minimise interpretation risk	Financial authorities should promote consistent understanding and minimise interpretation risk by providing an appropriate level of detail in setting reporting thresholds, using common terminologies and supplementing CIR guidance with examples
Promote timely reporting under materiality-based triggers	Financial authorities that use materiality thresholds should consider fine-tuning threshold language, or explore other suitable approaches, to encourage prompt reporting by FIs for material incidents

(continued)

Table 16.2 (continued)

Recommendations	Brief description
Supervisory activities and collaboration between authorities[6]	
Review the effectiveness of CIR and cyber incident response and recovery (CIRR) processes	Financial authorities should explore ways to review the effectiveness of FIs' CIR and CIRR processes and procedures as part of their existing supervisory or regulatory engagement
Conduct ad-hoc data collection	Financial authorities should explore ways to complement CIR frameworks with supervisory measures as needed and engage FIs in cyber incidents
Address impediments to cross-border information sharing	Financial authorities should explore methods for collaboratively addressing legal or confidentiality challenges relating to the exchange of CIR information on a cross-border basis
Industry engagement	
Foster mutual understanding of benefits of reporting	Financial authorities should engage regularly with FIs to raise awareness of the value and importance of incident reporting, understand possible challenges faced by FIs and identify approaches to overcome them when warranted
Provide guidance on effective CIR communication	Financial authorities should explore ways to develop toolkits and guidelines to promote effective communication practices in cyber incident reports
Capability development (individual and shared)	
Maintain response capabilities which support CIR	FIs should continuously identify and address any gaps in their cyber incident response capabilities which directly support CIR, including incident detection, assessment and training continuously
Pool knowledge to identify related cyber events and cyber incidents	Financial authorities and FIs should collaborate to identify and implement mechanisms to proactively share event, vulnerability and incident information amongst financial sector participants to combat situational
Protect sensitive information	Financial authorities should implement secure forms of incident information handling to ensure the protection of sensitive information at all times

Source Authors' adaptation from FSB (2023)

---

[6] An interesting example of collaboration between authorities is the Swiss Financial Sector Cybersecurity Centre (Swiss FS-CSC). Founded in April 2022, this association aims to facilitate the exchange of information between FIs, financial market players and authorities regarding policies and practices for cyber incident response and cyber crisis management. Membership of the Swiss FS-CSC association is open to all banks, insurance companies, financial market infrastructures and financial associations with their registered office in Switzerland and have been authorised by the Swiss Financial Market Supervisory Authority (FINMA), as well as subsidiaries and branches of

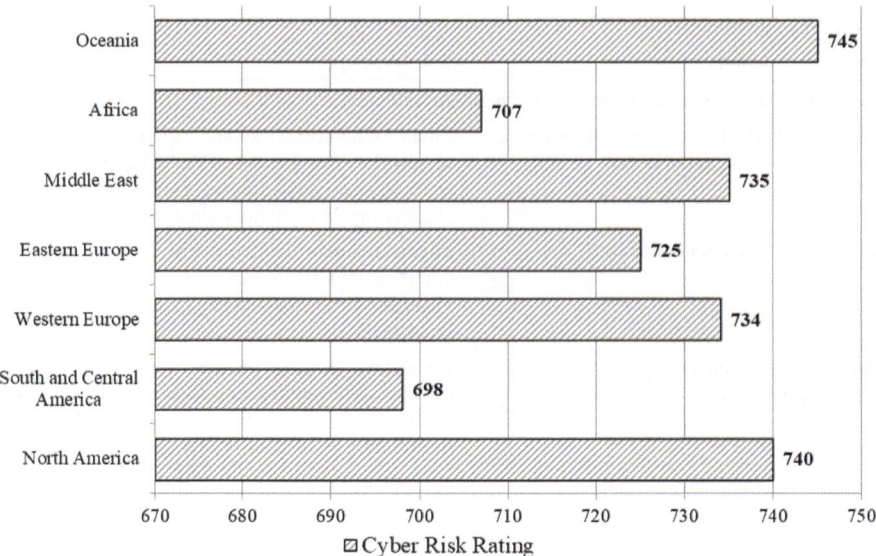

Fig. 16.2 Cyber risk rating for all banks worldwide (CRR as of 1 October, 2023). *Source* Authors' elaboration on ORBIS dataset

Fig. 16.3 Cyber risk rating for all global SIFIs (CRR as of 1 October, 2023). *Source* Authors' elaboration on ORBIS dataset

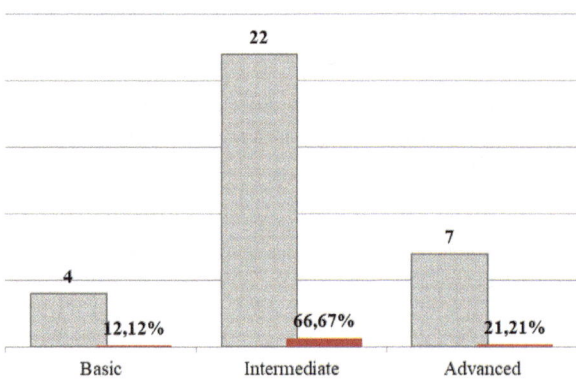

in Africa, Easter Europe and Middle East. The banks of Central and South America are in last place.

Figure 16.3 identifies the positioning of all Global Systemically Important Financial Institutions (SIFIs). This sub-sample of banks is also predominantly in the "Intermediate range" (22 SIFIs) with an average CRR equal to 694. Indeed, despite their systemic importance, only a minority of these institutions can boast advanced cyber risk management approaches (7 SIFIs), while 4 of them rank in the lower bracket.

---

foreign banks and insurance companies with FINMA authorisation. Currently, there are more than 80 founding members, including the Swiss National Bank.

Fig. 16.4 Cyber risk rating
for central banks (CRR as of
1 October, 2023). *Source*
Authors' elaboration on
ORBIS dataset

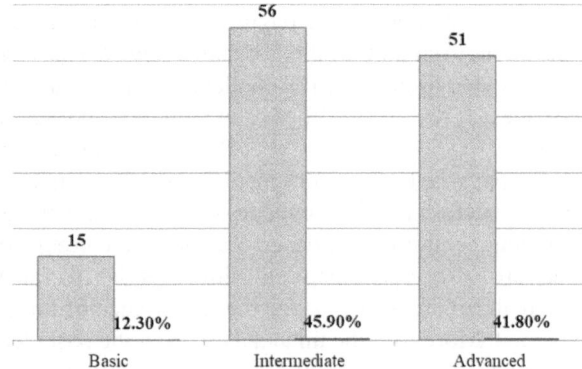

Table 16.3 The ranges of the cyber security rating

Categories	Rating ranges	Description	Distribution
Advanced	740–900	Strong security performance and lower risk	50% of entities
Intermediate	640–730	Fair security performance and moderate risk	45% of entities
Basic	250–630	Poor security performance and high risk	5% of entities

Finally, Fig. 16.4 focuses on Central Banks (CBs). Several authors document cyber-risk as an emerging threat for all types of financial institutions, including CBs as well as Fintech firms (Bouveret 2018). Figure 16.4 shows a balance between intermediate and advanced ranges of the CRR for CBs. Central Banks are thus one step ahead in cyber risk awareness and management, compared to the banks they supervise. This is an important aspect that bodes well for the further and imminent improvement of cyber risk management strategies by all financial institutions. A few days ago (3 January 2024), the European Central Bank (ECB) announced that it will conduct a cyber resilience stress test on 109 directly supervised banks in 2024. The exercise will assess how banks respond to a cyberattack, including activating emergency procedures, contingency plans and restoring normal operations. The findings of this predominantly qualitative exercise will be communicated in the summer of 2024.

(Non) Conclusive Remarks

Drawing conclusions about cyber risk in banks it is not currently possible because it is a risk on which we still have a lot to learn. It is certainly a complex risk to define and delineate as it is constantly evolving and requires the skills and expertise of different research fields (information technology, insurance, financial, etc.). In banks, cyber risk takes on even more severe connotations because it may very likely trigger negative systemic effects, putting the stability of the entire financial sector at risk. Hence supervisory authorities are paying increasing to this emerging risk, urging financial institutions to strengthen their organizational and governance controls as well as their level of cyber disclosure. The challenge is open: banks and other financial institutions have the task to provide effective and timely responses. The outlook seems favourable as the average level of the Cyber Risk Rating produced by Bitsight currently shows, banks appear to be in an intermediate position in managing cyber risk, i.e. both their security performance and cyber risk exposure are moderate. Supervisory authorities, including central banks, deserve some of the credit for this achievement. If banks today are better prepared to deal with cyber risk, it is also because the financial authorities realized at an early stage the importance of this new risk and its destabilizing potential. Moreover, the financial crises of the past (sub-prime crisis), even the most recent ones (Covid-pandemic crisis), have provided important lessons. The best way to manage a new risk, while minimizing its negative and systemic impacts, consists above all in creating diagnostic, measurement and containment tools. Recent initiatives by the banking regulator, including the launch of cyber stress tests and uniform cyber risk disclosure rules, are moving in this direction.

References

Adelmann, F., Elliott, J., Ergen, I., Gaidosch, T., Jenkinson, N., Khiaonarong, T., Morozova, A., Schwarz, N., Wilson, C.: Cyber risk and financial stability: it's a small world after all. In: IMF Staff Discussion Notes, No 20/07. Cyber Risk and Financial Stability: It's a Small World After All (imf.org) (2020). Accessed at 28 October (2023)

Agrafiotis, I., Nurse, J.R.C., Goldsmith, M., Creese, S., Upton, D.: A taxonomy of cyber-harms: defining the impacts of cyber-attacks and understanding how they propagate. J. Cybersecur. 4(1) (2018)

Akinbowale, O.E., Klingelhöfer, H.E., Zerihun, M.F.: The assessment of the impact of cyberfraud in the South African banking industry. J. Financ. Crime (2023). https://doi.org/10.1108/JFC-04-2023-0094

Akinbowale, O.E., Klingelhöfer, H.E., Zerihun, M.F.: Analysis of cyber-crime effects on the banking industry using balance scorecard: a survey of literature. 27(3), 945–958 (2020)

Aldasoro, I., Gambacorta, L., Giudici, P., Leach, T.: Operational and cyber risks in the financial sector. BIS Working Paper No. 840 (2020)

Aldasoro, I., Gambacorta L., Giudici P., Leach T.: The drivers of cyber risk. J. Financ. Stabil. 60, 100989 (2022). https://doi.org/10.1016/j.jfs.2022.100989

Allianz: Allianz Risk Barometer. Allianz-Risk-Barometer-2021.pdf. (2021). Accessed at 15 Oct (2023)

An, J., Duan, T., Hou, W., Liu, X.: Cyber risks and initial coin offerings: evidence from the world. Finance Res. Lett. 41, 101858 (2021). https://doi.org/10.1016/j.frl.2020.101858

Anand, K., Duley, C., Gai, P.: Cybersecurity and financial stability. Deutsche Bundesbank Discussion Paper No.08/2022 (2022)

Awiszus, K., Knispel, T., Penner, I., Svindland, G., Voß, A., Weber, S.: Modeling and pricing cyber insurance Idiosyncratic, systematic, and systemic risks. Eur. Actuar. J. 13, 1–53 (2023). https://doi.org/10.1007/s13385-023-00341-9

Bakker, T.G., Streff, K.: Accuracy of self-disclosed cybersecurity risks of large US banks. J. Appl. Econ. Bus. Stud. 18(3), 39–51 (2016)

Baldwin, A., Gheyas, I., Ioannidis, C., Pym, D., Williams, J.: Contagion in cyber security attacks. J. Oper. Res. Soc. **68**(7), 780–791 (2017)

Bank of England: Systemic Risk Survey Results—2023H2 (2023a)

Bank of England: Thematic findings from the 2022 cyber stress test, 29 March (2023b)

BCBS—Basel Committee on Banking Supervision: Cyber resilience: Range of practices. December (2018)

Berger, A., Curti, F., Mihov, A., Sedunov, J.: Operational risk is more systemic than you think: evidence from U.S. bank holding companies. J. Bank. Finance **143**, 106619 (2022)

Biener, C., Eling, M., Wirfs, J.H.: Insurability of cyber risk: an empirical analysis. Geneva Pap. Risk Insur. Issues Pract. **40**(1), 131–158 (2015). https://doi.org/10.1057/gpp.2014.19

Boungou, W.: Cyber-attacks and banking intermediation. Econ. Lett. **233**, 111354 (2023). https://doi.org/10.1016/j.econlet.2023.111354

Bouveret, A.: Cyber risk for the financial sector: a framework for quantitative assessment. IMF Working Paper No. WP/18/143 (2018). https://doi.org/10.5089/9781484360750.001

Boyer, M., Eling, M.: New advances on cyber risk and cyber insurance. Geneva Pap. Risk Insur. Issues Pract. **48**, 267–274 (2023). https://doi.org/10.1057/s41288-023-00294-w

Caruana, J.: Systemic Risk: How to Deal with it. Bank for International Settlements (2010)

Cebula, J.J., Popeck, M.E., Young, L.R.: A taxonomy of operational cyber security risks version 2. A Taxonomy of Operational Cyber Security Risks Version 2 (cmu.edu) (2014). Accessed at 20 Dec (2023)

Creado, Y., Ramteke, V.: Active cyber defence strategies and techniques for banks and financial institutions. J. Financ. Crime. **27**(3), 771–780 (2020)

Cremer, F., Sheehan, B., Fortmann, M., Kia, A.N., Mullins, M., Murphy, F., Materne, S.: Cyber risk and cybersecurity: a systematic review of data availability. Geneva Pap. Risk Insur. Issues Pract. **47**, 698–736 (2022). https://doi.org/10.1057/s41288-022-00266-6

Crosignani, M., Macchiavelli, M., Silva, A.F.: Pirates without borders: the propagation of cyber-attacks through firms' supply chains. Staff Report 937, Federal Reserve Bank of New York (2020)

Curti, F., Gerlach, J., Kazinnik, S., Lee, M.J., Mihov A.: Cyber risk definition and classification for financial risk management. J. Oper. Risk. **18**(2) (2023)

Deloitte: Cyber Value at Risk in the Netherlands (2016)

DTCC and Oliver Wyman: Large-Scale Cyber Attacks on the Financial System, March (2018)

DTCC: Systemic Risk Barometer Survey. 29873-Systemic_Risk-2024 (dtcc.com). Accessed at 18 Jan (2024)

Duffie, D., Younger, J.: Cyber Runs. Hutchins Center Working Paper 51. Brookings Institution (2019)

Dupont, B.: The cyber-resilience of financial institutions: significance and applicability. J. Cybersecur. **5**(1), 1–17 (2019)

Eisenbach, T.M., Kovner, A., Lee, M.J.: Cyber risk and the U.S. financial system: a pre-mortem analysis. J. Financ. Econ. **145**, 802–826 (2022)

Eling, M.: Cyber risk research in business and actuarial science. Eur. Actuar. J. **10**(2), 303–333 (2020)

Eling, M., Jung, K.: Heterogeneity in cyber loss severity and its impact on cyber risk measurement. Risk Manage. **24**, 273–297 (2022). https://doi.org/10.1057/s41283-022-00095-w

Eling, M., Wirfs, J.: What are the actual costs of cyber risk events? Eur. J. Oper. Res. **272**, 1109–1119 (2019). https://doi.org/10.1016/j.ejor.2018.07.021

Eling, M., McShane, M., Nguyen, T.: Cyber risk management: history and future research directions. Risk Manag. Insur. Rev. **24**(1), 93–125 (2021). https://doi.org/10.1111/rmir.12169

Eling, M., Schnell, W.: What do we know about cyber risk and cyber risk insurance? J. Risk Finance. **17**(5), 474–491 (2016). https://doi.org/10.1108/jrf-09-2016-0122

ESRB—European Systemic Risk Board: Systemic cyber risk, February (2020a)

ESRB—European Systemic Risk Board: The making of a cyber crash: a conceptual model for systemic risk in the financial sector. Occasional Paper Series No 16, May (2020b)

ESRB—European Systemic Risk Board: Mitigating systemic cyber risk, January (2022)

ESRB—European Systemic Risk Board: Advancing macroprudential tools for cyber resilience. February (2023)

Fell J., de Vette N., Gardó S., Klaus, B., Wendelborn J.: Towards a framework for assessing systemic cyber risk. Financ. Stab. Rev. Eur. Central Bank 2 (2022)

Firoozi, M., Mohsni, S.: Cybersecurity disclosure in the banking industry: a comparative study. Int. J. Discl. Gov. (2023). https://doi.org/10.1057/s41310-023-00190-8

Frost, J., Shapiro, J.: Cyber attacks 'the biggest risk in banking. Aust. Financ. Rev. (2021)

FSB—Financial Stability Board: Recommendations to Achieve Greater Convergence in Cyber Incident Reporting. Final Report. April (2023)

FSB—Financial Stability Board: FSB Financial Stability Surveillance Framework, September (2021)

FSI—Financial Stability Institute: Banks' cyber security—a second generation of regulatory approaches. Financial Stability Institute FSI Insights on policy implementation No 50 June (2023)

Gatzert, N., Schubert, M.: Cyber risk management in the US banking and insurance industry: a textual and empirical analysis of determinants and value. J. Risk Insur. 89, 725–763 (2022). https://doi.org/10.1111/jori.12381

Gulyás, O., Kiss, G.: Impact of cyber-attacks on the financial institutions. Procedia Comput. Sci. 219, 84–90 (2023). https://doi.org/10.1016/j.procs.2023.01.267

Harum, C.A., Gunadi, I.: Financial stability and systemic risk. In: Warjivo, P., Juhro, S.M. (eds.) Central Bank Policy Mix: Issues, Challenges, and Policy Responses. Springer, Singapore (2022). 978–981–16–6827–2.pdf (oapen.org)

Jamilov, R., Rey, H., Tahoun, A.: The anatomy of cyber risk. Institute for New Economic Thinking Working Paper Series No. 206 (2023)

Jin, J., Li, N., Liu, S., Nainar, S.M.K.: Cyber-attacks, discretionary loan loss provisions, and banks' earnings management. Finance Res. Lett. 54, 103705 (2023)

Kopp, E., Kaffenberger, L., Wilson, C.: Cyber risk, market failures, and financial stability, working paper. International Monetary Fund (WP/17/185) (2017)

Lagarde, C.: Estimating Cyber Risk for the Financial Sector. IMF Blog. June 22 (2018). https://blogs.imf.org/2018/06/22/estimating-cyber-risk-for-the-financial-sector/

Li, H., No, W.G., Wang, T.: SEC's cybersecurity disclosure guidance and disclosed cybersecurity risk factors. Int. J. Account. Inf. Syst. 30(C), 40–55 (2018)

Malavasi, M, Peters, G.W., Shevchenko, P., Trück, S., Jang, J. Sofronov, G.: Cyber risk frequency, severity and insurance viability. Insur. Math. Econ. 106, 90–114 (2022). https://doi.org/10.1016/j.insmatheco.2022.05.003

Mangala, D., Soni, L.: A systematic literature review on frauds in banking sector. J. Financ. Crime. 30(1), 285–301 (2023). https://doi.org/10.1108/JFC-12-2021-0263

Mazumder, M.M.M., Hossain, D.M.: Voluntary cybersecurity disclosure in the banking industry of Bangladesh: does board composition matter? J. Account. Emerg. Econ. 13(2), 217–223 (2023)

Mazumder, M.M.M., Sobhan, A.: The spillover effect of the Bangladesh bank cyber heist on bank's cyber risk disclosures in Bangladesh. J. Oper. Risk. 15(4), 53–76 (2021)

Mikhed, V., Vogan, M.: How data breaches affect consumer credit. J. Bank. Financ. 88, 192–207 (2018)

Nurse, J.R.C., Axon L., Erola A., Agrafiotis I., Goldsmith, M., Creese S.: The data that drives cyber insurance: a study into the underwriting and claims processes. In: 2020 International Conference on Cyber Situational Awareness, Data Analytics and Assessment (CyberSA), 15–19 June 2020

Pollmeier, S., Bongiovanni, I., Slapničar, S.: Designing a financial quantification model for cyber risk: a case study in a bank. Saf. Sci. 159, 106022 (2023). https://doi.org/10.1016/j.ssci.2022.106022

Ponemon Institute: Cost of Data Breach Study: Global Overview. https://www.ibm.com/security/data-breach (2020)

Radu, C., Smaili, N.: Board gender diversity and corporate response to cyber risk: evidence from cybersecurity related disclosure. J. Bus. Ethics **177**, 351–374 (2022). https://doi.org/10.1007/s10551-020-04717-9

Santucci, L.: Consumer Finance Institute discussion papers 18–3 Quantifying Cyber Risk in the Financial Services Industry. Federal Reserve Bank of Philadelphia (2018)

Sheehan, B., Murphy, F., Kia, A.N., Kiely, R: A quantitative bow-tie cyber risk classification and assessment framework. J. Risk Res. **24**(12), 1619–1638 (2021). https://doi.org/10.1080/13669877.2021.1900337

Sinanaj, G., Muntermann J.: Assessing corporate reputational damage of data breaches: an empirical analysis. In: Proceedings of the 26th International Bled eConference. Bled, 78–89 (2013)

Skinner, C.P.: Bank disclosure of cyber exposure. IOWA Law Rev. **105**, 239–281 (2019)

Smaili, N., Radu, C., Khalili, A.: Board effectiveness and cybersecurity disclosure. J. Manag. Gov. **27**, 1049–1071 (2023). https://doi.org/10.1007/s10997-022-09637-6

Statista: Global industry sectors most targeted by basic web application attacks from November 2020 to October 2021.https://www.statista.com/statistics/221293/cyber-crime-target-industries/. Last Accessed at 23 March (2023)

Strupczewski, G.: Defining cyber risk. Saf. Sci. **135**, 105143 (2021). https://doi.org/10.1016/j.ssci.2020.105143

Tian, S., Zhao, B., Olivares, R.O.: Cybersecurity risks and central banks' sentiment on central bank digital currency: evidence from global cyberattacks. Finance Res. Lett. **53**, 103609 (2023)

Uddin, M.H., Mollah, S., Ali, M.H.: Does cyber tech spending matter for bank stability? Int. Rev. Financ. Anal. **72**, 101587 (2020a). https://doi.org/10.1016/j.irfa.2020.101587

Uddin, M.H., Ali, M.H., Hassan, M.K.: Cybersecurity hazards and financial system vulnerability: a synthesis of literature. Risk Manage. **22**, 239–309 (2020b). https://doi.org/10.1057/s41283-020-00063-2

WEF—World Economic Forum: Understanding Systemic Cyber Risk. White Paper, October (2016)

WEF—World Economic Forum: The Global Risks Report 2020, 15th Edition (2020)

WEF and Deloitte: Partnering for Cyber Resilience Towards the Quantification of Cyber Threats, January (2015)

Welburn, J.W., Strong, A.M.: Systemic cyber risk and aggregate impacts. Risk. Anal. 42(8) (2022). https://doi.org/10.1111/risa.13715

Zhang, Q., Ou, Y., Chen, R.: Digitalization and stability in banking sector: a systemic risk perspective. Risk Manag. **25**(2), 1–29 (2023). https://doi.org/10.1057/s41283-023-00116-2

Chapter 17
The Dynamics of Crypto Markets and the Fear of Risk Contagion

Mauro Aliano, Massimiliano Ferrara, and Stefania Ragni

Abstract Decentralized finance has gained significance in recent years, as have concerns about the financial system's stability. Exchange mechanisms, such as those utilized on cryptocurrency platforms, enhance volatility, and transmit risk contagion to other financial actors globally, which may increase financial calamity. We propose a Susceptible-Infected-Recovered model with a time delay to examine the mechanism of risk contagion in the cryptocurrency markets during the last decade. The governance token prices of the main cryptocurrency exchange platforms, as well as their spillover effects, crash risks and indicators of people's attention, are assessed, and the obtained parameters are used in the Susceptible-Infected-Recovered model to replicate the dynamics of risk contagion in the examined crypto markets. Findings suggest high interconnection among crypto markets in short-run and the fear spread among people play an important contribution to financial risks. Under the new decentralized finance paradigm, predictive modeling of the temporal distribution of risk among cryptocurrencies may provide useful insights for policy and financial system stability, as well as for contagion risk.

Keywords Financial contagion · Financial crises · Crises' transmission channels · Tokenization

Introduction

In some ways, fear of infection and viral transmission is inherent in human nature, as well as in the financial market. The financial markets' reaction to contagion has implications similar to those that happened to medical disease during COVID-19, albeit with distinct features whether investors or stock traders characterize what

M. Aliano (✉) · S. Ragni
University of Ferrara, Ferrara, Italy
e-mail: mauro.aliano@unife.it

M. Ferrara
Mediterranea University of Reggio Calabria, Reggio Calabria, Italy

© The Author(s) 2025 323
V. Pacelli (ed.), *Systemic Risk and Complex Networks in Modern Financial Systems*,
New Economic Windows, https://doi.org/10.1007/978-3-031-64916-5_17

recently happened for Silicon Valley Bank, in the traditional banking system, but also for FTX, in the cryptocurrency sector.

We apply a Susceptible-Infected-Recovered (SIR) model to evaluate the effect of crashes on crypto platforms. To our best knowledge, this is the first attempt to apply this model to cryptocurrency platforms.

We suggest a structured approach that ensures readers can navigate through the details of our exploration seamlessly, gaining a holistic understanding of the nuanced interplay between SIR models, Granger Causality, Spillover effects, and the influence of fear on risk parameters in the dynamic realm of crypto platforms.

We show how risk contagion may evolve among Decentralized Finance (DeFi) exchanges using a dynamical method based on a SIR categorization. The SIR approach paradigm may be studied in the economic environment due to the parallels between financial systems and ecosystems.

While SIR models are not new, the application deserves to be highlighted due to the increased demand for financial services outside traditional schemes. We use the SIR approach to mimic risk contagion in the governance token market, which has recently been related to the most prominent cryptocurrency trading platforms. A governance token is a cryptocurrency token issued by a blockchain-based platform or protocol that allows its holders to participate in the governance of the platform. Holders of the governance token can vote on proposed protocol changes such as transaction fee changes or network infrastructure upgrades. In addition to voting, holders of governance tokens may have other benefits, such as earning a portion of the platform's revenue or suggesting changes themselves. Governance tokens are commonly utilized on DeFi platforms and protocols where the user community plays an important role in decision-making, according to Makridis et al. [1].

Given these token characteristics, which are analogous to equity instruments in certain ways, we evaluate governance price tokens and how they vary as a proxy for changing value for crypto platforms in our research. Furthermore, as a risk transmission channel method, we evaluate the information flow that we may capture in shifting price and risk for these governance tokens.

We investigate price movements for governance tokens in the context of market connectivity. In this regard, we conducted a causality analysis that revealed the presence of a link between the platform governance tokens in the data sample under consideration. We also make an original contribution by measuring the spillover effect and quantifying cross-platform contagion. The research on the spillover index computed from the price of the governance token shows how different platforms influence each other, which is known as interconnectivity. In this approach, risk contagion is related to information flow contained in the governance token.

The structure of the chapter can be outlined as follows: the second section extensively covers literature, the third section delves into data and methods, the fourth section presents findings and initiates discussion, and the final section addresses policy implications.

Literature Review

In the financial literature, contagion plays a key role in the so-called "systemic risk", in which both endogenous and exogenous events could determine a large cascade of crises (see [2]). According to the authors, starting from an outbreak with a domino effect, if a bank (or other financial intermediaries) moves toward a crisis or precrisis state, this could also cause a crisis or precrisis conditions for 50 other banks.

Financial network interdependencies, such as the interbank market, are one of the most important determinants of default propagation [3], and are considered a mechanism of contagion transmission. By using a probabilistic model, nodes, and Monte Carlo simulation, in Leonidova and Rumyantsev [4] the authors study the default contagion risk in the Russian interbank market. The use of network analysis in economic analysis has a long history, according to Callon [5], but it can also be used to explain financial crises.

For the banking sector, Babus [6] estimates the probability of systemic risk associated with a bank default in the interbank market, when it is at an equilibrium status. Financial networks are also analyzed, among other things, by Battiston and Caldarelli [7], who suggests that the interplay of network topology, capital requirements, and market liquidity are three important factors that could affect systemic risks. Under the liquidity risk perspective, Feinstein [8] outlines a model in which financial crisis propagation goes through illiquid assets and fire sales. Moreover, by accounting for the management effect, Caldarelli et al. [9] consider other networks: the board and director networks, price correlations, and stock ownership. A sort of spillover effect is used in Aït-Sahalia et al. [10], in which the authors propose a model to study the contagion jump process in different regions by studying the equity market and focusing their findings on the stock price propagation mechanism.

At the operative and costumers' level, the investigation in Barja et al. [11] exploits quarterly client's data from BBVA (i.e., a Spanish Bank) to study customer–supplier chain transactions. The authors consider a Susceptible-Infected-Susceptible (SIS) model to evaluate the patterns that are similar to the ones used in a spreading epidemic. Starting from catastrophic events, Torri et al. [12] evaluate the effects on non-life insurance by using balance sheet analysis. Default contagion and default degree in the capital chain are studied with the Copula metric for listed Chinese companies by Han [13]. Among the measures that can be applied by policymakers to contain the contagion effect on financial markets for banks, there is the short-selling ban for stocks and other financial instruments [14]. In the end, financial crises are boosted by psychological factors (see [15]) not only in their buildings but also in their spread in markets, instruments, and among economic players. For example, during the COVID-19 period, virus diffusion raised fear and uncertainty in the market [16].

Generally, under more uncertain scenarios, the behaviors of financial players are characterized by: (i) actions more sensitive to investment losses than gains [17],(ii) players triggering risks, and emotional or sentimental behavior that drives decisions [18], and (iii) mostly damaging investment decisions [19]. The so-called spillover effects are measured not only on the financial market, when the correlation between indices and stocks is analyzed, but also between firms. Among other contributions, Filbeck et al. [20] exploits event study methodologies to understand the stock reactions to disruption in the automobile industry supply chain, and measuring the contagion effect of the reduction of stock prices. The financial distress of a company caused by customer-supply chain relations is analyzed by Lian [21] from 1980 to 2014. The author finds that financial distress transfers from major customer firms to supplier firms, and the interfirm effect is persistent for up to two years. In Agca et al. [22] a credit default swap is used to evaluate credit shocks in the supply chain. Spatial analysis of the proximity effects of both financial distress and failure is considered in Barro and Basso [23]. Local factors as determinants of the default of a company also emerge in Barreto and Artes [24], Calabrese et al. [25], Maté-Sánchez-Val et al. [26]. Starting from commonalities in banks' balance sheets, Shi et al. [27], analyses China's banking system by considering the vulnerability of each bank according to some channels of transmission and complex relations among financial players.

According to Egloff et al. [28] there is another contagion mechanism through credit deterioration channels, in which the credit deterioration of a company could also deteriorate credit in other counterparties. Hertzel et al. [29] consider intra-industry bankruptcy and evaluate the consequences of distress both for customers and suppliers. They capture financial wealth effects on stock price reactions to distress and failures. Furthermore, Escribano and Maggi [30] analyze the default dependencies in a multisectoral framework starting from 1996 and evaluating the dot-com bubble and the global financial crisis (until 2015). The authors argue that the contagion effect between sectors manifests in two ways: (i) the "infectivity", or the degree of transmission of default among sectors, and (ii) the "vulnerability" of each sector. Moreover, Xie et al. [31] examine a dual-channel financing model in supply chain finance characterized by loans from the bank and trade credit from the manufacturer. In this chapter, credit risk is considered as a contagion channel from the supply chain perspective for small and medium enterprises (SMEs). Then, Calabrese [32] studies the contagion effects of UK small business failures and finds that the geographical location and the industry group are significant. In addition, the model in Fanelli and Maddalena [33] considers analogies between medical disease and credit risk contagion. It describes a nonlinear dynamic in the SIR framework and accounts for the transitory immunity time lag before a bank becomes defaultable. From another perspective, the authors in Xu et al. [34] study a contagion mechanism of associated credit risk with corporate senior executives' alertness; they exploit the SIR approach to construct the interaction model between the corporate senior executive alertness and the associated credit risk contagion in the network.

Concerns have developed in recent years about financial service providers who operate outside standard schemes or without Centralized Finance (CeFi), particularly in connection with cryptocurrencies. In contrast to the old financial system, the so-called DeFi phenomenon and automated smart contracts on the blockchain have expanded internationally in the crypto financial system.

Under a complexity and machine learning framework perspective, Ciano [35] forecasts the closing price of Bitcoins from the 61st day using a training dataset constructed from closing prices from the previous 60 days, emphasizing the market's significant volatility. The study dives into the association between cryptocurrencies, improving the analysis and providing insights into anticipating prices in this complex financial landscape.

Many concerns, such as leverage and liquidity mismatches, might be managed by policymakers and financial regulators from the standpoint of financial stability, according to Aramonte et al. [36]. The lack of internal shock absorbers during stressful times can be visible in the traditional financial system, such as liquidity issues, but without banks and central banks, it might lead to crypto runs. There are also concerns about consumer safety, ranging from operational platform failures to cyber-attacks, from volatility difficulties to the use of leverage [37].

Collaterals offered in stable coin issuance reflect liabilities; additionally, if values decrease owing to a bearish market, the value of collateral for crypto keepers falls. As a result, even for a less hazardous cryptocurrency like a stable coin, this procyclical system defines a liquidity mismatch due to a stable coin's liability-driven character, according to McLeay et al. [38].

Overcollateralization and high leverage worsen this procyclicality and spread risk contagion to other global financial actors. As an example, Three Arrows Capital (3AC) collapsed in June 2022 because of the failure of the so-called margin calls, and a few weeks later, in July 2022, another crash happened for Voyager Digital, the cryptocurrency broker that sold the 3AC bankruptcy action. The interconnectedness of cryptocurrencies, operators, and FinTech firms [39], as well as the influence of significantly over collateralization phenomena on the relationship between primary brokers and hedge funds borrowing [40], are just a few of the factors that can enhance the traditional contagion scheme in the crypto-financial system.

We employed an index of internet search traffic associated with a set of terms to estimate people's sentiment, building on previous studies [41, 42]. Many researchers utilize Internet search activity as a proxy for investor mood, demonstrating a correlation between people's attentiveness and stock volatility during the epidemic [43], [44].

Building on the insights garnered from the preceding literature review analysis, we construct our chapter employing a comprehensive conceptual framework. Initially, we study the existence of interconnection between platforms facilitated by governance tokens, aiming to capture this phenomenon through the application of VAR (Vector Autoregression) and Spillover methods. This approach allows us to assess the immediate impact in the short run. Subsequently, we incorporate risk measures designed to understand crisis conditions. Lastly, we apply SIR methods, utilizing parameters derived from VAR, Spillover, and risk analyses. Additionally, we factor in considerations of people and investor attention to formulate a thorough understanding of the long-run equilibrium.

Data and Methods

We examine governance tokens that have recently been related to the most popular cryptocurrency trading platforms. Using the Yahoo Finance data source, we retrieved the daily unbalanced values of each currency from 2017 to 2023. Our database is a hand-curated compilation of publicly available data, beginning with Yahoo Finance, where token prices are expressed in US dollars, and we begin our simulation with these numbers. We study potential connections across platforms and estimate the factors that are included in the SIR model by measuring the spillover impact, identifying the risk, and people's attention to understand how risk may spread throughout the market. The risk dynamics are then projected into the future and focused on long-term contagion in a stable state.

However, while Bitcoin came into existence in 2009, the emergence of crypto platforms, along with their associated tokens, has occurred more recently, particularly within the past years. This growth aligns with the increasing prevalence of cryptocurrencies, whether in the form of coins (including stable coins) or as investment assets (excluding stable coins). Table 17.1 outlines the inception year of the initial token emissions linked to crypto platforms, with our selection based on their trading volume as of early 2023 (https://coinmarketcap.com).

Table 17.1 Sample of tokens by launch year

Token	Year
Binance	2017
OKB	2019
FTX	2019
KuCoin Token	2017
Huobi Token	2020
Uniswap	2020
AAVE	2020
Compound	2020
Decentraland	2017
0x	2017
Decred	2017
Avalanche	2020
Bounce	2021
Ampleforth	2021
AntiMatter	2021
UNION Protocol	2020
Terra Classic	2019
Curve DAO Token	2022

Source Authors' elaboration

Table 17.2 Descriptive statistics for governance tokens

	Mean	Standard deviation	Skewness	Kurtosis
Binance	0.004	0.060	1.915	22.238
OKB	0.004	0.061	2.380	26.572
FTX	0.003	0.062	-0.963	32.848
KuCoin Token	0.004	0.072	3.065	31.576
Huobi Token	0.002	0.054	1.188	16.351
Uniswap	0.002	0.070	0.900	5.895
AAVE	0.003	0.070	0.328	2.422
Compound	0.003	0.081	4.711	70.363
Decentraland	0.006	0.092	5.637	84.238
0x	0.002	0.072	1.168	9.025
Decred	0.002	0.062	2.552	44.062
Avalanche	0.004	0.078	1.488	12.340
Bounce	0.001	0.080	1.451	11.140
Ampleforth	-0.002	0.083	2.984	32.366
AntiMatter	0.000	0.100	1.014	11.763
UNION Protocol	-0.001	0.102	2.157	20.447
Terra Classic	0.006	0.144	10.971	277.346
Curve DAO	-0.002	0.069	0.145	2.467

Source Authors' elaboration

For a deeper understanding of the tokens under consideration, Table 17.2 provides descriptive statistics on the daily returns within the sample utilized in this chapter.

VAR and Spillover

To support the interconnectedness theory, we begin by examining Granger Causality and the spillover impact among cryptocurrency platform tokens. The Granger Causality approach cited by Diebold et al. [45–47] allows for determining information flow among platforms because it is an efficient tool for determining whether the predicted distribution of one set of time series variables (i.e., cause variables, CV) has changed over time (i.e., effect variables, EV). The test examines the effect of the EV forecast on the mean squared error. To accomplish this purpose, the variables involved in the analysis must have stationary time series, otherwise, the data must be differenced. In this regard, we begin with the data and use the VAR model fit approach as well as the Akaike information criterion (AIC) to examine the Granger Causality among return series. We recorded the AIC score after testing the VAR model with lags ranging from 1 to 4 day and chose the VAR lag with the lowest AIC

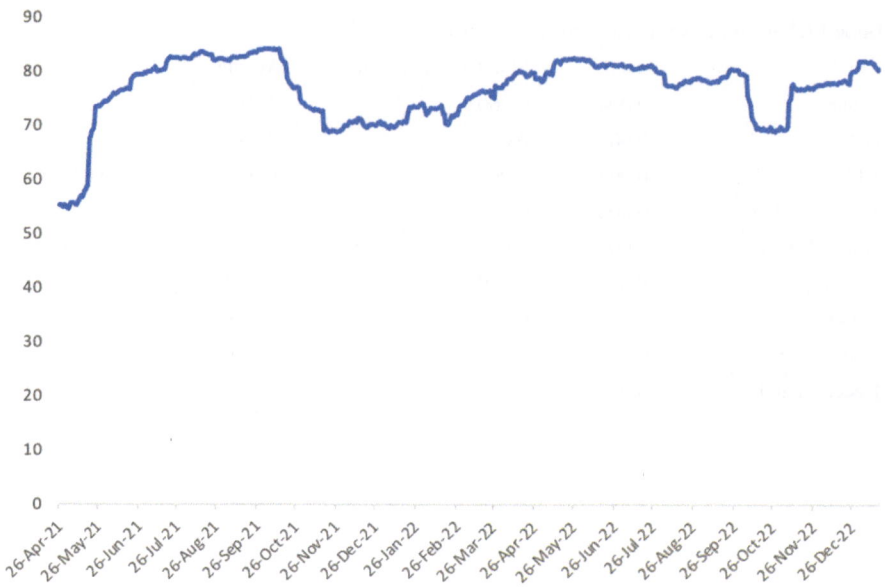

Fig. 17.1 Spillover Index

value. We ran a Granger Causality test on each variable and equation in the VAR system to see if one is the Granger causes of another. With the null hypothesis, we use Chi-Square to perform a leave-one-out Granger causality test.

We corroborate our initial interconnectedness intuition by conducting this causality analysis and conclude that there is a relevant interconnection in the platform governance tokens we account for. Then, considering the influence of spillover on risks and returns, we measure the extent of interconnection in platform governance tokens. In this regard, we create spillover indices using an extended decomposition of the forecast-error variance of the VAR model, as cited by Diebold et al. [45–47]. The net-spillover index (see also Fig. 17.1), as derived by the Diebold and Yilmaz technique, provides us with our answers. As a result, we may estimate the interconnection parameter involved in the prior SIR dynamics using this spillover analysis.

Risk Indicators

Daily prices are used to calculate risk indicators. Standard deviation, a well-known risk indicator for financial market analysis, serves as the initial measure. We also consider additional risk indicators, such as crash and idiosyncratic risks. After estimating token-specific daily returns, we compute risk indicators by taking into account the residual from regressing daily token returns in an enlarged index model, as proposed by Hutton et al. [48]:

$$r_{j,t} = \alpha_j + \beta_{a,j} r_{m,t-2} + \beta_{b,j} r_{m,t-1} + \beta_{c,j} r_{m,t} + \beta_{d,j} r_m + \beta_{e,j} r_{m,t+1} + \beta_{f,j} r_{m,t+2} + \varepsilon_{j,t}$$
$$(17.1)$$

where $r_{j,t}$ is the token return for daily t, and $r_{m,t}$ is the market index return for the same day (we use the S&P Cryptocurrency Broad Digital Market Index, as it has been recommended to utilize a worldwide index in the specific lack of a reference benchmark). We further insert forward and one- and two-day lagged market returns, per Hutton et al. [48]. We define $W_{j,t} = \ln(1 + \varepsilon_{j,t})$ to correct daily returns for the skewed residuals $\varepsilon_{j,t}$.

In our analysis of (1), we look at the next two risk metrics, which take idiosyncratic and crash risks into account.

The first one is a measure of crash risk that considers the negative conditional skewness of token-specific daily returns (cr1).

Cr1 measures token price up-movements [48] using an indicator of whether token-specific $W_{j,t}$ falls by more than 3.09 standard deviations above the average $W_{j,t}$ in that month:

$$cr1_{j,t} = \frac{\sum_{i=1}^n Risk_{j,i}}{n}$$
$$(17.2)$$

where, n in the number of observations and $Risk_{j,i}$ is:

$$Risk_{j,i} = \begin{cases} 1 \text{ if } \overline{W_{j,t}} * \sigma_w < -3.09 \\ 0 \text{ otherwise} \end{cases}$$
$$(17.3)$$

where $\overline{W_{j,t}}$ denotes the mean value of $W_{j,t}$, and σ_w denotes the standard deviation.

Our measure of risk cr2 is built over Dumitrescu et al., [49] and Habib et al., [50] and it represents the Negative Conditional Skewness (NCSKEW) of token returns. We calculate by taking the negative of the third moment of token-specific daily returns for each year and normalizing it by the standard deviation of daily returns raised to the third power. Specifically, cr2 is calculated as:

$$cr2_{j,t} = \frac{-\left[n(n-1)^{3/2} \sum W^3_{j,t}\right]}{\left[(n-1)(n-2)(\sum W^2_{j,t})^{3/2}\right]}$$
$$(17.4)$$

The third measure of crash risk is the down-to-up volatility measure (DUVOL) of the crash likelihood. For each token j over a fiscal year period τ, token-specific daily returns are separated into two groups: "down" days when the returns are below the annual mean, and "up" days when the returns are above the monthly mean. The standard deviation of token specific daily returns is calculated separately for each of these two groups. DUVOL is the natural logarithm of the ratio of the standard deviation in the "down" days to the standard deviation in the "up" days:

$$cr3_{j,t} = log\left((n-1)\frac{\sum_{down} W_{j,t}^2}{ndown-1}\sum_{up} W_{j,t}^2\right) \qquad (17.5)$$

As the fourth measure, we define "idion" in the chapter as an idiosyncratic risk applying the logistic transformation of R2 obtained in (1). According to Ferreira and Laux [51], we obtain the following measure:

$$idion_{j,t} = log\left(\frac{1-r2}{r2}\right) \qquad (17.6)$$

where r2 is the R-squared of model estimated in the Eq. (17.1). Jump measures token price up-movements [48] using an indicator of whether token-specific $W_{j,t}$ rises by more than 3.09 standard deviations above the average $W_{j,t}$ in that year (see cr1). According to previous scholars, investigating corporate governance and volatility risks [52] and financial return [49], we define Jump risk measures as one if a company or token record one or more $W_{j,t}$ 3.09 standard deviations above the mean value for that year, and zero otherwise:

$$Jump_{j,t} = \frac{\sum_{i=1}^{n} Risk_{j,i}}{n} \qquad (17.7)$$

with $Risk_{j,i}$ defined as:

$$Risk_{j,i} = \begin{cases} 1\ if\ \overline{W_{j,t}} * \sigma_w > 3.09 \\ 0\ otherwise \end{cases} \qquad (17.8)$$

where $\overline{W_{j,t}}$ denotes the mean value of $W_{j,t}$, and σ_w denotes the standard deviation.

On the one hand, we claim that "Susceptible" tokens are identified with platforms that have been quoted in the market; on the other hand, we need to develop a criterion for identifying "Infected" tokens and when they are "Recovered". In this regard, we assume that the infection will be evaluated in terms of crash risk (cr3). More particularly, we employ the risk assessment technique that estimates the Negative Conditional Skewness of the stock return variance, as cited in the literature [49, 50].

For the robustness check, we also consider other risk measures discussed in this paragraph, and the results do not change.

People and Investor Attention

We explore investor sentiment dynamics through a stock market-inspired lens, using internet research as a crucial tool. We argue that the Internet is a crucial avenue for gauging investor sentiment, especially during crisis periods when a significant portion of the population is confined to their homes, such as during the Covid-19 outbreak.

The Internet, acting as a primary information source, becomes instrumental for a diverse spectrum of investors, spanning institutional entities to individual households.

Our contribution aligns with the existing body of literature that delves into the intersection of online research activities and their intricate connection to risk considerations. Within this framework, we conceptualize online research as a manifestation of public interest, specifically geared toward concerns related to the crypto market—a surge in activity fueled by stakeholders (or token owners) seeking to stay abreast of patterns within the crypto landscape.

Drawing inspiration from the insights presented by Zhao et al. [53], our study posits that individuals engage in online research to evaluate endeavors and gauge public perceptions of performance behaviors. Mirroring the context of the stock market and beyond, stakeholders, including investors, leverage online research as a means of accessing information and asserting a form of regulatory oversight. This proactive engagement empowers stakeholders to exercise control, monitoring developments and staying well-informed about events and news that could potentially impact the companies and financial markets they are vested in.

To operationalize our approach, we draw on the foundations laid by previous literature [41, 42] and employ an internet search volume behavior index as a proxy for investor sentiment. This strategic choice allows us to establish a meaningful link between the attention parameter and token volatility, drawing upon insights from works such as Smales [43] and Tripathi and Pandey [44]. Our methodological approach enhances our understanding of the intricate interplay between online research, investor sentiment, and the dynamic landscape of the crypto market during unprecedented times.

We embark on the creation of a novel index, derived from the mean value of the Google Search Volume Index for key terms encompassing "crisis," "cryptocurrency," and "risk contagion." This innovative index serves as a foundational parameter in our exploration, enabling us to classify scenarios based on the level of contagion.

Our methodological approach involves the following steps. We aggregate the Google Search Volume Index for the specified keywords, calculating their mean value. This mean value, reflective of the collective online interest in crisis, cryptocurrency, and risk contagion, becomes a central parameter in our subsequent analysis. Moving forward, we employ this new index as a pivotal input for a SIR model. The SIR model, a widely used epidemiological framework, is adapted to our context, utilizing the derived index value as a crucial parameter. This strategic integration allows us to classify scenarios based on the influence of the aggregated online interest in crisis-related terms. By coupling the information gleaned from Google Search Volume with the SIR model, we establish a framework that discerns contagion scenarios in a nuanced manner. This innovative index not only reflects the collective attention on crisis, cryptocurrency, and risk contagion but also serves as a dynamic parameter guiding our classification of contagion intensity.

In essence, our approach amalgamates insights from online search behaviors with a robust epidemiological model, presenting a comprehensive strategy for classifying contagion scenarios. This innovative index stands as a testament to our commitment to leveraging diverse data sources and methodologies to enhance our understanding of contagion dynamics in the context of cryptocurrency and financial risk.

We employ this index, as elucidated in the subsequent paragraph, to moderate high and low risk levels and to delineate the designation of "Infected" within the SIR model.

SIR Model

We aim to model the spreading of financial risk among DeFi exchanges by employing a dynamical compartmental approach based on SIR classification. This mathematical tool was developed in the context of epidemiological models to analyze how an infectious illness spreads from its initial outbreak [54, 55]. Since it is possible to draw comparisons between financial systems and ecosystems, the SIR paradigm can be reviewed in the perspective of economics. To model crisis contagion, for example, the method is used to the banking network [33, 56], global financial crises across different nations [57], and credit risk contagion of peer-to-peer lending platforms on the Internet [58].

In our case, SIR approach is applied for modelling the contagion in terms of crash risk with low and high levels in a crypto market. The underlying concept is that platforms with higher risk are referred to as "Infected" because they have the potential to infect those with lower risk, which are referred to as "Susceptible". To refine our approach, we introduce a corrective measure for risk, incorporating the investor/people attention index. This correction factor is applied to distinguish between high and low risk for Infected tokens through the index established on GVSI (as detailed in Sect. 3.4). By doing so, we align high or low risk assessments with the prevailing perception among people, thereby calibrating our risk evaluations in accordance with public sentiment.

A portion of infected platforms become able to control and sustain a minimal degree of risk, making them no longer contagious. Consequently, these platforms are "Recovered" after healing and get a temporary financial immunity for a period of length $\tau > 0$. Under a mathematical viewpoint, this parameter τ represents a time delay involved in the dynamics. Over the time period τ, immunity ends, and some cryptocurrency platforms that have recovered may revert to come back susceptible compartment.

In this framework, risk contagion is described by modelling the dynamics of densities $S(t)$, $I(t)$ and $R(t)$ related to the susceptible, infected and recovered compartments, respectively, at each time $t \geq 0$. Moreover, by assuming that $0 < \delta < 1$ and $0 < \gamma < 1$, we account for the recovery rate δ from the high risk to the low risk and suppose that a portion of cryptocurrency platforms exits the market at any time according to the mortality rate γ. Under the previously stated reasoning, recovered cryptocurrency platform density evolves according to

$$\frac{dR}{dt} = \delta \cdot I(t) - \gamma \cdot R(t) - e^{-\gamma\tau}\delta \cdot I(t - \tau), \tag{17.9}$$

where the term $\delta \cdot I(t)$ corresponds to the portion of infected cryptocurrency platforms which is recovered, $\gamma \cdot R(t)$ is related to the cryptocurrency platforms which leave the market, while the portion $e^{-\gamma\tau}\delta \cdot I(t - \tau)$ reverts to susceptible compartment again. The previous equation can be integrated and density $R(t)$ can be evaluated once $I(t)$ is known. Therefore, we focus on the dynamics of the infected class which is closely related to the one of the susceptible compartments. In this respect, the strong analogy between any financial market and an ecosystem can be exploited to model the contagion among cryptocurrency platforms by employing Holling's response functions, which represent a common tool for studying population dynamics (for instance see [59, 60], [61], [62]. The extremely quick interconnectedness in cryptocurrency markets allows us to disregard the incubation period needed by a high-risk platform to process a susceptible one through infection. Therefore, we employ the type I response and model risk spread by the bilinear incidence term $a \cdot S(t) \cdot I(t)$, where $a > 0$ measures the interconnections among the exchanges and represents the removal rate due to contagion. Assuming further that new susceptible platforms enter the market at a given growth rate $b > 0$, the dynamics of the susceptible and the infected are described by the delay differential system.

$$\frac{dS}{dt} = b - \gamma \cdot S(t) - a \cdot S(t) \cdot I(t) + e^{-\gamma\tau}\delta \cdot I(t - \tau),$$

$$\frac{dI}{dt} = a \cdot S(t) \cdot I(t) - (\delta + \gamma) \cdot I(t), \tag{17.10}$$

which is completed by the following initial conditions:

$$S(0) = S_0,$$
$$I(s) = I_0(s) \geq 0 \, for \, all \, s \in [-\tau, 0], \, with \, I_0(0) > 0. \tag{17.11}$$

We assume that $I_0(\cdot)$ is a continuous function in the whole-time lag interval $[-\tau, 0]$ and determines the history of infected class before the initial time $t = 0$. Under this assumption, the fundamental theory of functional differential equations (see for instance [63] assures that the previous differential model admits a unique solution satisfying the initial conditions. It is not so difficult to prove that any solution gets positive values at any time.

The previous model has been proposed in the literature by Kyrychko and Blyuss, [64], for describing a disease transmission and an epidemic behavior. Here we apply this approach in the different framework of risk transmission through a cryptocurrency market.

The question of whether risk continues to exist in the cryptocurrency market over time requires careful consideration. In this respect, it is worthwhile to discuss the existence of different steady states. An important role is played by the basic reproduction number defined as

$$\rho_0 = \frac{ba}{\gamma(\gamma + \delta)}. \tag{17.12}$$

The model admits the risk-free steady state $E_0^* = (b/\gamma, 0)$ and another non-trivial equilibrium $E_\tau^* = (S_\tau^*, I_\tau^*)$ with

$$S_\tau^* = \frac{\gamma + \delta}{a}, I_\tau^* = \frac{\gamma(\gamma + \delta)}{a(\gamma + \delta - \delta e^{-\gamma\tau})}(\rho_0 - 1). \tag{17.13}$$

We notice that E_τ^* corresponds to an endemic or not-free-risk equilibrium. More-over, understanding long-term risk contagion requires a thorough analysis of steady state stability. According to the results provided in Kyrychko and Blyuss [64], the basic reproduction number ρ_0 has a cutoff value of 1 which marks the boundary between two distinct regions of stability. The first region corresponds to the case when $\rho_0 < 1$: the risk-free equilibrium E_0^* is locally asymptotically stable and no other equilibrium is feasible. In this stability region, risk contagion tends to vanish at the long run as the trajectories of the SIR system converge towards a risk-free situation corresponding to E_0^*. On the other hand, the second region corresponds to the opposite case when $\rho_0 > 1$: E_0^* is unstable, while the not-free-risk equilibrium E_τ^* becomes feasible. Without going into details, we state that condition $\rho_0 > 1$ can be enforced in order to guarantee the not-free-risk equilibrium point's stability both locally and globally.

Results

In this section, we endeavor to separate the results based on their short and long-term effects. The focus is on disentangling the outcomes derived from VAR analysis and spillover effects, which predominantly address short-term dynamics and their imme-diate impact on the behaviors of price tokens. Concurrently, the SIR results delve into the medium to long-term effects, providing insights into the overall structural dynamics of the token market. This dual perspective, examining both short-term volatility and the enduring impact on the broader market structure, aims to enhance the readability and comprehensive understanding of the results. By juxtaposing these two distinct aspects, we aim to offer a holistic view of the intricate dynamics shaping the cryptocurrency token landscape.

Short-Term Interconnection and Spillover Effect

In Fig. 17.1, the Spillover index is plotted for the last two years of a sample. We consider these two years according to Table 17.2 evaluating more tokens. Since the range goes from 0 to 100, the figure value shifts from 54 in April 2021 to 82 in October 2021. After this period, the Spillover index range is still high. The ascending phase of the Spillover index aligns with an upward trajectory in the valuation of cryptocurrencies, specifically referencing the S&P Cryptocurrency Broad Digital Market Index (USD). However, diverging from the latter, despite a substantial decline in cryptocurrency values commencing in November 2021 (with the pinnacle reached on November 10, 2021), the spillover index demonstrates a propensity to sustain elevated levels even in the latter period. This observation implies that the impact and interconnectedness across diverse platforms, notably heightened during the cryptocurrency boom, have endured beyond the reduction in cryptocurrency values. Consequently, in the short term, the interconnectivity between platforms has markedly surged, exhibiting a robust correlation with the people's attention metrics employed.

Simulated Dynamics Under the SIR Approach

Our numerical simulations illustrate the importance of time delay in establishing a long-term equilibrium for crypto platforms. Indeed, the results reveal that the shorter the financial immunity delay time, the sooner equilibrium is reached, and it is characterized by a low level of infection.

As a crucial observation, it's important to note that the time delay is not explicitly quantified in the available data. Nevertheless, we take a nuanced approach by considering various values for the time delay parameter (τ) to simulate diverse risk dynamics, accommodating different assumptions regarding the duration of financial immunity. Although precise measurements of the time delay are unavailable, we explore various values for τ to capture a spectrum of risk scenarios. Upon examining the data, we find that the minimum period between successive incidents of the same platform crashing spans three years throughout the sample. Consequently, we contend that τ does not surpass the threshold of 3. Given the recent surge in crypto platform crashes, it is reasonable to assume that, in the absence of policy and regulatory intervention, the immunity period is relatively short in comparison to this threshold. In this context, considering the swift pace facilitated by technology in market transactions, it becomes meaningful to compare the dynamics of very brief periods of immunity with longer intervals. To explore this, we delineate three distinct scenarios. Firstly, we assume a very brief temporary immunity, setting $\tau = 0.5$; subsequently, we consider a longer time delay, opting for $\tau = 2$. Finally, the third scenario involves a significantly extended financial immunity, setting $\tau = 3$. Numerical simulations of risk contagion are conducted, and the corresponding dynamics

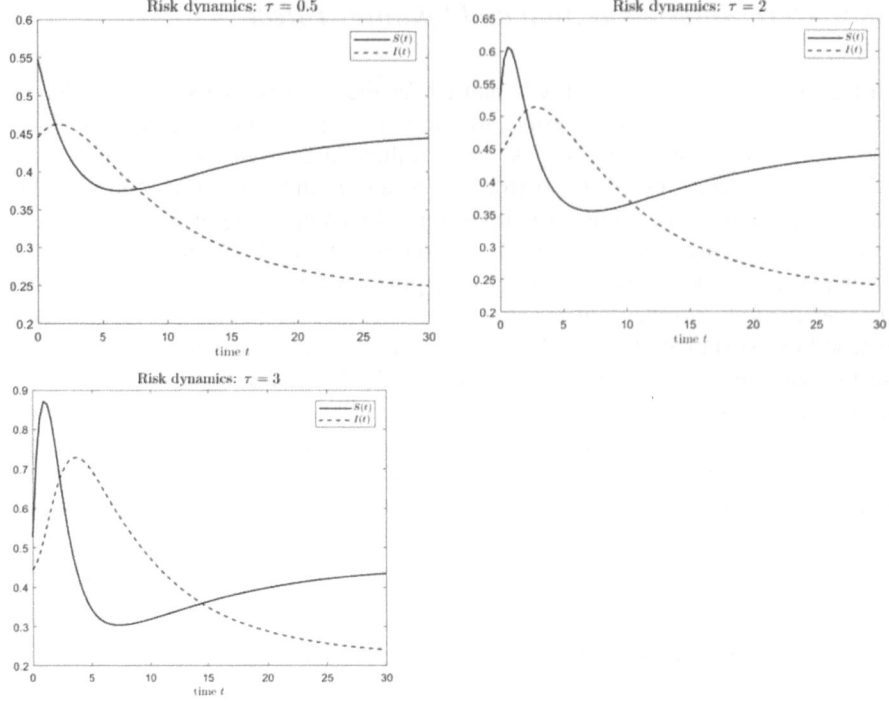

Fig. 17.2 SIR dynamics in terms of Susceptible and Infected densities

are illustrated in Fig. 17.2. The results for $\tau = 0.5$, $\tau = 2$, and $\tau = 3$ are displayed. This comprehensive exploration allows us to assess and compare the implications of different time delay scenarios on the simulated dynamics of risk contagion.

The not-free-risk equilibrium endemically attracts the trajectories of SIR solution in each case connected to the varied time delays under study. From a financial standpoint, this means that risk infection will stay prevalent in the market among crypto platforms in the long term. However, it is reassuring that at steady state, the number of vulnerable platforms surpasses the level of infected platforms by roughly 30%. Additionally, it is possible to note that as time delay τ lengthens, then the level of infected cryptocurrency platforms is lower at the steady state.

Conclusions and Policy Implications

In summary, our study highlights the growing concerns regarding contagion risk and the performance of interconnected assets, especially within cryptocurrency platforms. The significance of these findings is underscored by the implication of a potentially endemic condition, suggesting a scenario with low or virtually no risk. Turning our attention to the global landscape, the policy implications aimed at reducing the period of financial vulnerability present a formidable challenge. While the role of over-indebtedness and collateral assets is evident in platform failures and contagion spread, the ultimate challenge lies in the implementation of effective global policies. It is crucial to maintain awareness of the contrast between cryptocurrencies viewed as a form of investment and those regarded strictly as money, especially when interpreting the outcomes of risk contagion simulations. The stark difference in interpretation requirements for these two aspects of cryptocurrencies highlights the necessity for a currency, whether considered an investment or medium of exchange, to possess a maximum level of confidence to achieve a risk-free equilibrium, irrespective of trading circuit failures.

The lack of boundaries within the crypto asset ecosystem limits the effectiveness of national regulatory efforts (see to [65] and emphasizes the importance of international collaboration. The lack of boundaries within the crypto asset ecosystem hampers the effectiveness of national regulatory efforts and underscores the need for international collaboration. For instance, the European Union's Market-in-Crypto-Asset (MiCA) regulatory framework aims to address this issue by promoting legislative uniformity across member states. A worldwide framework could increase collaboration across platforms and users, therefore encouraging the adoption of risk-prevention and risk-containment strategies. Adhering to the constraints and control mechanisms stated in international laws may encourage cryptocurrency platforms to embrace better ethical standards, particularly when it comes to client relations. The success of these rules is dependent on collective collaboration and conformity to the global framework.

Our analysis marks an initial effort to address and mitigate risk within a specific crypto market. While the dynamics proposed by the SIR model are a valuable aspect of future research, it is essential to acknowledge the limitations of our study. Overlooking factors such as the high degree of interconnectedness among platforms and the rapid spread of effects through technology, our analysis lays the groundwork for further exploration and a more comprehensive understanding of risk dynamics in the cryptocurrency landscape. Future work should also take into account the selection of governance tokens, the types of trade assets, and the rights associated with each contract. On the other hand, this work serves as the foundation for an analysis that can be strengthened by incorporating cooperative activities to reduce opportunistic behavior and control risk contagion.

References

1. Makridis, C.A., Frowis, M., Sridhar, K., B.hme, R.: The rise of decentralized cryptocurrency exchanges: Evaluating the role of airdrops and governance tokens. J. Corp. Finance **79**, 102358 (2023)
2. Haldane, A.G., May, R.M.: Systemic risk in banking ecosystems. Nature **2011**(469), 351–355 (2011)
3. Roukny, T., Bersini, H., Pirotte, H., Caldarelli, G., Battiston, S.: Default cascades in complex networks: Topology and systemic risk (2013)

4. Leonidova, A.V., Rumyantsevb, E.L.: Default contagion risks in Russian interbank market. Phys. A Stat. Mech. Its Appl. **2016**, 451 (2016)
5. Callon, M.: Techno-economic Networks and Irreversibility. Sociol. Rev. **1990**(38), 132–161 (1990)
6. Babus, A.: The formation of financial networks. Rand J. Econ. **2016**(47), 239–272 (2016)
7. Battiston, S.; Caldarelli, G.: Systemic risk in financial networks. J. Financ. Manag. Mark. Inst. **1**, 129–154 (2013)
8. Feinstein, Z.: Financial contagion and asset liquidation strategies. Oper. Res. Lett. **2017**(45), 109–114 (2017)
9. Caldarelli, G., Battiston, S., Garlaschelli, D., Catanzaro, M.: Emergence of Complexity in Financial Networks; Ben-Naim, E., Frauenfelder, H., Toroczkai, Z. (eds.) Springer: Berlin/Heidelberg, Germany, pp. 399–423 (2004)
10. Aït-Sahalia, Y., Cacho-Diaz, J., Laeven, R.J.A.: Modeling financial contagion using mutually exciting jump processes. J. Financ. Econ. **2015**(117), 585–606 (2015)
11. Barja, A., Martínez, A., Arenas, A., Fleurquin, P., Nin, J., Ramasco, J.J.: Tomás, E. Assessing the risk of default propagation in interconnected sectoral financial networks. EPJ Data Sci. **8**, 32 (2019)
12. Torri, G., Radi, D., Dvořáčková, H.: Catastrophic and systemic risk in the non-life insurance sector: A micro-structural contagion approach. Financ. Res. Lett. **2022**(47), 102718 (2022)
13. Han, L.: Controlling Default Contagion Through Small-World Networks. Procedia Comput. Sci. **2019**(154), 47–53 (2019)
14. Previati, A.D., Aliano, M., Galloppo, G., Paimanova, V.: Why do banks react differently to short-selling bans? Evidence from the Asia-Pacific area and the United States. Q. Rev. Econ. Financ. **2021**(80), 144–158 (2021)
15. Akerlof, G.A., Shiller, R.J.: Animal Spirits: How Human Psychology Drives the Economy, and Why It Matters for Global Capitalism; Princeton University: Princeton, p. 2009. NJ, USA (2009)
16. Adekoya, O.B., Oliyide, J.A.: Commodity and financial markets' fear before and during COVID-19 pandemic: Persistence and causality analyses. Resour. Policy **2022**(76), 102598 (2022)
17. Giot, P.: Relationships between implied volatility indexes and stock index returns. J. Portfolio Manag. **2005**(31), 92–100 (2005)
18. Economou, F., Panagopoulos, Y., Tsouma, E.: Uncovering asymmetries in the relationship between fear and the stock market using a hidden-co-integration approach. Res. Int. Bus. Financ. **2018**(44), 459–470 (2018)
19. Chen, X., Chiang, T.C.: Empirical investigation of changes in policy uncertainty on stock returns- Evidence from China's market. Res. Int. Bus. Financ. **2020**(53), 101183 (2020)
20. Filbeck, G., Kumar, S., Liu, J., Zhao, X.: Supply chain finance and financial contagion from disruptions evidence from the automobile industry. Int. J. Phys. Distrib. Logist. Manag. **2016**(46), 414–438 (2016)
21. Lian, Y.: Financial distress and customer-supplier relationships. J. Corp. Financ. **2017**(43), 397–406 (2017)
22. Agca, S., Babich, V., Birge, J.R., Wu, J.: Credit shock propagation along supply chains: Evidence from the CDS market. Manag. Sci. **2021**(68), 6506–6538 (2021)
23. Barro, D., Basso, A.: Credit contagion in a network of firms with spatial interaction. Eur. J. Oper. Res. **2010**(205), 459–468 (2010)
24. Barreto, G., Artes, F.: Spatial correlation in credit risk and its improvement in credit scoring. Eur. J. Oper. Res. **2016**(249), 517–524 (2016)
25. Calabrese, R., Andreeva, G., Ansell, J.: 'Birds of a feather' fail together: Exploring the nature of dependency in SME defaults. Risk Anal. **2019**(39), 71–84 (2019)
26. Maté-Sánchez-Val, M., Lóopez-Hernandez, F., Rodriguez Fuentes, C.C.: Geographical factors and business failure: An empirical study from the Madrid metropolitan area. Econ. Model. **2018**(74), 275–283 (2018)

27. Shi, Q., Sun, X., Jiang, Y.: Concentrated commonalities and systemic risk in China's banking system: A contagion network approach. Int. Rev. Financ. Anal. **2022**(83), 102253 (2022)
28. Egloff, D., Leippold, M., Vanini, P.: (2007) A simple model of credit contagion. J. Bank. Financ. **31**, 2475–2492 (2007)
29. Hertzel, M.G., Li, Z., Officer, M.S., Rodgers, K.J.: Inter-firm linkages and the wealth effects of financial distress along the supply chain. J. Financ. Econ. **2008**(87), 374–387 (2008)
30. Escribano, E., Maggi, M.: (2016) Intersectoral default contagion: A multivariate Poisson autoregression analysis. Econ. Model. **82**, 376–400 (2016)
31. Xie, X., Shi, X., Gu, J., Xu, X.: Examining the Contagion Effect of Credit Risk in a Supply Chain under Trade Credit and Bank Loan Offering. Omega **2023**(115), 102751 (2023)
32. Calabrese, R.: Contagion effects of UK small business failures: A spatial hierarchical autoregressive model for binary data. Eur. J. Oper. Res. **2022**(305), 989–997 (2022)
33. Fanelli, V., Maddalena, L.: A nonlinear dynamic model for credit risk contagion. Math. Comput. Simul **2020**(170), 45–58 (2020)
34. Xua, K., Qianb, Q., Xiec, X., Zhoud. Z.: Study on the contagion mechanism of associated credit risk with corporate senior executives' alertness. Procedia Comput. Sci. **199**, 207–214 (2022)
35. Ciano T.: Bitcoin price prediction and machine learning features: New financial scenarios. In: Encyclopedia of Monetary Policy, Financial Markets and Banking (2023)
36. Aramonte, S., Huang, W. Schrimpf, A.: DeFi risks and the decentralisation illusion. BIS Q. Rev. (2021)
37. Bains, P., Ismail, A., F., M. Sugimoto, N.: Regulating the crypto ecosystem: The case of unbacked crypto assets. FinTech Notes No 2022/007. Int. Monet. Fund (2022)
38. McLeay, M., Radia, A. Thomas, R.: Money creation in the modern economy. Bank England, Q. Bull. (2014)
39. Bazan-Palomino, W.: Interdependence, contagion and speculative bubbles in cryptocurrency markets. Finance Res. Lett. **49**, 103132 (2022)
40. Kruttli, M.S., Monin Sumudu, F., Watugala, W.: The life of the counterparty: Shock propagation in hedge fund-prime broker credit networks. J. Financial Econ. **146**(3), 965–988 (2022)
41. Gao, Z., Ren, H., Zhang, B.: Googling investor sentiment around the world. J. Financ. Quant. Anal. **55**(2), 549–580 (2020)
42. Hsu, Y.L., Tang, L.: Effects of investor sentiment and country governance on unexpected conditional volatility during the COVID-19 pandemic: Evidence from global stock markets. Int. Rev. Financ. Anal. **82**, 102186 (2022)
43. Smales, L.A.: Investor attention and global market returns during the COVID-19 crisis. Int. Rev. Financ. Anal. **73**, 101616 (2021)
44. Tripathi, A., Pandey, A.: Information dissemination across global markets during the spread of COVID-19 pandemic. Int. Rev. Econ. Financ. **74**, 103–115 (2021)
45. Diebold, F.X., Yilmaz, K.: Better to give than to receive: Predictive directional measurement of volatility spillovers. Int. J. Forecast. **28**, 57–66 (2012)
46. Diebold, F.X., Yilmaz, K.: On the network topology of variance decompositions: measuring the connectedness of financial firms. J. Econom. **182**, 119–134 (2014)
47. Diebold, F. X., Yilmaz, K.: Trans-Atlantic equity volatility connectedness: U.S. and European financial institutions. J. Financial Econom. **14**, 81–127 (2016)
48. Hutton, A.P., Marcus, A., Tehranian, H.: Opaque financial reports, R2, and crash risk. Can. Entomo. **94**, 67–86 (2009)
49. Dumitrescu, A., Zakriya, M.: Governance, information flow, and stock returns. J. Corp. Finance **72**, 102168 (2022)
50. Habib, A., Hasan, M.M., Jiang, H.: Stock price crash risk: review of the empirical literature. Account. & Finance **58**, 211–251 (2018)
51. Ferreira, M. A., Laux, P. A.: Corporate governance, idiosyncratic risk, and information flow. J. Financ. 62(2), 951–989 (2007)
52. Tadele, H., Ruan, X., Li, W.: Corporate governance and firm-level jump and volatility risks. Appl. Econ. **54**(22), 2529–2553 (2022)

53. Zhao, X., Fang, L., Zhang, K.: Online search attention, firms' ESG and operating performance. Int. Rev. Econ. Financ. **88**(2023), 223–236 (2023)
54. Kendall, D.G.: Deterministic and stochastic epidemics in closed populations. Proc. Third Berkeley Symp. Math. Stat. Probab.: Contrib. Biol. Probl. Health **4**(1956), 149–165 (1956)
55. Kermack, W.O., McKendrick, A.G.: A contribution to the mathematical theory of epidemics. Proc. R. Soc. London. Ser. A, Contain. Pap. Math. Phys. Character **115**(772), 700–721 (1927)
56. Cao, H.H., Zhu, J.M.: Research on banking crisis contagion dynamics based on the complex network of system engineering. Syst. Eng. Procedia **5**(2012), 156–161 (2012)
57. Garas, A., Argyrakis P., Rozenblat, C., Tomassini, M., Havlin, M.S.: Worldwide spreading of economic crisis. New J. Phys. **12**(11), 30–43 (2010)
58. Zhao, C., Li, M., Wang, J., Ma, S.: The mechanism of credit risk contagion among internet P2P lending platforms based on a SEIR model with time-lag. Res. Int. Bus. Financ. **57**(2021), 101407 (2021)
59. Holling, C.S.: The components of predation as revealed by a study of small-mammal predation of the European sawfly. Can. Entomo. **91**(1959), 293–320 (1959a)
60. Holling, C.S.: Some characteristics of simple types of predation and parasitism. Can. Entomo. **91**(1959), 385–398 (1959b)
61. Baker, C.M., Diele, F., Marangi, C., Martiradonna, A., Ragni, S.: (2018) Optimal spatiotemporal effort allocation for invasive species removal incorporating a removal handling time and budget. Nat. Resour. Model. **31**(4), e12190 (2018)
62. Marangi, C., Martiradonna, A., Ragni, R.: Optimal resource allocation for spatiotemporal control of invasive species. Appl. Math. Comput. **439**(2023), 127614 (2023)
63. Hale, J.: 1977. Theory of Functional Differential Equations. Springer-Verlag, Heidelberg (1977)
64. Kyrychko, Y.N., Blyuss, K.B.: (2005) Global properties of a delayed SIR model with temporary immunity and nonlinear incidence rate. Nonlinear Anal. RWA **6**(3), 495–507 (2005)
65. IMF: Elements of effective policies for crypto assets. Policy Paper No. 2023/004 (2023)

Chapter 18
Cryptocurrencies and Systemic Risk. The Spillover Effects Between Cryptocurrency and Financial Markets

Vincenzo Pacelli, Caterina Di Tommaso, Matteo Foglia, and Stefania Ingannamorte

Abstract This research delves into the intricate relationship between cryptocurrencies and systemic risk within the framework of global financial markets. Utilizing a comprehensive dataset that amalgamates relevant indices from the cryptocurrency market along with global equity indexes from Europe, the United States, and China, the study employs a VAR for VaR model. This approach allows for the computation of spillover effects at different risk quantiles, offering insights into both downside and upside risk scenarios. The analysis underscores the notable spillover between cryptocurrency and traditional financial markets, revealing a complex interplay of risk factors that are not confined to geographical or asset-class boundaries. Our findings suggest that these interconnections could have far-reaching implications for global financial stability, regulatory policies, and risk management practices. By shedding light on these underexplored dimensions of financial markets, this study contributes to a deeper understanding of the systemic risks introduced by the growing prominence of cryptocurrencies.

Keywords Cryptocurrency market · Spillover effects · Equity markets

V. Pacelli (✉) · C. Di Tommaso · M. Foglia · S. Ingannamorte
University of Bari Aldo Moro, Bari, Italy
e-mail: vincenzo.pacelli@uniba.it

C. Di Tommaso
e-mail: caterina.ditommaso@uniba.it

M. Foglia
e-mail: matteo.foglia@uniba.it

S. Ingannamorte
e-mail: stefania.ingannamorte@uniba.it

© The Author(s) 2025
V. Pacelli (ed.), *Systemic Risk and Complex Networks in Modern Financial Systems*,
New Economic Windows, https://doi.org/10.1007/978-3-031-64916-5_18

Introduction

The recent defaults of banks, including Silicon Valley Bank and Silvergate Bank, have raised concerns about the potential correlation between cryptocurrencies and systemic risk in the financial system. The relationship between the cryptocurrency market and the traditional financial market is multifaceted. While cryptocurrencies can be seen as disruptive, they are also prompting traditional banks to adapt, innovate, and explore opportunities in the evolving digital financial landscape. The regulatory environment, customer demand, and technological advancements will continue to shape the nature of this relationship in the years to come.

Overall, this chapter adds valuable insights to the growing body of literature that seeks to understand the intricate connections between cryptocurrencies and traditional financial markets (Cao and Xie 2022; Aharon et al. 2023; Ugolini et al. 2023; among others), considering both the variety of cryptocurrencies and the different market conditions. This research can be crucial for investors, policymakers, and academics seeking to navigate the evolving landscape of cryptocurrencies and their impact on the broader financial ecosystem. While previous papers have studied the reaction of the cryptocurrency market in bearish market conditions (Conlon and McGee 2020; Corbet et al. 2020), we explore the spillover effects between the cryptocurrency market and the broader financial market under both bearish (downward-trending) and bullish (upward-trending) conditions. This analysis can shed light on how cryptocurrencies impact traditional financial markets during different economic situations. Furthermore, this essay takes a more comprehensive approach, considering a cryptocurrency index, which includes multiple cryptocurrencies whereas much of the existing literature often concentrates on a single cryptocurrency and, specifically on Bitcoin or Ethereum (Patel et al. 2022; Corbet et al. 2020; Bouri et al. 2018). This broader perspective allows for a more holistic understanding of how various cryptocurrencies collectively affect financial markets. Finally, this chapter evaluates the interconnection between cryptocurrency and the financial market in different geographical areas. We focus on the U.S., Europe, and China to understand if the response of the financial market may vary depending on country factors or cryptocurrency market development.

The findings from this research shed light on the potential contagion effects and vulnerabilities that cryptocurrencies may pose to the financial system, particularly in the wake of cryptocurrency defaults. By analyzing the correlation patterns and considering the dynamics of risk transmission, this essay aims to provide insights for regulators, policymakers, and financial institutions in managing and mitigating potential systemic risks associated with cryptocurrencies.

Furthermore, this research will contribute to a better understanding of the risk-return characteristics of cryptocurrencies and their relationship with the stability of the financial markets. By considering the bearing and the bullish market conditions as critical events, we assess the systemic risk implications of cryptocurrencies in a real-world context, helping investors and market participants make more informed decisions regarding portfolio diversification and risk management strategies.

Overall, this chapter seeks to bridge the gap between cryptocurrencies and systemic risk, providing valuable insights into the potential interdependencies and risk factors within the financial system. Understanding the correlation between cryptocurrencies and systemic risk in the context of recent cryptocurrency defaults is crucial for ensuring financial stability and resilience in an increasingly digital and interconnected financial landscape.

Literature Review

The academic discourse surrounding cryptocurrencies is rapidly expanding, enveloping a diverse range of methodologies and presenting a variety of results that highlight the intricate dynamics of cryptocurrency markets.

Several prominent studies have utilized a range of models to analyze the behavior and relationships of cryptocurrencies with traditional financial assets. However, most of the recent empirical studies (Cao and Xie 2022; Zhang et al. 2021; Wang et al. 2022; Bendob et al. 2022) highlight the interconnectedness between cryptocurrencies and traditional financial markets, calling attention to the importance of monitoring and managing risks associated with this relationship.

Several research endeavors have harnessed various analytical models to scrutinize the link between cryptocurrencies and conventional financial assets. The prevailing trend in recent empirical investigations underscores the intricate interdependence between cryptocurrencies and traditional financial markets (Wang et al. 2022; Bendob et al. 2022; Cao and Xie 2022; Zhang et al. 2021; Scagliarini et al. 2022, among others), underscoring that there exists a complex relationship between the two markets, and it is imperative for vigilance and effective risk management to study this dynamic relationship.

The very recent literature applying different models to study the relationship between cryptocurrency and the stock market highlights that the relationship is not stable over time (Wang et al. 2022; Bendob et al. 2022) with a notable positive correlation during extreme market events such as Covid-19 and varies across regions in the case of Arab countries (Bendob et al. 2022). Some other studies, such as that of Cao and Xie (2022), reveal that the relationship between cryptocurrency and the stock market is complex because of an asymmetric and time-varying risk spillover effect. Specifically, studying China's financial market they point out that the spillover from cryptocurrencies to China's financial market is stronger than the reverse. Focusing on the relationship between Bitcoin and the U.S. dollar index, MSCI world equity index, S&P Goldman Sachs Commodity Index (GSCI), and PIMCO Investment Grade Corporate Bond Index ETF, Zhang et al. (2021) show the existence of downside risk spillover between Bitcoin and four assets, and they emphasize the necessity of dynamic risk management strategies.

Wang et al. (2022) employed the CVaR and ADCC-GARCH model to investigate the linkage between Bitcoin and traditional financial assets, uncovering a notable positive correlation between Bitcoin and risk assets, which intensifies during extreme market events such as the COVID-19 outbreak in 2020. Similarly, Bendob et al. (2022) used the DCC-GARCH model to explore the dynamic correlations between

Bitcoin, gold, oil, and stock market indices in selected Arab countries, highlighting Bitcoin's significantly varying behavior based on geographic regions and market conditions.

Cao and Xie (2022) utilized the TVP-VAR model to reveal an asymmetric and time-varying risk spillover effect between cryptocurrencies and China's financial market, wherein the spillover from cryptocurrencies to China's financial market is stronger than the reverse. Zhang et al. (2021) further investigated the risk spillover between Bitcoin and conventional financial markets, emphasizing the necessity of dynamic risk management strategies.

Scagliarini et al. (2022) ventured into an unprecedented pathway, leveraging the Granger causality and O-information methodology to scrutinize the high-order dependencies in the cryptocurrency trading network. The researchers emphasized the inherent unpredictability and extreme risks involved in the crypto market, advocating for policies grounded in tangible values to foster market stability.

In a study by Hassan et al. (2022), the researchers explored the dynamic spillover of cryptocurrency environmental attention across various assets using Continuous Wavelet Transforms (CWT), uncovering a negative impact of the ICEA index on the WTI and soybean indices.

Various other studies have brought forth insights into the dynamics of cryptocurrency markets. For instance, Johnson (2019) reflected on JPY/BTC trading behavior, suggesting a more domestic than international market for JPY/BTC trading. Furthermore, Charfeddine et al. (2020) investigated the relationship between cryptocurrencies and conventional assets, suggesting new opportunities for diversification. Soloviev et al. (2020) affirmed the applicability of the Random Matrix Theory in the early diagnosis of crisis phenomena in financial systems.

Jeris et al. (2022) took a meta-analytical approach to evaluate the evolution and current state of research on the relationship between cryptocurrency and stock markets, identifying critical areas and future research avenues.

Studies such as those by Hung (2021) using DECO-GARCH and Omane-Adjepong et al. (2021) employing the CSAD approach have investigated correlations and causality in emerging markets. In advanced economies, Isah and Raheem (2019) and Erdas and Caglar (2018) have used predictive models and asymmetric causality tests, respectively, to study the U.S. market.

The COVID-19 pandemic has catalyzed a surge of research reevaluating these interconnections. Jeribi et al. (2021) employed the NARDL technique to claim changing dynamics during the pandemic, while Mariana et al. (2021) used DCC-GARCH to analyze volatility. Studies like those by Kumah et al. (2021) and Lahmiri and Bekiros (2021) have further examined the pandemic's effects on market interdependence and long-term memory.

Other explorations extend into unconventional markets, such as Islamic stock indices, as well as focus on the impact of specific global events, including the COVID-19 pandemic (Umar et al. 2020; Grobys 2021; Caferra and Vidal-Tomás 2021).

Collectively, these studies indicate an increasingly interconnected financial landscape. They suggest that the relationship between cryptocurrencies and stock markets is neither static nor confined to traditional financial paradigms, but is rather influenced by global events and market sentiments.

This literature review stands as a testament to the vibrant and ever-evolving landscape of cryptocurrency research, drawing from a rich array of methodological approaches and findings. It underscores the crucial role of grounded policy interventions in fostering stability in a market characterized by volatility and intrinsic unpredictability.

Methodology

This chapter studies the extreme spillover effect between the cryptocurrency and equity markets. We use the VAR for VaR model (White et al. 2015) to compute the spillover effect at different quantiles (0.05 and 0.95), which represent the downside and upside risk, respectively. This method allows us to identify the degree and direction of the spillover effect across quantiles, which can help us to better understand the transmission mechanism of risk between the crypto world and equity markets. The VAR for VaR model is defined as follows:

$$q_{1t} = c_1(\theta) + a_{11}(\theta)r_{1(t-1)} + a_{12}(\theta)r_{2(t-1)}$$
$$+ b_{11}(\theta)q_{1(t-1)} + b_{12}(\theta)q_{2(t-1)} \tag{18.1}$$

$$q_{2t} = c_2(\theta) + a_{21}(\theta)r_{1(t-1)} + a_{22}(\theta)r_{2(t-1)}$$
$$+ b_{21}(\theta)q_{1(t-1)} + b_{22}(\theta)q_{2(t-1)} \tag{18.2}$$

where q_{1t} (q_{2t}) is the level θ conditional quantiles; $r_{1(t-1)}$ ($r_{2(t-1)}$) stands for the return; c_1 (c_2) is the constant term, while a_{ij} and b_{ij} are the coefficient terms. The (b_{ij}) coefficient captures the degree of risk, while q_{1t} and q_{2t} are the VaRs of the two-time series which are defined as:

$$q_{1,t} = VaR_{1,t} = -Q_\theta(F_{t-1}) = -inf\{q \in R|Pr(F_{t-1}) \geq \theta\} \tag{18.3}$$

$$q_{2,t} = VaR_{2,t} = -Q_\theta(F_{t-1}) = -inf\{q \in R|Pr(F_{t-1}) \geq \theta\} \tag{18.4}$$

where Q_θ is the quantile function at confidence interval $\theta \in (0,1)$. F_{t-1} denotes the information set available at the time $t - 1$.

To examine the effects of a one-off shock on the conditional quantiles of asset returns, we analyze the pseudo quantile impulse response functions (QIRFs) derived from the vector autoregressive (VAR) model for value at risk (VaR). The QIRFs allow us to investigate how negative (positive) shocks in one market spread to another market and how long it takes for the markets to absorb these shocks. The complete absorption occurs when the pseudo QIRFs converge to zero. The QIRFs are defined as follows: $\Delta_{is}(\tilde{r}_{1,t}) = \tilde{q}_{i,t+s} - q_{i,t+s}$, where $\tilde{q}_{i,t+s}$ represents the θth conditional quantile, while $q_{i,t+s}$ is the θth conditional quantile of unaffected return. We calculate the pseudo impulse response, starting with $i = 1$, to capture the response of the equity market to the shock in the crypto and then for $i = 2$, i.e., the response of the crypto market to the shock in the equity ones. Mathematically:

$$\Delta(\tilde{r}_{1,t}) = a_{11}(\tilde{r}_{1,t} - r_{i,t}) + a_{12}(\tilde{r}_{2,t} - r_{2,t}) \text{for } s = 1$$
$$\Delta(\tilde{r}_{1,t}) = b_{11}\Delta_{1,s-1}(\tilde{r}_{1,t}) + b_{12}\Delta_{2,s-1}(\tilde{r}_{1,t}) \text{for } s > 1 \tag{18.5}$$

for $i = 2$, i.e.,

$$\Delta(\tilde{r}_{2,t}) = a_{21}(\tilde{r}_{1,t} - r_{i,t}) + a_{22}(\tilde{r}_{2,t} - r_{2,t}) \text{for } s = 1$$
$$\Delta(\tilde{r}_{2,t}) = b_{21}\Delta_{1,s-1}(\tilde{r}_{2,t}) + b_{22}\Delta_{2,s-1}(\tilde{r}_{2,t}) \text{for } s > 1 \tag{18.6}$$

Data

In this section, we present an overview of the data that forms the foundation of our analytical framework in this chapter. Our dataset comprises crucial components that help us gain a comprehensive understanding of market dynamics. To capture the cryptocurrency market, we rely on the BITWISE 10 Crypto Index, which provides a thorough representation of the cryptocurrency landscape. To represent the stock market, we incorporate key global equity indexes from significant regions. For the United States (U.S.), we utilize the S&P 500, a prominent benchmark for the U.S. equity market (Jeris et al. 2022; Wang et al. 2022). To gain insights into the European market (EU), we consider the EuroStoxx 600 index (Bua et al. 2022; Stolowy and Paugam 2018). Following Li et al. (2023), Lao et al. (2018) and Jiang et al. (2020), for the Chinese market (CN), we rely on the Shanghai Stock Exchange Composite Index (Shanghai SE). This diverse and extensive dataset equips us with the tools necessary for a comprehensive analysis. By examining the interplay between the cryptocurrency and global equity markets across these distinct regions, we aim to have deep insights that contribute to a clear understanding of market dynamics and trends.

Table 18.1 presents the summary statistics of our sample. From Table 18.1, it is evident that the crypto market demonstrates the highest levels of volatility. Furthermore, the kurtosis values suggest that none of the series adheres to Gaussian distributions, a conclusion supported by the Jarque–Bera (JB) test results. Besides, the outcomes of the Elliott, Rothenberg, and Stock (ERS) test confirm that there is no presence of a unit root, meeting the stationarity prerequisite for our VAR modeling.

The trend of cryptocurrencies and global equity indexes is shown in Fig. 18.1. Figure 18.1 notably emphasizes the heightened volatility characteristic of the cryptocurrency market. One remarkable observation is that the cryptocurrency market became increasingly attractive amid the COVID-19 pandemic. This can be attributed to a number of factors, including the perception of cryptocurrencies as digital gold or a store of value, their decentralized nature, and their potential to generate profits during economic instability. However, the collapse of FTX in 2022 had a cascading impact on the entire cryptocurrency ecosystem. The failure of this major exchange led to a sudden and significant decline in market confidence, which ushered in the so-called

Table 18.1 Summary statistics

	Mean	Std. Dev	Skewness	Ex.Kurtosis	JB	ERS
BITWISE	0.02	0.04	−1.09	9.74	7230.32***	−11.83***
S&P 500	0.04	0.01	−0.86	16.30	19,475.03***	−9.28***
STOXX EUROPE 600	0.01	0.01	−1.36	18.13	24,355.26***	−9.15***
SHANGHAI SE	0.01	0.01	−0.64	6.09	2812.69***	−7.31***

The table reports the summary statistics of the cryptocurrency market (BITWISE 10 Crypto Index) and the global equity indexes (S&P500, Stoxx Europe 600, and Shanghai SE). We report the mean, standard deviation, Skewness, Kurtosis, Jarque–Bera (JB), and Elliott, Rothenberg, and Stock (ERS) tests

Note *** significant at the 1% level; ** significant at the 5% level; * significant at the 10% level

"Crypto Winter". During this period, cryptocurrency prices and market capitalizations plummeted, resulting in a colossal loss of value estimated to be around USD 2 trillion (Arner et al. 2023). Furthermore, it is noteworthy that the U.S. and EU equity indexes exhibit a parallel trend, both experiencing a decline in 2019. In contrast, the Chinese market follows a divergent path, with a decline observed at the outset of 2018 and again in late 2022.

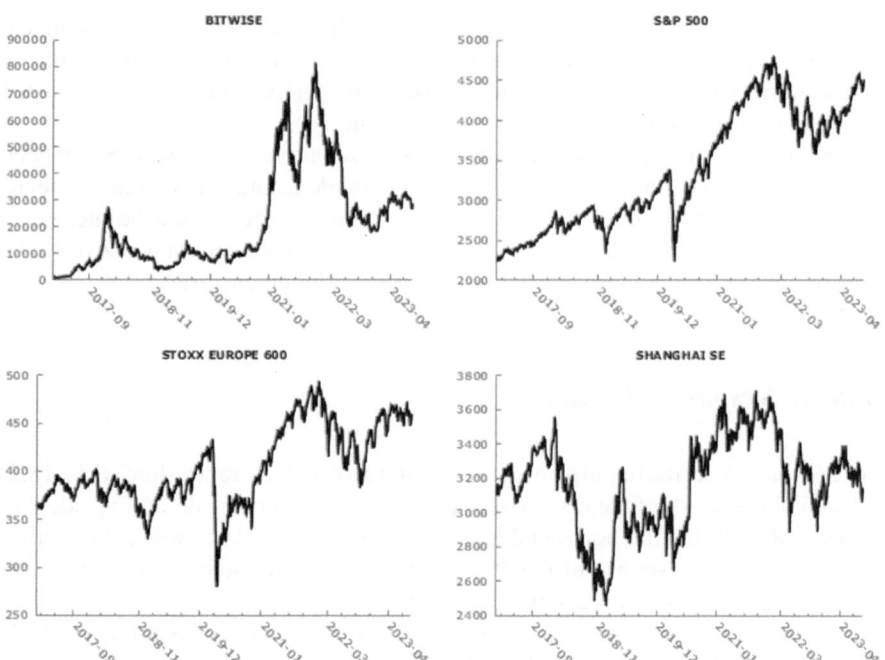

Fig. 18.1   Trend of BITWISE 10 Crypto Index, S&P 500 index, EuroStoxx 600 index, and Shanghai Stock Exchange Composite Index

Results

In Table 18.2, we present the results and standard errors of the VAR for VaR model, categorized by country (U.S., Europe, and China). Panel A presents findings related to downside quantile risks, specifically focusing on $\theta = 0.05$, which represents bearish market conditions. Conversely, Panel B showcases results pertaining to upside quantile risks, specifically targeting $\theta = 0.95$, symbolizing bullish market conditions.

The results for bearish market conditions in Panel A suggest an inverse correlation between returns and risks in both the cryptocurrency and equity markets. This implies that an increase in returns in either the crypto or equity markets corresponds to a reduction in the associated market risks. Similarly, the findings for bullish market conditions in Panel B also align with this direction, indicating that increased returns in these markets are associated with decreased risks.

Examining the parameters b_{11} and b_{22}, which capture the lagged values of risk, we observe their significance across all series and under various financial market conditions. The positive sign of these parameters indicates that the current level of risk is positively impacted by its past value, suggesting a persistence in risk dynamics. Analyzing these effects within specific markets, we identify heterogeneous outcomes. Notably, the U.S. equity market plays a substantial role in influencing the risk dynamics of the cryptocurrency realm. Both the downside and upside risks of cryptocurrencies appear to be influenced by the performance of the U.S. market (as indicated by coefficient a_{12}) as well as its own risk level (as indicated by coefficient b_{12}). This relationship, however, does not hold true for other markets. Interestingly, crypto returns emerge as a determinant of risk, both negative and positive, within the European (EU) market (as suggested by coefficient a_{21}).

In summary, our findings suggest that an increase in risk within the U.S. financial market can exert an influence on the risk in other markets, indicative of the presence of risk spillover effects. From an investor's standpoint, these results can be interpreted as reflecting "flight to quality" effects, where investors seek to divest perceived risky assets in favor of safer alternatives during periods of heightened risk.

Impulse Response Results

This section reports the impulse response results to a cryptocurrency shock (Fig. 18.2) and to a shock in the financial market (Fig. 18.3) in a downside and upside risk spillover. Recall that the horizontal axis represents time in days, while the vertical axis measures the response of the 1% quantiles of crypto (stock market), i.e., the percentage of returns, to an equity (crypto) shock.

Focusing on the reaction of the financial market to a cryptocurrency shock, Fig. 18.2 illustrates the reactions of the United States (U.S.), the European Union (EU) and China (CN) financial markets to a cryptocurrency shock in a downside (left-side) and upside (right-side) risk spillover. The results reveal that all three regions' stock markets—the EU, the U.S., and China—significantly respond to bearish conditions

Table 18.2 VAR for VAR results

Panel A: downside quantile ($\theta = 0.05$)

U.S. market vs crypto market

Crypto	c_1	a_{11}	a_{12}	b_{11}	b_{12}
	−0.001	−0.038	−0.512***	0.975***	−0.251***
	(0.001)	(0.029)	(0.228)	(0.015)	(0.105)
U.S	$c2$	a_{21}	a_{22}	b_{21}	b_{22}
	−0.001	0.008	−0.431***	0.001	0.783***
	(0.001)	0.008	(0.095)	(0.008)	(0.061)

EU market vs crypto market

Crypto	c_1	a_{11}	a_{12}	b_{11}	b_{12}
	−0.001	−0.028	−0.051	0.984***	−0.044
	(0.003)	(0.025)	(0.409)	(0.016)	(0.309)
EU	$c2$	a_{21}	a_{22}	b_{21}	b_{22}
	−0.003***	0.022***	−0.342***	−0.009	0.739***
	(0.001)	(0.007)	(0.101)	(0.009)	(0.079)

Chinese market vs crypto market

Crypto	c_1	a_{11}	a_{12}	b_{11}	b_{12}
	0.006	−0.039*	−0.044	0.939***	0.011
	(0.006)	(0.019)	(0.265)	(0.010)	(0.157)
CN	$c2$	a_{21}	a_{22}	b_{21}	b_{22}
	−0.001	0.038	−0.124***	0.003	0.925***
	(0.001)	(0.036)	(0.039)	(0.002)	(0.021)

Panel B: upside quantile ($\theta = 0.95$)

U.S. market vs crypto market

Crypto	c_1	a_{11}	a_{12}	b_{11}	b_{12}
	0.008*	0.147***	−0.115*	0.824***	−0.003
	(0.005)	(0.032)	(0.062)	(0.087)	(0.075)
U.S	$c2$	a_{21}	a_{22}	b_{21}	b_{22}
	0.001	−0.005	0.233***	0.001	0.869
	(0.001)	(0.005)	(0.038)	(0.007)	(0.022)

EU market vs crypto market

Crypto	c_1	a_{11}	a_{12}	b_{11}	b_{12}
	0.001	0.135***	0.001	0.911***	−0.007
	(0.001)	(0.039)	(0.007)	(0.022)	(0.012)
EU	$c2$	a_{21}	a_{22}	b_{21}	b_{22}
	0.009*	−0.157***	0.155***	0.013	0.786***
	(0.006)	(0.053)	(0.042)	(0.112)	(0.121)

Chinese market vs crypto market

Crypto	c_1	a_{11}	a_{12}	b_{11}	b_{12}
	0.001*	0.151***	0.043	0.821***	−0.082
	(0.000)	(0.028)	(0.144)	(0.074)	(0.116)

(continued)

Table 18.2 (continued)

CN	c2	a_{21}	a_{22}	b_{21}	b_{22}
	0.001	−0.004	0.073***	0.001	0.969***
	(0.001)	(0.005)	(0.015)	(0.005)	(0.001)

The Table reports the results of the VAR model. Panel A and Panel B report the results divided into downside ($\theta = 0.05$) and upside ($\theta = 0.95$) quantiles, respectively. We categorized the results by country (U.S., EU, and Chinese). Standard errors are reported in parenthesis. ***, **, and * indicate statistical significance at the 1%, 5% and 10% levels, respectively.

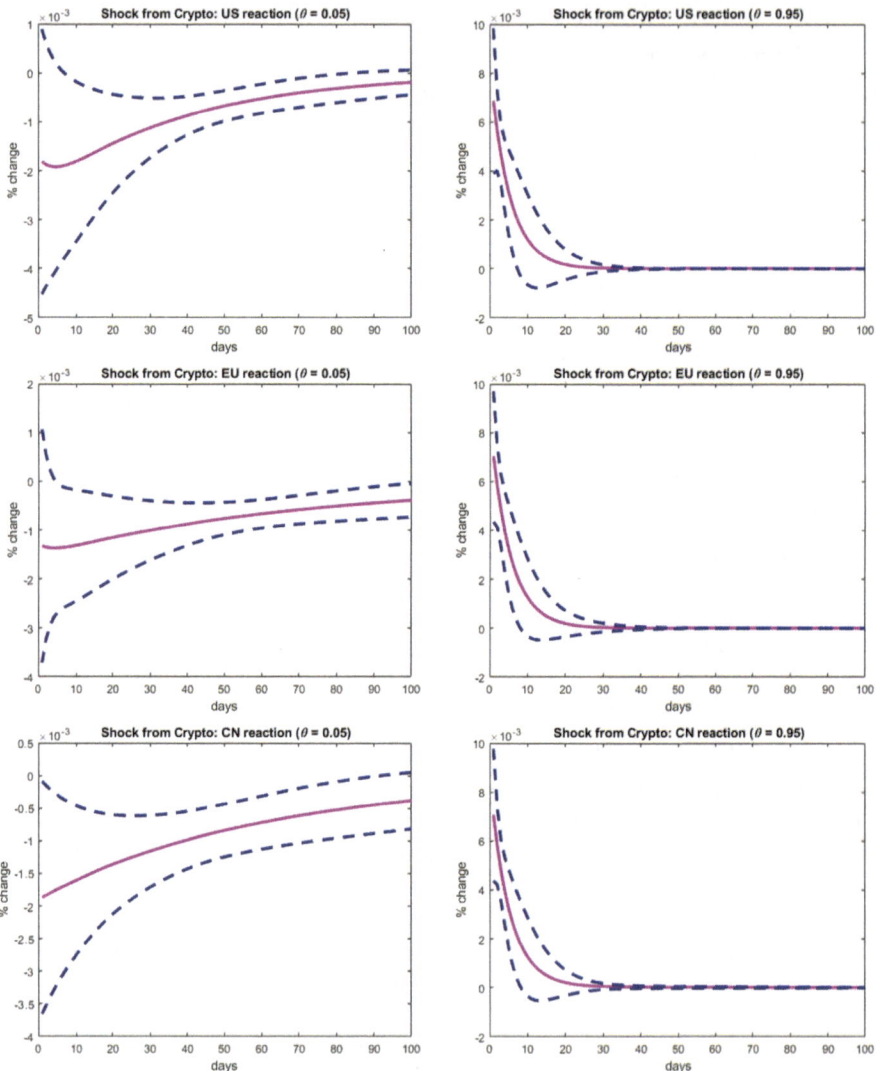

Fig. 18.2 The impulse response of the U.S., EU, and CN markets to a cryptocurrency shock in a downside (left-side) and upside (right-side) risk spillover

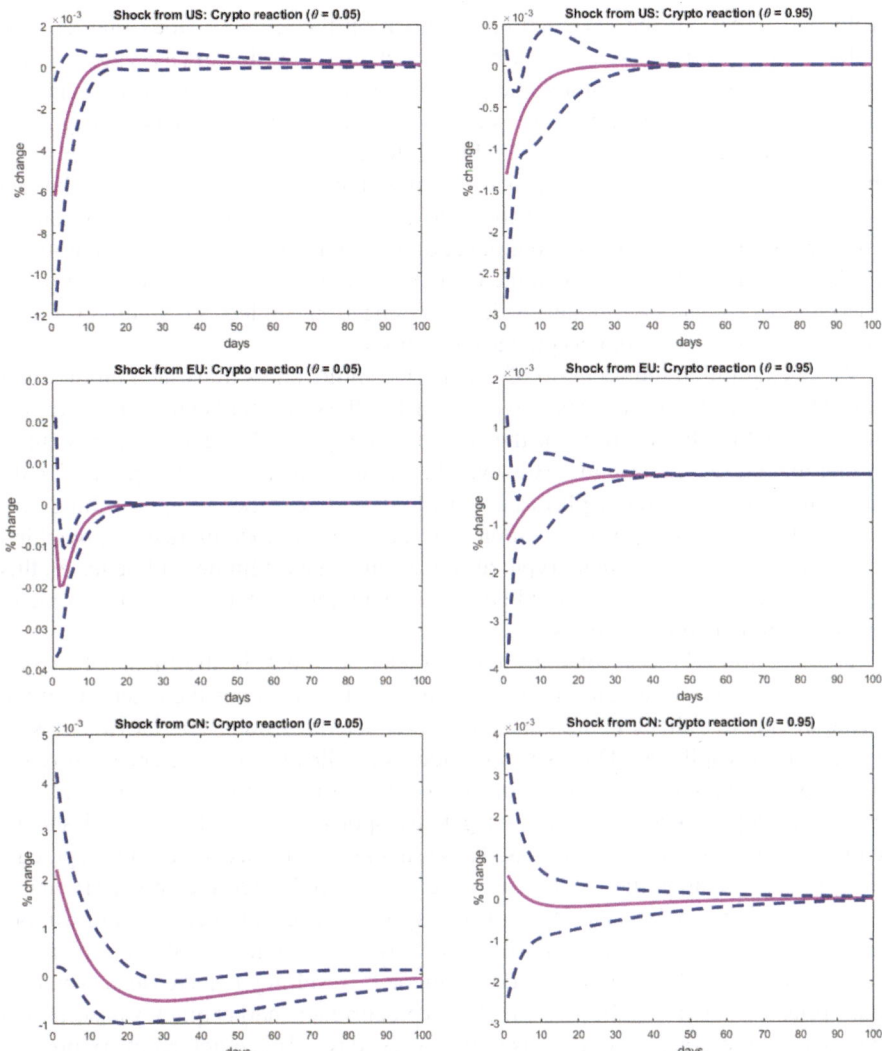

Fig. 18.3 The impulse response of the cryptocurrency market to the U.S., EU, and CN to a shock in the financial market in a downside (left-side) and upside (right-side) risk spillover

in the cryptocurrency market. However, the degree of these responses varies significantly. The Chinese stock market exhibits the most pronounced reaction, experiencing a substantial decline of -2.5%. The U.S. stock market falls in between, with a reduction of roughly -2%. In contrast, the EU stock market records a more moderate reduction, approximately -1%. Furthermore, it is worth noting that the Chinese stock market takes a longer time to return to equilibrium, where the percentage change reaches 0, in contrast to the EU and U.S. stock markets. This suggests a prolonged impact of the cryptocurrency shock on the Chinese stock market. The prolonged impact witnessed in the Chinese stock market following a cryptocurrency shock is the result of a complex interplay of various factors. These factors encompass investor sentiment, regulatory actions, liquidity conditions, market integration with cryptocurrencies, economic variables, information dissemination, and external shocks. However, China's stringent regulatory stance on cryptocurrencies, known for its strict measures, further contributes to an extended impact. Regulatory actions have the power to sway investor confidence and disrupt market stability. Furthermore, the degree of integration between the Chinese stock market and the cryptocurrency markets is a critical determinant. A deeper connection can lead to a more profound and prolonged impact from cryptocurrency shocks.

When examining the scenario of upside risk spillover, as illustrated on the right side of Fig. 18.2, we observe a positive response of the stock market to cryptocurrency shocks. Notably, the reactions of the stock markets in the EU and the U.S. exhibit similar magnitudes and timing. However, in line with Kostika and Laopodis (2020), the Chinese stock market displays a notably low positive response, approaching zero in magnitude. The low positive response of the Chinese stock market, approaching zero magnitude, suggests that cryptocurrencies may have a limited influence on this particular stock market. This could be due to regulatory factors, investor sentiment, or other regional considerations.

The reaction of the cryptocurrency market to a shock in the financial market varies from region to region. Figure 18.3 specifically illustrates the reactions of the crypto market to a shock in the financial market in a downside (left-side) and upside (right-side) risk spillover. During a downside risk spillover event, the cryptocurrency market demonstrates varying reactions to shocks in different financial markets. The cryptocurrency market experiences a negative response when confronted with shocks in the EU and U.S. financial markets. These shocks result in a decrease in cryptocurrency prices. Notably, the reaction is more pronounced when the shock originates from the U.S. financial market, with the cryptocurrency market showing a more substantial drop of approximately 6% in the days surrounding the shock. The negative reaction to shocks in the EU and U.S. markets, resulting in decreased cryptocurrency prices, is relatively short-lived. The cryptocurrency market quickly absorbs and incorporates these negative impacts within a few days. This quick recovery process brings the cryptocurrency market back to equilibrium, with its reaction approaching zero. This suggests resilience and adaptability within the cryptocurrency market to external financial market shocks. In contrast, the cryptocurrency market exhibits a positive reaction when a shock occurs in the CN financial market. This positive reaction may indicate that, during certain conditions, the cryptocurrency market views shocks from the Chinese financial market as opportunities or as less detrimental to its overall health. The positive reaction to shocks in the CN market, resulting in increased cryptocurrency prices, is relatively long-lived. The cryptocurrency market shows an increase in prices in the 15/20 days surrounding the shock.

When examining upside risk spillovers, the findings consistently reveal contrasting reactions within the cryptocurrency market when exposed to shocks from different financial markets. While the cryptocurrency market responds negatively to shocks originating in the EU and U.S. financial markets, it exhibits a positive response to shocks from the CN financial market. Notably, the response remains more accentuated when the shocks originate in the U.S. financial market (see e.g., Ji et al. 2020; Smales 2022). Importantly, in all instances, the reactions observed are of a transient nature, characterized by their short-lived duration (Li et al. 2023).

These different reactions can be attributed to several factors. For instance, market sentiment and perception of cryptocurrencies can vary from one region to another. In China, for example, there may be a different perception of cryptocurrencies and their potential, leading to a positive market reaction in response to a shock. The positive reaction to CN shocks may reflect positive sentiment in the region or the perception that cryptocurrency is a safe haven. However, it is important to underline that a crucial role is played by the Regulatory Environment. Regulatory clarity or the lack thereof can impact investor behavior and market reactions. Positive regulatory developments in CH, for example, could lead to a more favorable market response. It is important to note that China has had a historically strict regulatory stance on cryptocurrencies, including banning initial coin offerings (ICOs) and cryptocurrency exchanges. This strong regulatory environment can create a different market response compared to regions with more lenient or evolving regulations.

Conclusion

This essay investigates the relationship between the cryptocurrency market and the global equity indexes by considering the bearish and bullish market conditions in both markets. We employ a VAR for VaR model and an impulse response function to investigate how global financial markets react to the cryptocurrency market and vice versa.

Our results can be summarized as follows. First, the inverse correlation between returns and risks suggests that market participants often exhibit risk-averse behavior. In bearish market conditions ($\theta = 0.05$), investors tend to avoid risky investments and instead seek out safer, lower-risk assets. This is because they are more cautious and worried about the potential for further losses. As a result, they are willing to accept lower returns in exchange for the stability of less risky investments. On the other hand, in bullish market conditions ($\theta = 0.95$), investors are more optimistic about the future of the market and are willing to take on more risk in order to secure higher returns. This is because they believe that the potential for profit outweighs the risk of losses. As a result, they are willing to invest in riskier assets, even if these assets offer lower returns in the short term. This inverse relationship between risk and return, observed in both bearish and bullish market conditions, highlights the risk-return trade-offs. Investors' willingness to accept risk is intricately linked to their expectations of potential returns. These findings align with the concept of investors' sentiment-driven decisions and the pursuit of safer assets during turbulent times, i.e., the "flight to quality" effect. The results are in line with the work of Wang et al. (2022), which found this asymmetric effect between returns and risks in the context of oil markets. Second, the significance of the parameters b_{11} and b_{22}, indicating the persistence of risk dynamics, underscores the interconnectedness of financial markets. The results suggest that risk in one market is influenced by its own past risk levels, demonstrating that market conditions are not isolated events. The fact that these relationships are more pronounced in the U.S. market highlights its global importance and influence on risk dynamics (see e.g,. Ji et al. 2020; Smales 2022). Third, the influence of the U.S. market on both cryptocurrency and equity market risks is noteworthy. This highlights the role of the U.S. market as a global financial hub and indicates that events or changes in the U.S. market can have ripple effects in other financial markets worldwide. On the other hand, the fact that cryptocurrency returns are a determinant of risk in the European market underscores the growing influence of digital assets in the global financial landscape (OECD 2022).

The cryptocurrency market's sensitivity to traditional financial markets indicates a level of integration with the broader financial system. This can influence the behavior of cryptocurrency prices, potentially leading to correlations with traditional assets during certain market conditions. The pronounced reaction of the cryptocurrency market to shocks in the U.S. financial market, as well as its quick recovery, underscores the high volatility of cryptocurrencies. This can create opportunities for traders but also presents significant risks for investors. Cryptocurrency market reactions can influence investor sentiment and behavior. Sudden price drops in response to traditional market shocks may trigger panic selling or speculative buying, leading to amplified market dynamics.

The implications of the cryptocurrency market's responses to downside risk spillovers emphasize the need for careful risk management, highlight the potential for diversification challenges, and underscore the importance of understanding the dynamic and evolving nature of the cryptocurrency market. These implications are relevant to a wide range of stakeholders, from investors to regulators and researchers.

References

Aharon, D.Y., Butt, H.A., Jaffri, A., Nichols, B.: Asymmetric volatility in the cryptocurrency market: new evidence from models with structural breaks. Int. Rev. Financ. Anal. **87**, 102651 (2023)

Arner, D.W., Zetzsche, D.A., Buckley, R.P., Kirkwood, J.M.: The Financialization of Crypto: Lessons from FTX and the Crypto Winter of 2022–2023. Available at SSRN 4372516 (2023)

Bendob, A., Othman, A., Sirag, A.: Understanding the dynamic correlation between Bitcoin, gold, oil, and stock market indices in the selected Arab countries: application of the DCC-GARCH model to the banking and insurance sectors. Arab Monetary Fund **2022**, 25 (2022)

Bouri, E., Das, M., Gupta, R., Roubaud, D.: Spillovers between Bitcoin and other assets during bear and bull markets. Appl. Econ. **50**(55), 5935–5949 (2018)

Bua, G., Kapp, D., Ramella, F., Rognone, L.: Transition versus physical climate risk pricing in European financial markets: a text-based approach (2022)

Caferra, R., Vidal-Tomás, D.: Who raised from the abyss? A comparison between cryptocurrency and stock market dynamics during the COVID-19 pandemic. Financ. Res. Lett. **43**, 101954 (2021)

Cao, G., Xie, W.: Asymmetric dynamic spillover effect between cryptocurrency and China's financial market: evidence from TVP-VAR based connectedness approach. Financ. Res. Lett. **49**, 103070 (2022)

Charfeddine, L., Benlagha, N., Maouchi, Y.: Investigating the dynamic relationship between cryptocurrencies and conventional assets: implications for financial investors. Econ. Model. **85**, 198–217 (2020)

Conlon, T., McGee, R.: Safe haven or risky hazard? Bitcoin during the COVID-19 bear market. Financ. Res. Lett. **35**, 101607 (2020)

Corbet, S., Larkin, C., Lucey, B.: The contagion effects of the COVID-19 pandemic: evidence from gold and cryptocurrencies. Financ. Res. Lett. **35**, 101554 (2020)

Erdas, M.L., Caglar, A.E.: Analysis of the relationships between Bitcoin and exchange rate, commodities and global indexes by asymmetric causality test. Eastern J. Eur. Stud. **9**(2) (2018)

Grobys, K.: When Bitcoin has the flu: on Bitcoin's performance to hedge equity risk in the early wake of the COVID-19 outbreak. Appl. Econ. Lett. **28**(10), 860–865 (2021)

Hassan, M.K., Hasan, M.B., Halim, Z.A., Maroney, N., Rashid, M.M.: Exploring the dynamic spillover of cryptocurrency environmental attention across the commodities, green bonds, and environment-related stocks. North Am. J. Econ. Finance **61**, 101700 (2022)

Hung, N.T.: Bitcoin and CEE stock markets: fresh evidence from using the DECO-GARCH model and quantile on quantile regression. Eur. J. Manag. Bus. Econ. **30**(2), 261–280 (2021)

Isah, K.O., Raheem, I.D.: The hidden predictive power of cryptocurrencies and QE: evidence from US stock market. Physica A **536**, 121032 (2019)

Jeribi, A., Jena, S.K., Lahiani, A.: Are cryptocurrencies a backstop for the stock market in a COVID-19-led financial crisis? Evidence from the NARDL approach. Int. J. Financ. Stud. **9**(3), 33 (2021)

Jeris, S.S., Chowdhury, A.N.U.R., Akter, M.T., Frances, S., Roy, M.H.: Cryptocurrency and stock market: bibliometric and content analysis. Heliyon (2022)

Ji, Q., Liu, B.Y., Cunado, J., Gupta, R.: Risk spillover between the US and the remaining G7 stock markets using time-varying copulas with Markov switching: evidence from over a century of data. North Am. J. Econ. Finance **51**, 100846 (2020)

Jiang, Y., Tian, G., Mo, B.: Spillover and quantile linkage between oil price shocks and stock returns: new evidence from G7 countries. Financ. Innov. **6**(1), 1–26 (2020)

Johnson, J.: JPY/BTC Trading Behaviour: A Reflection of the Japanese Economy or Due to the Construction of the Bitcoin Price Index. Available at SSRN 3378873 (2019)

Kostika, E., Laopodis, N.T.: Dynamic linkages among cryptocurrencies, exchange rates and global equity markets. Stud. Econ. Financ. **37**(2), 243–265 (2020)

Kumah, S.P., Abbam, D.A., Armah, R., Appiah-Kubi, E.: African financial markets in a storm: cryptocurrency safe havens during the COVID-19 pandemic. J. Res. Emerg. Markets **3**(2), 60–70 (2021)

Lahmiri, S., Bekiros, S.: The effect of COVID-19 on long memory in returns and volatility of cryptocurrency and stock markets. Chaos, Solitons Fractals **151**, 111221 (2021)

Lao, J., Nie, H., Jiang, Y.: Revisiting the investor sentiment–stock returns relationship: a multi-scale perspective using wavelets. Physica A **499**, 420–427 (2018)

Li, Z., Mo, B., Nie, H.: Time and frequency dynamic connectedness between cryptocurrencies and financial assets in China. Int. Rev. Econ. Financ. **86**, 46–57 (2023)

Mariana, C.D., Ekaputra, I.A., Husodo, Z.A.: Are Bitcoin and Ethereum safe-havens for stocks during the COVID-19 pandemic? Financ. Res. Lett. **38**, 101798 (2021)

OECD: Environmental impact of digital assets. OECD Business and Finance Policy Papers, December (2022)

Omane-Adjepong, M., Paul Alagidede, I., Lyimo, A.G., Tweneboah, G.: Herding behaviour in cryptocurrency and emerging financial markets. Cogent Econ. Finance **9**(1), 1933681 (2021)

Patel, R., Migliavacca, M., Oriani, M.E.: Blockchain in banking and finance: a bibliometric review. Res. Int. Bus. Financ. **62**, 101718 (2022)

Scagliarini, T., Pappalardo, G., Biondo, A. E., Pluchino, A., Rapisarda, A., Stramaglia, S.: Pairwise and high-order dependencies in the cryptocurrency trading network. Sci. Rep. **12**(1), 18483 (2022)

Smales, L.A.: Spreading the fear: the central role of CBOE VIX in global stock market uncertainty. Glob. Financ. J. **51**, 100679 (2022)

Soloviev, V., Yevtushenko, S., Batareyev, V.: Comparative analysis of the cryptocurrency and the stock markets using the Random Matrix Theory (2020)

Stolowy, H., Paugam, L.: The expansion of non-financial reporting: an exploratory study. Account. Bus. Res. **48**(5), 525–548 (2018)

Ugolini, A., Reboredo, J.C., Mensi, W.: Connectedness between DeFi, cryptocurrency, stock, and safe-haven assets. Financ. Res. Lett. **53**, 103692 (2023)

Umar, M., Hung, N.T., Chen, S., Iqbal, A., Jebran, K.: Are stock markets and cryptocurrencies connected? Singap. Econ. Rev. 1–16 (2020)

Wang, P., Liu, X., Wu, S.: Dynamic linkage between Bitcoin and traditional financial assets: a comparative analysis of different time frequencies. Entropy **24**(11), 1565 (2022)

Wen, D., Wang, G.J., Ma, C., Wang, Y.: Risk spillovers between oil and stock markets: a VAR for VaR analysis. Energy Econ. **80**, 524–535 (2019)

White, H., Kim, T.H., Manganelli, S.: VAR for VaR: measuring tail dependence using multivariate regression quantiles. J. Econometrics **187**(1), 169–188 (2015)

Zhang, Y.J., Bouri, E., Gupta, R., Ma, S.J.: Risk spillover between Bitcoin and conventional financial markets: an expectile-based approach. North Am. J. Econ. Finance **55**, 101296 (2021)

Chapter 19
Financial Challenges and Threats of Circular Economy Logistics

Claudia Capozza, Anatoliy Mokiy, Iryna Zvarych, Olha Ilyash, and Ivan Vankevych

Abstract This chapter explores the economic aspects related to circular economy logistics and highlights the key financial challenges and threats that organizations may face when transitioning to a circular economy business model. We developed a methodological approach for researching financial challenges and threats, which consisted of 4 steps. In the 1st stage of the research, we identified the leader in the trade of processed raw materials—Germany and characterized the factors why this country is the leader. In the next stage, the authors reasonably identified possible financial threats that prevent the effective development of circular economy logistics. In the third stage, the relationship between the rate of reuse of materials and the exchange rate, interest rates, and inflation rates was investigated using the multiple regression method. Then, based on the results obtained in the first stage, we investigated which companies carry out their activities in the field of circular economy logistics. The hypothesis of the chapter is that the transition to circular economy logistics presents several financial problems and threats that must be carefully considered and resolved. By recognizing these challenges and adopting appropriate strategies, organizations can navigate financial challenges and unlock the economic benefits associated with the circular economy, promoting sustainability and resource efficiency.

Keywords Circular economy · Circular economy logistics · Financial challenges · Financial threats · Reverse supply chains

C. Capozza
University of Bari Aldo Moro, Bari, Italy

A. Mokiy
Institute of Regional Studies Named After M.I. Dolishny, National Academy of Sciences of Ukraine, Lviv, Ukraine

I. Zvarych · I. Vankevych
West Ukrainian National University, Ternopil, Ukraine

O. Ilyash (✉)
National Technical University of Ukraine "Igor Sikorsky Kyiv Polytechnic Institute", Kyiv, Ukraine
e-mail: oliai@meta.ua

© The Author(s) 2025
V. Pacelli (ed.), *Systemic Risk and Complex Networks in Modern Financial Systems*,
New Economic Windows, https://doi.org/10.1007/978-3-031-64916-5_19

Introduction

In recent years, the transition to a circular economy has attracted considerable attention due to its potential to solve environmental problems and promote sustainable development. Circular economy logistics plays a key role in ensuring the efficient flow of resources and materials within the circular economy. However, as with any transformational process, there are several financial challenges and threats that need to be addressed in order to successfully implement circular economy logistics.

In order to overcome financial challenges and threats, policymakers and businesses must adopt a holistic approach that combines financial incentives, regulatory frameworks, and supporting infrastructure, highlighting the aim of this article.

Encouraging collaboration between the public and private sectors, providing financial support and incentives, and promoting knowledge sharing and capacity building can contribute to the successful integration of circular economy logistics. In addition, the development of standardized indicators and evaluation methods, specific to the practice of a closed economy, will allow a better assessment of financial efficiency and contribute to the adoption of informed decisions.

Literature Review

The logistics of a circular economy encompass the entire lifecycle of products, from design and manufacturing to end-of-life recovery and recycling. Research in this area focuses on developing efficient and sustainable logistics strategies to support the circular flow of materials. Scholars investigate the optimization of supply chains, transportation networks, and storage facilities to minimize waste and energy consumption. Key considerations include the integration of digital technologies, such as blockchain and the Internet of Things (IoT), to enhance transparency and traceability within circular supply chains.

In spite of recognizing the advantages of a circular economy (CE), many industries continue to adhere to the linear economy model, characterized by the processes of taking, making, using, disposing, and polluting, as outlined by Zhang et al. (2021).

As per the Ellen MacArthur Foundation (2015), the emphasis should be on prioritizing re-use over remanufacturing, and in turn, prioritizing remanufacturing over recycling. Essentially, there is a structured hierarchy for handling the end-of-life processes of products, aiming to preserve the invested effort in the original product and the energy embedded in its material composition.

The systematic planning, execution, and control of the optimal movement of raw materials, finished goods, in-process inventories, and related information from the consumption point back to the origin to recapture value is termed as reverse logistics (Rogers and Tibben-Lembke 2001).

In recent years, reverse logistics has emerged as a crucial tool in establishing a circular economy. Reverse logistics is described as the logistics process occurring in the opposite direction, as highlighted by Makarova et al. (2018) and Guide and Van Wassenhove (2009).

In their study (Butt et al. 2023), the connection between reverse logistics and the circular economy (CE) was investigated. The findings revealed that effective reverse logistics operations can significantly contribute to the shift towards a circular economy.

In their research (Fernando et al. 2023), it was contended that reverse logistics (RL) provides a resolution for the dynamic interaction between committing to sustainable resources and achieving financial performance. The financial well-being of many companies is influenced by economic downturns, and RL contributes by generating additional revenues to bolster the company. This underscores the advantage of incorporating circular economy principles and processes, which can generate value. Hence, the circular economy facilitates the establishment of circular supply chains and promotes environmentally friendly practices.

According to Rémy Le Moigne (2020), reverse logistics is characterized as the systematic movement of goods from their consumption point to a central consolidation point, aiming to extract value or ensure appropriate disposal. This process involves the gathering of goods, transportation to a central facility, and categorization based on their ultimate destination, such as refurbishment, reuse, or recycling.

A more comprehensive perspective on reverse logistics involves minimizing materials within the forward system to decrease the flow of materials back, enabling the possibility of material reuse, and facilitating recycling (Guide et al. 2003).

It is a procedure through which companies can enhance their environmental efficiency by engaging in recycling, reusing, and minimizing the quantity of materials employed (Alghababsheh et al. 2022).

Ellen MacArthur Foundation (2017) contended that the circular economy functions as a feedback system aiming to reduce the input of resources (both biological and technical) and the creation of waste that escapes into the environment. This involves extending the principles of reverse logistics into a broader framework, which consists of two subsystems: one associated with biological goods (e.g., food) and the other with technical goods (products).

According to Lee and Klassen (2008), circular economy logistics is characterized as environmentally conscious supply chain management employed by a company or organization. This approach considers environmental concerns and incorporates them into supply chain management to influence the environmental performance of both suppliers and customers.

Since the 1970s, there has been great interest in the study of reverse logistics due to the possibility of restoring the value of old items. A study by (Fleishman et al. 2004) compared reverse and forward logistics strategies using quantitative models. Areas such as distribution planning, inventory control and production planning were considered as parts of reverse logistics. Carter and Elram (1998) investigated aspects of reverse logistics, focusing on reverse allocation and resource minimization. They pointed to cost reductions caused by government regulation and environmental issues in reverse logistics. Supply chain coordination and reverse logistics have already been used in previous studies. Green logistics activities involve assessing the environmental consequences of diverse distribution strategies, mini-mizing energy consumption in logistic operations, decreasing waste volume, and overseeing its disposal methods (Sibihi et al. 2010). Tang et al. (2009) addressed the value of joint cycle time for the economic lot planning problem of multiple products, including new and remanufactured return items. The authors focused on a specific study of a company that manufactures and processes various goods on a single production line, determining the optimal batch sizes and production sequence for each product.

Moreover, reverse logistics creates opportunities for establishing new businesses due to the recyclable nature of the original items (Farooque et al. 2019a). Conversely, a circular economy (CE) promotes logistics and reverse manufacturing, yielding social, environmental, and economic advantages, including job generation and environmental preservation (Zhang et al. 2021).

Reverse logistics can play a pivotal role in facilitating the transition to a circular economy by completing the cycle of product life cycles (Makarova et al. 2018). The significance of recovering and recycling a product is widely acknowledged as a key factor in waste reduction, and effective reverse logistics has the potential to minimize waste from the initial stages of product design to the conclusion of the production process, thereby contributing to a reduced carbon footprint for a brand (Burke et al. 2021).

However, during the review of the literature on the selected research topic, the authors did not find any materials, reports, or articles related to financial challenges and threats that could potentially interfere with the implementation of the circular economy and that the logistics of the circular economy may face on the way to implementation.

Research Methodology

To conduct a study on the assessment of financial risks, the authors offer a step-by-step analysis, which is divided into the following steps:

1 Step. To start with Eurostat, we took an indicator in which logistics plays a key role—this is the trade of recycled raw materials (RRW) in the countries of the Euro-pean Union (27). Data for 2021 was used, as it was the most recent relevant informa-tion. All raw data were collected in Microsoft Excel and with the help of a filter, we managed to single out the top 5 exporters and importers of processed raw materials (see Table 19.2). First, information was presented in general for all industries, and then separately for each industry, to highlight the leaders in each industry. Then we took the example of Germany, the leader in almost every industry, and analyzed the trends of export and import of recycled raw materials.

Table 19.1 Characteristics of circular logistics

Efficient circular logistics	Logistics is essential in managing the reverse flow of products from consumers back to manufacturers. This involves the collection, transportation, and processing of returned goods, enabling the remanufacturing or refurbishment of products for resale
	Supports closed-loop supply chains by efficiently transporting materials from end-of-life products to recycling facilities. This includes the transportation of recyclable materials like paper, plastic, metal, and electronic waste
	Plays a key role in optimizing transportation routes and distribution networks to reduce energy consumption and minimize environmental impact. This contributes to resource efficiency and sustainability
	Due to strategic location of manufacturing, distribution, and recycling facilities can minimize transportation distances, reducing energy consumption and emissions. This helps create a more sustainable and circular supply chain
	Practices contribute to the reduction of packaging waste and the overall environmental impact of transportation. This involves optimizing packaging design, using reusable packaging, and minimizing unnecessary handling

2 Step. The Authors Identified Financial Risks that May Interfere with the Functioning of Effective Circular Logistics:

Exchange rate risks;

Changes in interest rates;

Economic instability;

Credit risks;

Tariffs and duties;

Supply Chain Management Risks.

Each risk was characterized in detail and explained why the authors consider it a challenge of circular economy logistics.

3 Step. The authors used the circular material waste (CMW) indicator for countries of European Union (27) in order to assess how these risks affect the reuse of materials. Using the multiplicative regression method, it was possible to investigate the relationship between the variable y and x_1, x_2, x_3, where.

y—Circular material waste;

x_1—exchange rates;

x_2—interest rates;

x_3—inflation rate.

The result was obtained (R = 0.43364795), which indicates that there is a weak relationship between the variables.

4 Step. Based on the results obtained in step 1 regarding the leaders (Germany, the Netherlands, Belgium) in the trade of processed raw materials, the search for information on startups that contribute to the functioning of efficient logistics of the circular economy was additionally carried out.

Table 19.2 Top 5 traders of recycled raw material

Branch	Top 5 exporters	Thousand Euro	Top 5 importers	Thousand Euro
Total	Germany	4 359 822,35,875	Germany	11 986 598,38,166
	Netherlands	3 334 864,93,817	Belgium	8 126 277,47,178
	Belgium	2 559 257,7039	Italy	5 655 257,63,751
	Spain	1 376 669,77,658	Netherlands	3 864 317,81,344
	Italy	1 345 889,54,228	Spain	3 657 651,65,281
Paper and cardboard	Top 5 exporters	Thousand Euro	Top 5 importers	Thousand Euro
	Italy	168 502,608	Germany	915 920,228
	Netherlands	140 808,52	Netherlands	343 311,18
	Spain	123 999,098	Austria	317 395,889
	Ireland	75 994,864	Spain	267 007,171
	France	69 977,34	France	131 415,648
Plastics	Top 5 exporters	Thousand Euro	Top 5 importers	Thousand Euro
	Netherlands	96 733,955	Netherlands	213 420,303
	Germany	82 701,23	Germany	150 429,421
	Belgium	55 267,485	Belgium	90 882,962
	Italy	19 815,526	Austria	73 676,274
	Spain	16 629,049	Italy	69 545,088
Rubber	Top 5 exporters	Thousand Euro	Top 5 importers	Thousand Euro
	Germany	25 125,764	Germany	38 649,056
	Netherlands	17 590,312	Netherlands	22 475,18
	Belgium	16 693,589	Spain	18 046,465
	Spain	13 812,704	France	9 623,654
	Italy	6 114,422	Poland	5 074,382
Wood	Top 5 exporters	Thousand Euro	Top 5 importers	Thousand Euro
	Portugal	15 045,6289	Spain	47 689,11,581
	Germany	9 225,32,075	Germany	40 836,58,166
	Sweden	7 717,80,643	France	33 297,98,782
	Latvia	2 936,19,044	Portugal	28 667,6989
	France	2 703,54,017	Austria	23 382,46,033
Textiles	Top 5 exporters	Thousand Euro	Top 5 importers	Thousand Euro
	Germany	165 520,277	Netherlands	146 602,539
	Belgium	152 709,592	Poland	123 741,222
	Poland	143 524,934	Germany	81 501,385
	Italy	121 999,657	Belgium	63 897,752
	Netherlands	121 655,708	Italy	62 805,2
Glass	Top 5 exporters	Thousand Euro	Top 5 importers	Thousand Euro
	Germany	9 709,613	Netherlands	45 905,225

(continued)

Table 19.2 (continued)

Branch	Top 5 exporters	Thousand Euro	Top 5 importers	Thousand Euro
	Belgium	6 805,433	Germany	20 976,586
	Ireland	5 971,061	Czechia	14 984,208
	France	5 606,473	Italy	14 366,653
	Netherlands	4 336,731	Portugal	9 380,396
Organic	Top 5 exporters	Thousand Euro	Top 5 importers	Thousand Euro
	Netherlands	325 984,516	Germany	976 034,168
	Denmark	249 492,771	France	869 625,224
	Germany	244 387,192	Netherlands	677 956,692
	France	159 777,54	Belgium	582 947,489
	Romania	140 336,88	Italy	426 991,088
Mineral	Top 5 exporters	Thousand Euro	Top 5 importers	Thousand Euro
	Spain	271 683,352	Belgium	344 774,342
	Germany	33 762,807	Germany	339 370,81
	Netherlands	23 834,325	Netherlands	169 691,464
	France	22 627,782	Sweden	149 963,158
	Greece	21 049,066	Spain	148 660,099
Metal—ferrous	Top 5 exporters	Thousand Euro	Top 5 importers	Thousand Euro
	Netherlands	1 829 712,33	Belgium	2 806 040,016
	Belgium	1 162 778,459	Italy	2 225 134,396
	Germany	544 987,929	Germany	1 902 362,411
	Romania	480 981,9	Netherlands	1 435 138,452
	Denmark	465 856,806	Spain	1 338 375,74

Source Calculated by author

Results

Logistics plays a crucial role in the circular economy by facilitating the efficient and sustainable movement of goods, materials, and resources within closed-loop systems (Fernando et al. 2019). The circular economy is an alternative economic model that aims to minimize waste and make the most of resources by promoting the reuse, remanufacturing, recycling, and refurbishment of products (Alghababsheh et al. 2022). Here are several ways in which logistics contributes to the circular economy (Table 19.1).

Logistics plays a crucial role in the trade of recycled raw materials for several reasons (Tang et al. 2009). First, it ensures efficient use of resources. Good logistics services help to reduce the costs of transportation, storage, and other operations. Secondly, logistics increases the competitiveness of enterprises. Enterprises that have efficient logistics can offer their customers more competitive prices and better service conditions. Thirdly, logistics contributes to the development of the economy. It creates jobs, increases labor productivity, and promotes the growth of trade.

Step 1

To conduct a qualitative study, the authors used statistical data from Eurostat on trade in recycled raw materials (RRM) for all countries of the European Union for 2021 as a basis. This choice is explained by the fact that logistics is an important component of trade in recycled raw materials. It ensures continuous movement of goods from the producer to the consumer. Logistics services include transportation, storage, packaging, sorting, and inspection of goods. Thus, we want to show how the logistics system of transporting recycled waste affects the circular economy and what financial threats it faces or may arise in a few years.

The initial data for graph modeling were collected in a Microsoft Excel spreadsheet. We received information on how European countries exported and imported recycled raw materials in general and for each branch over the past 15 years. As already mentioned, we only take data for 2021 to reflect the current situation in Europe.

Using the filter, we were able to get the top 5 exporters and importers of recycled raw materials in general and by each individual industry (see Table 19.2).

Table 19.2 shows the results of the analysis. As you can see, Germany ranks first in the volume of export and import of processed raw materials. It is also worth noting that the same Germany occupies a leading position in exports and imports in each industry. This shows that the German government pays a lot of attention to the reprocessing of raw materials. This is due to the fact that the German government seeks to strengthen the role of secondary processing of raw materials in the country's industry in order to increase the reliability of the supply of metals and industrial minerals. The created platform for dialogue between industry, scientists, and civil society presented its final report (more security of supply through recycling metals and industrial minerals) to the Ministries of Economy and Environment in order to achieve this goal. They developed recommendations for improving the recycling of raw materials, which include the implementation of product design that promotes recycling; clear guidelines and framework conditions; enshrining the circular economy in legislation; and using the full potential of digitization, in particular in the collection, recording, and sorting of materials.

Analyzing Fig. 19.1, we can conclude that during 2011–2021, Germany imported more raw materials that can still be processed and reused. This shows that Germany has a very well-established waste processing system. The German government understood the importance of reducing dependence on raw material exporting countries. For countries with a limited amount of their own raw materials, such as Germany, there are two options—processing already available raw materials on the market for reuse and importing the same raw materials for processing to achieve a closed loop.

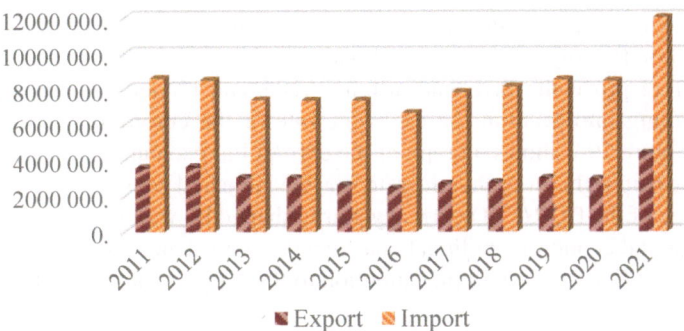

Fig. 19.1 The volume of trade of Germany in recycled raw materials for 2011–2021 years

This explains why Germany imports more than export. To achieve this, it is important to systematically recover secondary raw materials from industrial and domestic waste, treating waste not as a material to be disposed of, but as a source of raw materials. Ideally, this should cover all categories of waste, from old packaging to waste electrical and electronic equipment.

Germany has numerous seaports, such as the port of Bremen and Hamburg, which are important for foreign trade and transportation of raw materials.

Germany is one of the leading countries in the world in the processing of raw materials and production of goods. The export and import of processed raw materials are important to ensure a constant flow of raw materials for German industry.

German companies are famous for their high-quality products and innovation. This makes their products competitive on world markets and provides demand for processed raw materials.

Logistics and trade in recycled raw materials play an important role in the economy of Germany, which is considered one of the largest and most developed countries in the world. Germany is one of the leading countries in the field of industry and exports, and therefore logistics is a key element of its economic system.

Germany has a developed transport infrastructure, including road, rail, water and air transport. Thanks to this, transport logistics allows you to efficiently move processed raw materials to all regions of the country and abroad.

In Germany, much attention is paid to the management of the supply chain to ensure the efficiency and reliability of the supply of raw materials for production.

Step 2

There are several financial challenges that can affect the logistics and trade of recycled raw materials in:

Exchange rate risks: Changes in exchange rates can affect the cost of raw materials and transportation costs, as many trade transactions are conducted in foreign currencies. Unwanted changes in exchange rates can lead to losses for companies that depend on the import or export of goods.

In order to prove the fact that currency risks can hinder efficient logistics, it is worth taking into account the exchange rate of the euro/dollar during the last 15 years. Figure 19.2 shows a graph of the euro/dollar exchange rate over 25 years. It should be said right away that the rate was unstable and constantly changing. The performance of the EUR/USD currency pair is shaped by various factors that mirror the fundamental economic, political, and social circumstances in both the eurozone and the United States. We will pay attention to the biggest changes in the exchange rate.

From its introduction on January 1, 1999, the euro has been valued below a dollar for less than two years. The highest point was reached on April 22, 2008, when the exchange rate peaked at $1.60. A high euro-to-dollar exchange rate implies that your dollar could buy more in the European Union, whereas a low rate indicates that you would acquire less there. Investors initially believed that the subprime mortgage crisis would be confined mainly to the United States. This caused the euro to strengthen, but when it became clear that the recession would be global, the euro fell to $1.39. The next sharp decline in the exchange rate began in 2014, caused by Russia's occupation of the eastern regions of Ukraine. Political division over joining the European Union or unification with Russia has caused a crisis in Ukraine. In 2015, the euro experienced a decline to $1.12 following the announcement by the European Central Bank (ECB) that it would commence monthly purchases of 60 billion euros in euro-denominated bonds starting in March. At the beginning of 2020 the euro was

Fig. 19.2 Exchange rates Euro US dollar for 25 years. *Source* Trading Economics

priced at $1.12, but by mid-year, it had dropped to $1.06, coinciding with the severe impact of the coronavirus pandemic on Europe In 2022, the war in Ukraine caused another exchange rate slump and led to the euro reaching parity with the dollar.

Payment for logistics transportation is made at the exchange rate between the euro and the dollar, which is unstable and constantly changing, as shown in Fig. 19.2. This justifies this financial risk of circular economy logistics.

Changes in interest rates: Changes in the level of interest rates can affect the cost of financing for logistics operations and trade operations. An increase in interest rates can lead to an increase in the cost of borrowing capital.

Figure 19.3 shows the graph of changes in ECB interest rates over 15 years. In July 2022, the European Central Bank (ECB) raised its fixed interest rate to 0.5 percent, marking the initial hike since March 2016. Following this, the ECB continued to incrementally raise the fixed interest rate nearly every month. By October 2023, the rate reached 4.5 percent, marking the highest level since the global financial crisis of 2007 and 2008. The ECB's interest rate represents the rate at which the ECB provides overnight loans to banks. Commercial banks utilize these loans to maintain short-term liquidity.

The European Central Bank (ECB) kept interest rates at record levels for a second consecutive meeting and signaled it intended to quickly end its last remaining bond-buying scheme as part of measures to combat high inflation. The prime refinancing rate remained at 4.5%, the highest in 22 years, while the deposit rate remained at a record 4%. The ECB also announced that full reinvestment under the PEPP will end on June 30, and the portfolio will shrink by €7.5 billion per month until the end of 2024. Politicians have also pledged to keep rates at a fairly restrictive level as long as necessary. The ECB forecasts that inflation will average 5.4% in 2023, 2.7% in 2024, 2.1% in 2025 and 1.9% in 2026. Core inflation is projected to be slightly higher at 5.0% in 2023, 2.7% in 2024, 2.3% in 2025 and 2.1% in 2026 (Farooque et al. 2019b).

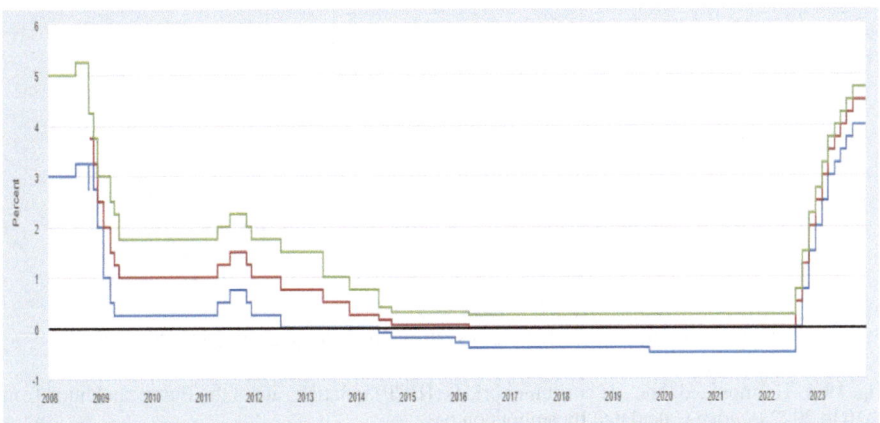

Fig. 19.3 European Central Bank interest rates. *Source* European Central Bank (2023)

Having analyzed the trends in setting the level of ECB interest rates, we can conclude that they are constantly changing. The circular economy aims to minimize waste and resource consumption by promoting closed-loop systems where materials are reused, recycled, and repurposed. However, like any economic system, it's not immune to external factors like interest rates. Higher interest rates make it more expensive to borrow money for circular economy initiatives, like building recycling facilities or developing new reuse technologies. This can discourage investment and slow down the transition to a circular model. Logistics within the circular economy often involve collecting, sorting, and transporting materials. Higher interest rates can increase the cost of these activities, making it more expensive to operate circular businesses.

Higher interest rates can dampen consumer spending, potentially reducing demand for recycled or reused products. This can make it harder for circular businesses to generate revenue and stay afloat.

When interest rates rise, businesses might prioritize short-term profits over long-term investments in sustainability initiatives like the circular economy. This can further slowdown the adoption of circular practices.

Economic instability: Economic hardship or recession may result in reduced demand for goods and services, including recycled raw materials. This can affect trading volumes and lead to lower profits for companies. In order to show how economic instability affects the logistics of the circular economy in this case, we will use the inflation rate indicator in the European Union during 2010–2022. The raw data for the Harmonized Index of Consumer Prices was compiled in Microsoft Excel and interpreted in the form of a graph, which you can see in Fig. 19.4.

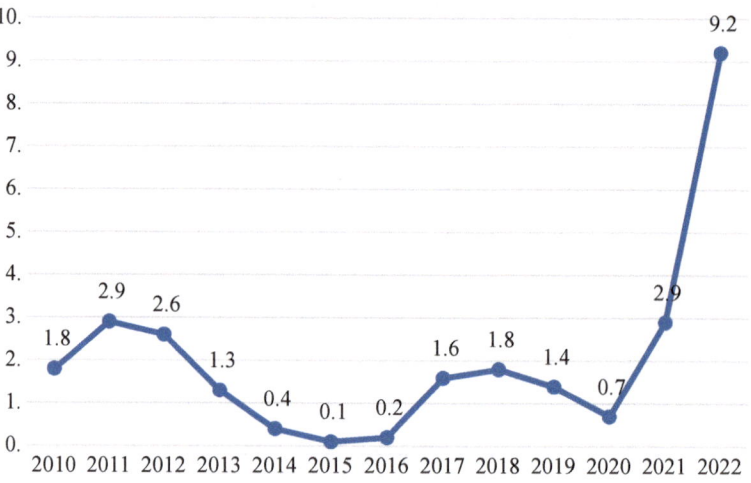

Fig. 19.4 Harmonized index of consumer prices (HICP) inflation rate of the European Union from 2010 to 2022. *Source* Calculated by authors on base

As can be seen from the broken line graph, the inflation rate in the countries of the European Union increased sharply in 2021. This phenomenon was caused by various factors that simultaneously led to a rise in prices. Even as market demand quickly recovered from the COVID-19 restrictions in 2020, global supply chains have not yet fully recovered. In particular, this led to an increase in the prices of energy resources and food products, especially after the Russian invasion of Ukraine in February 2022.

Without overstating, it can be asserted that elevated inflation posed a worldwide challenge in 2022. The United States, for instance, encountered its most substantial inflation rate in 40 years in March 2022, reaching 8.5 percent. Similarly, the United Kingdom witnessed a nine percent inflation rate in April 2022, fueled by escalating energy and housing expenses. If elevated inflation persists and intersects with sluggish economic growth and elevated unemployment, concerns arise that the existing crisis might evolve into a period of stagflation, reminiscent of the early 1980s.

Additionally, Fig. 19.5 shows the level to which prices for specific goods or services have risen.

The rise in the EU's figures was significantly influenced by a notable 18% increase in consumer prices for housing, water, gas, and other fuels over the course of a year. Transport saw a rise of 12.1%, and food and non-alcoholic beverages increased by 11.9%, trailing housing expenses. In 2022, every other major category covered by the HICP experienced an increase, except for a marginal 0.1% decrease in communication consumer prices.

As researched, inflation has led to higher prices for fuel and other transportation costs, making it more expensive to move materials and products in a circular economy. In turn, this prevented businesses from participating in circular initiatives, especially for low-value materials.

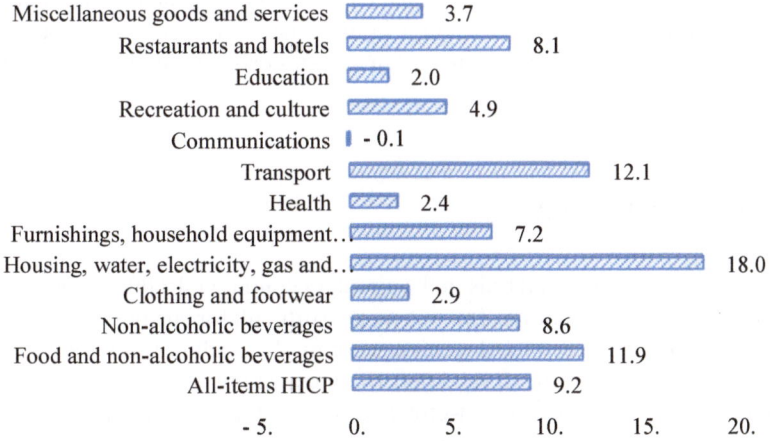

Fig. 19.5 Inflation rates of individual items included in Harmonised Index of Consumer Prices (HICP). *Source* Calculated by authors on base (Trading Economy 2023)

A high Inflation rate can disrupt the delicate balance of supply and demand in a circular economy, where materials are constantly being reused or recycled. This can lead to shortages of certain materials or difficulties finding markets for recycled goods, impacting the overall efficiency of the system.

Inflation can put a strain on household budgets, potentially leading consumers to prioritize cheaper, non-circular products over more sustainable options. This can make it harder for circular businesses to compete and scale up their operations.

Credit risks: A deterioration in the creditworthiness of partners or customers can affect financial flows and lead to losses for companies. The inability to pay for goods or services may arise due to financial difficulties in partner companies.

The influence of credit risks on the logistics of a circular economy is a complex and multifaceted issue, with potential impacts across various stages of the closed-loop system. Here's a breakdown of some key points to consider:

Increased Financial Risks

Lending hesitancy: Banks and financial institutions might be hesitant to provide loans to businesses operating in the circular economy due to perceived higher credit risks associated with novel business models, unproven revenue streams, and potential dependence on external factors like recycling infrastructure. This can hinder access to capital needed for investment in circular logistics infrastructure and operations.

Debt burden: Businesses in the circular economy may face higher debt burdens due to the upfront costs associated with setting up reverse logistics systems, investing in reusable materials, and managing complex product lifecycles. This can lead to cash flow constraints and limit their ability to absorb unexpected financial shocks.

Supply chain disruptions: Credit risks can also be amplified by disruptions in the circular supply chain, such as delays in the collection, sorting, or processing of used materials. This can lead to increased inventory costs, production stoppages, and ultimately, financial losses.

Operational Challenges

The reliance on recycled materials in the circular economy introduces uncertainty in terms of quality and availability. This can make it difficult for businesses to accurately forecast demand and plan their logistics accordingly, leading to inefficiencies and potential financial losses due to stockouts or overstocking.

Implementing efficient and cost-effective reverse logistics systems can be challenging, especially for complex products with multiple components or those requiring specialized processing. This can lead to higher operational costs and reduced profitability, impacting the overall creditworthiness of businesses in the circular economy.

The success of the circular economy hinges on consumer acceptance of recycled products and their willingness to participate in reverse logistics systems. Low consumer engagement can lead to decreased demand for recycled materials and reduced revenue for businesses, impacting their ability to meet their financial obligations.

Tariffs and duties: The introduction of new tariffs or changes to existing ones may affect the cost of customs duties when importing and exporting goods. The European Commission engages in trade negotiations by directly interacting with other countries or regions and by participating in the World Trade Organization (WTO). The WTO, the sole international organization addressing multinational trade issues and establishing global trade rules among nations, primarily aims to facilitate smooth, predictable, and unrestricted trade. The General Agreement on Tariffs and Trade (GATT) specifically addresses international trade in goods.

EU trade agreements enhance the competitiveness of European businesses, allowing them to increase exports to countries and regions beyond the EU. These agreements also provide improved access to essential raw materials and components for importers within the EU, offering consumers a broader range of products. Additionally, these trade pacts may necessitate partner governments to safeguard human rights, labor rights, and the environment. For instance, addressing workplace safety or promoting gender equality could be among the stipulations.

The EU enjoys the advantage of being among the most open economies globally, with approximately 71% of its imports entering the EU at zero tariffs. In 2022, Fig. 19.1 illustrates the proportion of EU-imported goods from selected partners, indicating various tariff levels. Notably, China had the lowest share of zero-tariff imports at 45%, while Nigeria had the highest at 98%, and the United States stood at 72%.

Figure 19.6 shows the import restrictions of the European Union countries from different countries. Based on this, we can conclude that customs barriers will stand in the way of exporting processed raw materials.

Tariffs and duties raise the price of imported materials and goods, including recycled content, making circular economy solutions less competitive compared to virgin materials. This can discourage businesses and consumers from adopting circular practices.

Trade barriers can disrupt the flow of materials within and between countries, hindering the efficient movement of waste, recyclables, and repaired/remanufactured goods. This can lead to logistical bottlenecks and inefficiencies.

High tariffs and duties can make it difficult for circular economy businesses to export their products and services, limiting their growth potential and market reach.

Supply Chain Management Risks: Failure to effectively manage the supply chain can cause delivery delays, additional costs, and lost opportunities.

Unreliable suppliers, inefficient reverse logistics, and unforeseen interruptions in processing facilities can lead to delays and shortages of recovered materials, hampering production planning and product availability.

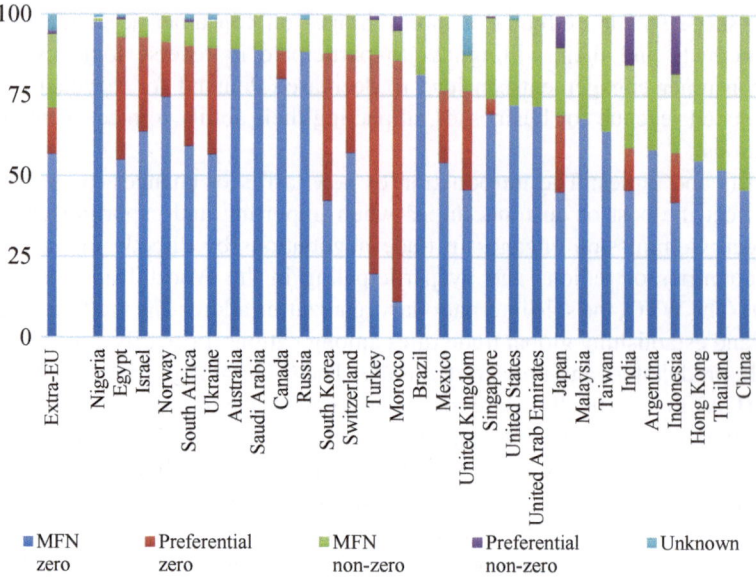

Fig. 19.6 Imports of selected EU partners by tariff regime, 2022 (%). *Source* World Integrated Trade Solution (WITS) (European Union 2023)

The quality of recovered materials can vary greatly depending on collection and sorting processes. Ineffective risk management might result in inconsistent material inputs, leading to production problems and reduced product quality.

Reverse logistics often involve transporting bulky or hazardous materials, with higher associated costs and environmental risks. Poor routing, inefficient transport modes, and lack of infrastructure can significantly escalate logistics costs and carbon footprint.

Inaccurate or incomplete information about material availability, location, and quality can lead to mismatched supply and demand, missed opportunities, and inefficient resource allocation. Lack of transparency and collaboration throughout the supply chain can further amplify these risks.

As circular economy relies heavily on data sharing and digital tools, there's an increased risk of cyberattacks. Data breaches can disrupt operations, compromise sensitive information, and lead to financial losses.

Step 3

As part of the study of the financial challenges of the circular economy, it is important to follow the impact of the challenges proposed by the authors in step 2 on the reuse of materials, which is the basis of the logistics of the circular economy.

For this, data on the reuse of materials in the countries of the European Union for 2021 from Eurostat were downloaded to Microsoft Excel. We marked y—reuse of materials, × 1—exchange rate, × 2—interest rate, × 3—inflation rate. Data for 2021 were used. Based on these data, we performed a multiple regression analysis in the STATISTICA program. The following results were obtained (see Table 19.3).

	Summary Statistics; DV: Circular material use rate
Statistic	Value
Multiple R	0,433,647,952
Multiple R?	0,188,050,546
Adjusted R?	0,0,773,301,664
F(3,22)	1,69,842,757
p	0,196,503,013
Std.Err. of Estimate	6,21,151,131

Source Calculated by authors

The conducted analysis made it possible to estimate the average correlation coefficient between u and three other variables ($R = 0.4336$). This means that there is a weak relationship between all variables.

	Current Status of Sweep Matrix; DV: Circular material use rate			
N = 26	exchange rate	interest rate	inflation rate	Circular material use rate
Exchange rate	−104,092	−0,03,165	−0,20,467	−0,001,445
Interest rate	−0,03,165	−100,151	−0,02,954	0,391,914
Inflation rate	−0,20,467	−0,02,954	−104,079	−0,176,894

(continued)

Table 19.3 Multiply regression

Regression Summary for Dependent Variable: Circular material use rate

N = 26	b*	Std.Err. of b*	b	Std.Err. of b	t(22)	p-value
Intercept			−753,304	8,001,903	−0,941,406	0,356,720
Exchange rate	−0,001,445	0,196,002	−0,3941	5,343,543	−0,007,374	0,994,183
Interest rate	0,391,914	0,192,256	1,173,431	5,756,335	2,038,503	0,053,696
Inflation rate	−0,176,894	0,195,990	−0,9216	102,114	−0,902,568	0,376,532

(continued)

	Current Status of Sweep Matrix; DV: Circular material use rate			
Circular material use rate	−0,00,145	0,39,191	−0,17,689	0,811,949

	Variables currently in the Equation; DV: Circular material use rate (Spreadsheet1)							
Variable	b* in	Partial Cor	Semipart Cor	Tolerance	R-square	t(22)	p-value
Exchange rate	−0,001,445	−0,001,572	−0,001,417	0,960,692	0,039,308	−0,007,374	0,994,183
Interest rate	0,391,914	0,398,593	0,391,620	0,998,497	0,001,503	2,038,503	0,053,696
Inflation rate	−0,176,894	−0,188,961	−0,173,394	0,960,811	0,039,189	−0,902,568	0,376,532

Source Calculated by authors

Based on the multiplicative regression analysis, we obtained correlation coefficients that show the relationship between the variables (see Table 19.4). It should be noted that no strong relationship was found, as the correlation coefficients are smaller than the modulus of 0.5. However, the coefficient with the largest positive value between the reuse of materials and interest rates stands out ($R = 0.396074$). As interest rates rise, the reuse of materials also increases. This indicates that there is a direct, albeit weak, relationship between these indicators. It should also be noted that there is an inverse relationship between the rate of inflation and the reuse of materials. This is evidenced by the negative sign of the correlation coefficient ($R = -0.185743$). So, when the rate of inflation decreases, the reuse of materials increases.

Figure 19.7 clearly demonstrates the direct interdependence between interest rates and circular material use (CMU).

Rising interest rates don't automatically boost circular material use. Higher borrowing costs can make it pricier for businesses to invest in long-term circular projects, potentially slowing their adoption. However, the impact varies across industries and regions, and other factors like environmental concerns can also influence the shift towards circular practices.

Figure 19.8 shows how inflation rates affect circular material use. When the rate of inflation decreases, the circular material use.

Having received such results, a conclusion must be drawn. Inflation can have various impacts on circular material use, which refers to the practice of using and

Table 19.4 Correlation coefficients

	Correlations		
Variable	Exchange rate	Interest rate	Inflation rate
Circular material use rate	0,023,089	0,396,074	−0,185,743

Source Calculated by authors on base

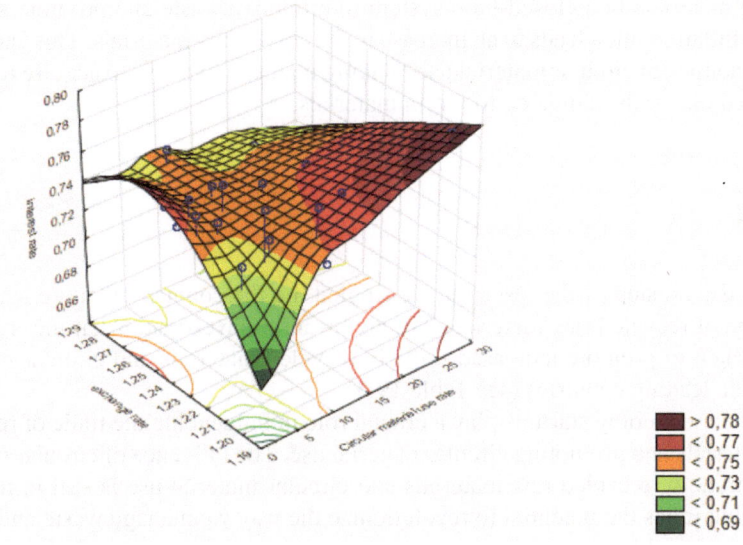

Fig. 19.7. 3D Surface: Circular material use rate vs. exchange rate vs. interest rate. *Source* calculated by authors

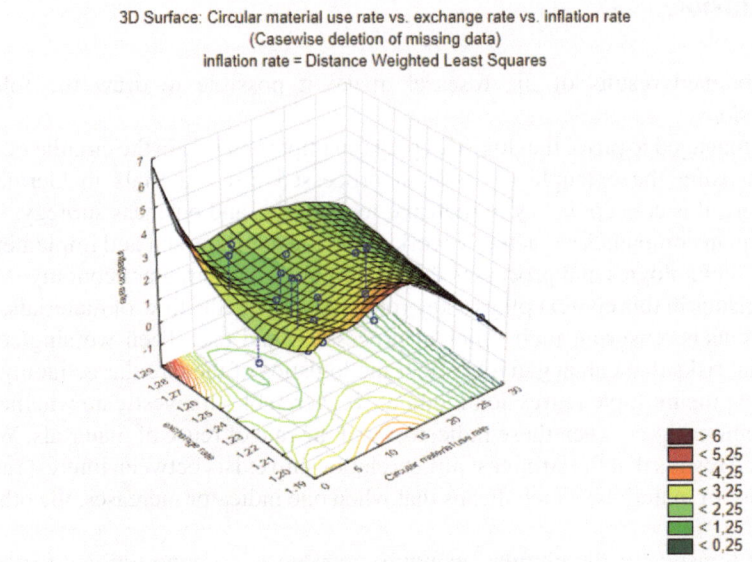

Fig. 19.8. 3D Surface: Circular material use rate vs. exchange rate vs. inflation rate. *Source* Calculated by authors

reusing materials in a closed-loop system to minimize waste and promote sustainability. Inflation often leads to an increase in the cost of raw materials. This can affect the economics of circular material use, making it more expensive to acquire recycled or reused materials compared to virgin materials.

Step 4

During the 1st step of the research, we singled out the countries that are leaders in the trade of recycled raw materials. We now want to investigate what measures are being taken to promote increased trade in recycled materials and circular material use in the leading countries (see Table 19.5).

Circular economy startups play a crucial role in influencing the trade of recycled raw materials and promoting circular material use. The influence of circular startups on the trade of recycled raw materials and circular material use is still in its early stages, but it has the potential to revolutionize the way we manage waste and create a more sustainable future.

Conclusion

The obtained results of the research make it possible to draw the following conclusions.

We managed to prove that logistics plays an important role in the circular economy system using the example of trade in processed raw materials in Germany. In Germany, this concept is key in industry, technology, and business strategy.

German companies are actively working on the development and implementation of new technologies and processes that contribute to the circular economy.

Logistics in this context plays a key role in managing the flow of materials, waste, and products, ensuring their efficient processing and use. Then we singled out 6 financial risks that can negatively affect the logistics of the circular economy.

Using the multiple regression method, we were able to investigate whether there is a relationship between these indicators and the rate of reuse of materials. We have obtained results that the strongest direct relationship exists between interest rates and the reuse of materials, which means that when one indicator increases, the other also increases.

The logistics of the circular economy open up wide opportunities for creating more sustainable and efficient resource management systems. However, along with this, it also brings its own financial challenges and threats.

Table 19.5 Circular economy startups

No	The name of the startup	Country of Origin	Description of the startup
1	Aurubis	Germany	Aurubis, a German company, specializes in the processing of copper and other metals. The company's main plant in Lünen, Germany, uses only recycled raw materials to produce high-quality copper cathodes. Aurubis describes these cathodes as identical to those obtained from primary copper production. Sources of secondary raw materials include copper cable waste, electronic waste such as printed circuit boards, and industrial waste and sludge. They are processed, smelted and refined using Kayser's processing system
2	Numi.circular	Germany	Numi.circular, a German startup, is the creator of numi.platform, a software designed to promote the circular economy. This innovative platform facilitates product return programs, allowing brands to minimize waste and explore additional revenue streams. Through the numi.platform, customers can return used products and earn recovery points. The system developed by numi.circular automates the sorting and distribution of these returned goods. In addition, it carefully tracks all return transactions and generates impact reports, allowing brands to quantify their environmental initiatives
3	Circular logistics	Germany	German startup Circular Logistics produces the BikeBox, an eco-friendly transport package for bicycles made of durable polypropylene. This foldable box can shrink to 1/8 of its original size, encouraging constant reuse and eliminating the need for cardboard, reducing waste. The use of polypropylene ensures that the box can be used many times and is 100% recyclable at the end of its life cycle. The BikeBox design allows for easy assembly, reminiscent of the convenience of traditional cardboard boxes. Circular Logistics manages box returns after delivery, promoting a closed loop system that minimizes waste

(continued)

Table 19.5 (continued)

No	The name of the startup	Country of Origin	Description of the startup
4	Circular in Motion	Netherlands	Dutch startup Circular in Motion is the creator of Cirinmo, an e-commerce platform for buying and selling certified materials. Using blockchain certificates built into the system, Cirinmo offers proof of origin for cyclical products and supports transparent chain-of-custody data for products with higher recycled content throughout the supply chain. In this way, Circular in Motion promotes secure digital connections between companies, helping to create businesses and create value chains within the circular economy
5	Sykell	Germany	German startup Sykell produces bisphenol A (BPA)-free reusable containers using a monocomposite material that can be recycled. In order to effectively monitor the movement of containers, the company has implemented a reusable platform as a service in accordance with the reusable obligations set out in §33 VerpackG. This platform creates an open and centrally managed system for returning and cleaning containers, ensuring transparent inventory management. By offering these reusable containers, Sykell is replacing single-use packaging, helping to promote circular packaging practices

(continued)

Table 19.5 (continued)

No	The name of the startup	Country of Origin	Description of the startup
6	EIT RawMaterials	Germany	Aiming to drive innovation across the raw materials and advanced materials value chain, EIT RawMaterials, the world's largest network in the field, has made an investment in Circular Silicon Europe GmbH. Based in Braunschweig, Germany, this innovative cleantech startup has received seed capital from EIT's RawMaterials mission. This investment enables the company to create the first industrial recycling line with a capacity of 140 tons, designed to regenerate silicon from discarded solar panels. This pioneering method of recycling an important raw material marks a significant step in moving the solar industry towards a circular economy
7	Recycllux	Belgium	The company is developing a system that uses Earth observation data and artificial intelligence to detect plastic debris in natural bodies of water, making it easier to collect. In addition, the platform intends to use blockchain technology to establish links between plastic producers and waste collectors

Source Formed by authors

References

Alghababsheh, M., Abu Khader, D.E., Butt, A.S., Moktadir, M.A.: Business strategy, green supply chain management practices, and financial performance: A nuanced empirical examination. J. Clean. Prod. 380(2): 134865 (2022). https://doi.org/10.1016/j.jclepro.2022.134865

Burke, H., Zhang, A., Wang, J.X.: Integrating product design and supply chain management for a circular economy. Prod. Plan. & Control. 1–17, (2021). https://doi.org/10.1080/09537287.2021.1983063

Butt, A., Ali, I., Govindan, K.: The role of reverse logistics in a circular economy for achieving sustainable development goals: A multiple case study of retail firms. Prod. Plan. & Control. (2023). https://doi.org/10.1080/09537287.2023.2197851

Carter, C.R., Ellram, L.M.: Reverse logistics: A review of the literature and framework for future investigation. J. Bus. Logist. 19 (1998)

Guide, V.D.R., Jr., Jayaraman, V., Linton, J.D.: Building Contingency Planning for Closed-Loop Supply Chains with Product Recovery. J. Oper. Manag. 21(3), 259–279 (2003). https://doi.org/10.1016/S0272-6963(02)00110-9

Ellen MacArthur Foundation, The Circular Economy and Supply Chains. (2023). https://www.ellenmacarthurfoundation.org/circular-supply-chains

European Central Bank: Key ECB interest rates. Retrieved September 20, 2023 (2023) from https://www.ecb.europa.eu/home/html/index.en.html

European Union: Eurostat Database. Eurostat (2023). https://ec.europa.eu/eurostat/data/database

Farooque, M., Zhang, A., Thurer, M., Qu, T., Huisingh, D.: Circular Supply Chain Management: A Definition and Structured Literature Review. J. Clean. Prod. **228**, 882–900 (2019a). https://doi.org/10.1016/j.jclepro.2019.04.303

Farooque, M., Zhang, A., A, and Y. Liu.: Barriers to Circular Food Supply Chains in China. Supply Chain. Manag.: Int. J. **24**(5), 677–696 (2019b). https://doi.org/10.1108/SCM-10-2018-0345

Fernando, Y., Jose Chiappetta Jabbour, C., Wah, W.-X.: Pursuing green growth in technology firms through the connections between environmental innovation and sustainable business performance: Does service capability matter? Resources. Conserv. Recycl. **141**, 8–20 (2019), ISSN 0921–3449, https://doi.org/10.1016/j.resconrec.2018.09.031

Fernando, Y., Shaharudin, M.S., Abideen, A.Z.: Circular economy-based reverse logistics: dynamic interplay between sustainable resource commitment and financial performance. Eur. J. Manag. Bus. Econ. **32**(1) (2023). https://www.emerald.com/insight/content/doi/https://doi.org/10.1108/EJMBE-08-2020-0254/full/html

Fleischmann, M., Bloemhof-Ruwaard, J.M., Beullens, P., Dekker, R.: Reverse Logistics Network Design. In: Reverse Logistics, pp. 65–94. Berlin; Heidelberg: Springer (2004). https://doi.org/10.1007/978-3-540-24803-3_4

Lee, S.Y., Klassen, R.D.: Drivers and enablers that foster environmental management capabilities in small- and medium-sized suppliers in supply chains. Prod. Oper. Manag. **17**, 573–586 (2008). https://doi.org/10.3401/poms.1080.0063

Makarova, I., Shubenkova, K., Pashkevich, A.: The Concept of the Decision Support System Is to Plan the Reverse Logistics in the Automotive Industry. In: 2018 26th International Conference on Software, Telecommunications and Computer Networks (SoftCOM), 1–6 (2018). IEEE. https://doi.org/10.23919/SOFTCOM.2018.8555760

Rémy Le Moigne: Reverse Logistics, The Circular Economy Weakest Link. (2020). https://www.renewablematter.eu/articles/article/reverse-logistics-the-circulareconomy-weakest-link

Rogers, D.S., Tibben-Lembke, R.: An Examination of Reverse Logistics Practices. J. Bus. Logist. **22**(2), 129–148 (2001). https://doi.org/10.1002/j.2158-1592.2001.tb00007.x

Tang, O., Teunter, R.: Economic Lot Scheduling Problem with Returns. Prod. Oper. Manag. **15**(4), 488–497 (2009). https://doi.org/10.1111/j.1937-5956.2006.tb00158.x

Trading Economy: Interest Rates of ECB. Euro Area Interest Rate (2023). https://tradingeconomics.com/euro-area/interest-rate

World Bank: World Integrated Trade Solution (2023). https://wits.worldbank.org/

Zhang, A., Wang, J.X., Farooque, M., Wang, Y., Choi, T.M.: MultiDimensional Circular Supply Chain Management: A Comparative Review of The State-of-the-Art Practices and Research. Transp. Res. Part E: Logist. Transp. Rev. **155**, 102509 (2021). https://doi.org/10.1016/j.tre.2021.102509

Chapter 20
Systemic Risks to Capital Investment Flows in the Post-crisis Economy of Ukraine

Alessandro Rubino, Anatoliy Mokiy, Mariya Fleychuk, Viktoriia Khaustova, and Tetiana Salashenko

Abstract The study investigates systemic risks affecting capital investment flows in Ukraine's post-crisis economy, employing VAR models and the Kalman filter. It analyzes the relationship between investment flows and key macroeconomic indicators. Utilizing neural network tools, the research identifies pivotal factors influencing investment processes amidst wartime. The study delineates primary strategies for risk mitigation, including the adoption of modern warfare economics principles, combating systemic dysfunctions such as corruption, and legitimizing property rights through targeted investments. Furthermore, in the financial sector, there is a notable emphasis on the imperative: to enhance commercialization in banking, align regulations with EU standards, develop money, bond, and securities markets, facilitate small business financing and financial inclusivity, and mobilize external financing to sustain financial stability.

Keywords Systemic risks · Investment flows · Modern warfare economics

A. Rubino
University of Bari Aldo Moro, Bari, Italy
e-mail: alessandro.rubino@uniba.it

A. Mokiy
Institute of Regional Studies Named After M.I. Dolishny, National Academy of Sciences of Ukraine, Lviv, Ukraine
e-mail: amokiy320@ukr.net

M. Fleychuk
Stepan Gzhytskyi National University of Veterinary Medicine and Biotechnologies, Lviv, Ukraine
e-mail: fleychukm@lvet.edu.ua

V. Khaustova (✉) · T. Salashenko
Research Center for Industrial Problems of Development of the National Academy of Sciences of Ukraine, Kharkiv, Ukraine
e-mail: ndc_ipr@ukr.net

T. Salashenko
e-mail: tisandch@gmail.com

© The Author(s) 2025
V. Pacelli (ed.), *Systemic Risk and Complex Networks in Modern Financial Systems*,
New Economic Windows, https://doi.org/10.1007/978-3-031-64916-5_20

Focus and the Novelty of the Research

The focus of the research is to analyze systemic risks impacting capital investment flows within Ukraine's post-crisis economy. This involves employing advanced
econometric techniques such as VAR models and the Kalman filter to understand the
relationship between investment flows and key macroeconomic indicators. Additionally, the study utilizes neural network tools to identify critical factors influencing
investment processes during times of conflict.

The novelty of the research lies in its comprehensive approach to understanding
and addressing systemic risks in the context of post-crisis economic conditions and
conflict environments. By integrating various analytical methods and focusing on
specific aspects such as modern warfare economics principles and combating corruption, the research offers insights into effective strategies for risk mitigation and
sustainable economic development. This multidimensional approach contributes to
the advancement of knowledge in both theoretical and practical domains, offering
valuable implications for policymakers, investors, and other stakeholders involved
in Ukraine's economic reconstruction and stabilization efforts.

Literature Review

At this stage, there is an urgent need to mobilize substantial amounts of capital assets
for the reconstruction and development of Ukraine during the post-war and concurrently crisis-ridden period for the socio-economic system following the large-scale
invasion of Ukraine by the Russian Federation. In the case of Ukraine's victory over
Russia, this, on the one hand, presents significant opportunities for Ukraine's development through the military and financial-economic support of a coalition of states
led by the United States. However, on the other hand, with the influx of substantial
amounts of capital assets for the reconstruction and development of Ukraine, there
exist numerous external and internal risks and threats that could potentially mitigate
the positive effects of such capital investments not only for Ukraine as a state but
also for external and internal investors.

Vasyechko in this context investigates critical components of protecting foreign
investors during wartime in Ukraine, discusses the volatility of Foreign Direct Investment (FDI) due to market riskiness, highlights how investment decisions rely on
investor aversion and the host country's investment climate, emphasizes the uncertainty and extreme risks of investing in war-torn countries, proposing systemic risk
management approaches (Vasyechko 2023). Lino expands on Stiglitzs work, focusing
on new discontents and protectionist movements, addresses the negative impacts of
globalization on developed nations, reflecting political unrest and inequality (Lino
2019).

Nell warns about potential economic setbacks due to slow and unpredictable donor disbursements, highlights risks to macroeconomic stability, currency weakening, and inflation without upfront donor commitments, advocates for clear commitments from donors to prevent economic setbacks during the war (Nell et al. 2022). Hryhoriev analyzes post-war economies with significant war expenditures and effects of sovereign debt restructuring, utilizes system dynamics modeling to demonstrate scenarios of external debt minimization, emphasizes the need for radical decisions like sovereign debt cancellation to stabilize the economy (Hryhoriev 2023).

Hohg and others highlight the importance of sustainable public finances, sound monetary policy, and flexible labor markets for successful reconstruction, discusses the challenges faced by Ukraine in generating revenue and the policy mix required for stable macroeconomic growth, and propose platform creation for dialogue and policy discussion on Ukraine's reconstruction (Hong 2023).

Research Methodology

For comprehensive examination of the systemic risks of capital investments in the economy of a country experiencing conflict, it is advisable to employ the Vector Autoregression (VAR) methodology (Lam 2000) and the Kalman filter (Lacey 1998), both of which offer several advantages for the subject of our study:

1. **Modeling Multidimensional Dependencies**: VAR enables the consideration of interrelations among various economic variables, allowing an analysis of how different factors influence capital investments during conflicts.
2. **Forecasting and Evaluation**: VAR can be utilized for predicting future economic changes and assessing their impact on capital investments during conflicts.
3. **Adaptation to Changes**: The Kalman filter permits the adjustment of forecasts and assessments to new information available in real-time, a crucial aspect in conflict situations where conditions swiftly evolve.
4. **Accounting for Instability**: Both methods enable the consideration of instability and unpredictability amid conflict conditions, making risk assessments more adaptive.
5. **Impact Assessment on Markets and Investment Strategies**: Employing these methodologies in risk analysis allows a deeper comprehension of conflict's influence on markets and investment decisions, aiding in the development of portfolio management strategies and risk mitigation.

Incorporating sustainable development into VAR and Kalman filter methodologies also allows for the consideration of ecological, social, and other pertinent aspects crucial for long-term and stable economic development amidst conflict.

A generalized vector autoregression methodology appears as follows (Eberly College of Science 2023):

VAR Model Specification: The VAR model is conceived as a system of equations where each variable can be explained by its lagged values and the lagged values of other variables in the model:

$$Y_t = A_1 mathrm{Y}_{t-1} + A_2 Y_{t-2} + \ldots A_p Y_t - p + u_t \tag{20.1}$$

where Y_t—vector of endogenous variables; $A1$, $A2$, ..., Ap—parameter matrices; u_t—vector of standard errors.

Parameter estimation: The least squares method is used.

Kalman filter in this context is also an effective method for real-time state estimation of dynamic systems, particularly in the presence of noise and uncertainties in measurements and the model. Utilizing this method allows for the assessment of systemic risks associated with capital investments in a country amidst conflict. The outlined methodology involves the following stages:

1. **Outline of the forecasted value**:

$$\hat{x}_{\overline{k}} = F_k \hat{x}_{k-1} + B_k u_k \tag{20.2}$$

where: $\hat{x}_{\overline{k}}$—is the forecasted state value at time k prior to updating; F_k—denotes the transition matrix modeling the system's dynamics; $\hat{x}_{k-1}$ is the estimation of the system's state at time $k - 1$; B_k—represents the control matrix (if control is considered); u_k—stands for the control vector (given control is considered).

2. **Update of the forecast covariance**:

$$P_{\overline{k}} = F_k P_{k-1} F_k^T + Q_k \tag{20.3}$$

where $P_{\overline{k}}$—presents the forecasted state covariance at time k, before updating; $P_k - 1$—is the state covariance at time $k - 1$; Q_k—stands for the covariance matrix of model disturbances.

3. **Determination of measurement disturbances**:

$$K_k = P_{\overline{k}} H_k^T (H_k P_{\overline{k}} H_k^T + R_k)^{-1} \tag{20.4}$$

where K_k—is the Kalman matrix determining the importance of each measured signal for updating the estimate; H_k—is the measurement matrix reflecting how measurements represent the system state; R_k is the covariance matrix of measurement noise.; H_k—is the measurement matrix reflecting how measurements represent the system state; R_k—is the covariance matrix of measurement noise.

4. **State estimate update**:

$$\hat{x}_k = \hat{x}_{\overline{k}} + K_k (z_k - H_k \hat{x}_{\overline{k}}) \tag{20.5}$$

where z_k—measurement vector at time k.

5. **Update of the state estimation covariance**:

$$\mathbf{P_k} = (\mathbf{I} - \mathbf{K_k H_k})\mathbf{P_{\bar{k}}} \qquad (20.6)$$

where I—identity matrix.

For a deeper understanding of investors' concerns regarding capital investments in Ukraine during and after the war, it is also advisable to employ neural network tools within the context of content analysis (Beck 2018) of internet sources related to this topic. The overall methodology involves the following steps: (1) *data collection and processing* (collecting information from various web sources such as news portals, blogs, social media, analytical reports, etc.); (2) *preliminary data processing* (the collected data undergoes preliminary processing, including text tokenization using Python modules (segmentation into individual words or phrases related to investment risks associated with the Ukrainian economy during wartime, '**tokens = text. split**', followed by cleansing from unnecessary characters, lemmatization (reducing words and phrases to their base form (using the NLTK library)); removing stop words, etc.); *vector representation of text* (transforming text into vectors or numerical representations understandable for neural networks using Word Embeddings (Word2Vec) and TF-IDF (Term Frequency-Inverse Document Frequency): the formula for computing the TF-IDF value for a word: TF-IDF(t,d) = TF(t,d) × IDF(t), where TF denotes the word or phrase frequency in the document, IDF represents the inverse frequency of occurrence in the document corpus); neural network construction (creating a neural network that analyzes and learns textual data to identify patterns, connections, and risk assessments. In our study, a Recurrent Neural Network (RNN) was utilized); *model training and testing* (the neural network is trained on the collected data and tested for accuracy on test data. This enables the model to learn complex word dependencies and identify potential risks and patterns in texts related to capital investments during wartime); *results analysis* (after training the model, an analysis of the obtained results is conducted, identifying connections between identified risks and specific events in the context of war.

This approach enables investors and the state to additionally evaluate potential consequences, risks, and challenges for capital investments and make informed decisions based on the obtained forecasts. Here are the general steps used to build the neural network during the content analysis of internet sources concerning systemic investment risks in and from the Ukrainian economy during wartime:

1. **Word Embeddings (Word2Vec)**:

Vector representation of a word (phrase):

$$\mathbf{V(w_i)} = \frac{\sum_{j=1}^{N} \mathbf{f}(\mathbf{w_i}, \mathbf{w_j}) \mathbf{V(w_j)}}{\sum_{j=1}^{N} \mathbf{f}(\mathbf{w_i}, \mathbf{w_j})}, \qquad (20.7)$$

where $V(w_i)$—represents the vector representation of the word (phrase) w_i, $f(w_i, w_j)$—stands for the similarity function between words (phrases) w_i and w_j, N—N is the number of context words (phrases).

2. **LSTM (Long Short-Term Memory)** (Table 20.1):

whereas, here is the decoding of variables used in the formulas: $V(w_i)$—vector representation of a word (phrase) w_i in vector space; $f(wi, wj)$—signifies the similarity function (or distance) between words (phrases) wi and wj; N stands for the number of contextual words in the vector representation of a word; i_t, f_t, $\tilde{c}_t$, c_t, o_t, h_t—denote different layers and components in the operation of LSTM (Long Short-Term Memory)—a recurrent neural network with long and short-term memory; $Wii, Whi, bii, bhi, Wif, Whf, bif, bhf, Wic, Whc, bic, bhc, Wio, Who, bio, bho$—refer to the weights and biases used in various layers of LSTM, optimized during the neural network training; x_t—is the input vector; σ—represents the activation function (e.g., sigmoid or hyperbolic tangent); tanh—refers to the hyperbolic tangent function.

Developed by authors, using (Beck 2018).

The general architecture of the neural network for text analysis consists of an input layer (vectorized text associated with potential investors in Ukraine's economy or representatives of Ukraine's government), a hidden layer (LSTM) (for sequential text processing), and an output layer (prediction of systemic investment risks concerning Ukraine during wartime).

In our research we also use cluster analyses to cluster countries based on indicators of investment process shadowing risk (considering Corruption Perception Index, signs of offshore jurisdictions, and GDP per capita. This methodology involves the following stages:

1. **Selection of the number of clusters**: initially determining the number of clusters we wish to identify among the set of studied objects—based on our prior assessments, we chose 5 clusters.
2. **Calculation of similarity between objects**: for this purpose, we utilized the Euclidean distance between vectors representing the objects.
3. **Grouping of objects**: Based on the computed similarity, the algorithm groups objects into clusters so that objects within the same cluster exhibit high similarity.

Table 20.1 The structure of the LSTM stage in utilizing neural networks for content analysis

№	Stages of construction and utilization of a neural network	Formula
1	Input update	$i_t = \sigma(W_{ii}x_t + b_{ii} + W_{hi}h_{t-1} + b_{hi}$
2	Update for memory	$f_t = \sigma(W_{if}x_t + b_{if} + W_{hf}h_{t-1} + b_{hf})$
3	Memory update	$\tilde{c}_t = \tanh(W_{ic}x_t + b_{ic} + W_{hc}h_{t-1} + b_{hc})$
4	Memory state update	$c_t = f_t \bullet c_{t-1} + i_t \bullet \tilde{c}_t$
5	Output update	$o_t = \sigma(W_{io}x_t + b_{io} + W_{ho}h_{t-1} + b_{ho})$
6	LDTM Output Update	$h_t = o_t \bullet \tanh(c_t)$

4. **Evaluation of clustering quality**: following the formation of clusters, it's necessary to assess their quality, which was done by evaluating intra-cluster similarity and inter-cluster distinctiveness (using Statistica 10.0 software package).

Within the context of cluster analysis, the following methodological approach was employed:

1. **Calculation of the arithmetic mean (average value)**:

$$\text{Average} = \frac{1}{n} \sum_{i=1}^{n} x_i, \tag{20.8}$$

where x_i—sample value; n—number of samples.

2. **Variance**:

$$\text{Variance} = \frac{1}{n} \sum_{i=1}^{n} (x_i - \mu)^2 \tag{20.9}$$

where x_i—sample value; μ—average.

3. **Covariance between two variables (X I Y)**:

$$\text{Covariance}\,(X, Y) = \frac{1}{n} \sum_{i=1}^{n} (x_i - \mu X) \cdot (y_i - \mu Y) \tag{20.10}$$

where x_i, y_i—sample value; $\mu X, \mu Y$—the respective variable means.

Results

Risk disclosure encompasses deliberate endeavors aimed at detailing and conveying to stakeholders the risks that have been successfully addressed, along with the devised strategies to handle prospective future risks. The act of disclosing risks holds importance as it sheds light on how the governmental leadership and investor representatives steer through these risks, offering insights into their implications for the ongoing sustainability of the system in question. Clearly, effective risk management requires a meticulously crafted approach inclusive of an extensive range of analyses, guiding principles, strategies, rationales, and measures geared towards offering suitable responses to intricate high-risk scenarios (Hong 2023).

Note that experts from the World Bank have stated that Ukraine requires over 400 billion USD for post-war recovery. This estimation of damages was made based on an analysis of Ukrainian infrastructure destruction during the year of conflict up to February 24, 2023. However, these assessments have not yet accounted for the extensive technological catastrophe—the sabotage of the Kakhovka Hydroelectric Power

Station by Russian forces (Ukrinform 2023) and others. According to the Prime Minister of Ukraine, D. Shmyhal, as of September 2023, the estimated expenses for the country's reconstruction amounted to 750 billion USD. The President of the European Investment Bank, W. Hoyer, indicated a sum exceeding 1 trillion USD. Evidently, due to the ongoing war, the extent of damages will continue to escalate. M. A. Green, the Chairman of the Woodrow Wilson International Science Center, specified that in its calculations, the World Bank did not consider the costs for the restoration of territories occupied by the Russian Federation since 2014, including the Crimean Peninsula and parts of the Donetsk and Luhansk regions (Sorokin 2023). In this context, it is imperative to conduct a comprehensive assessment, taking into account both external and internal risks and perils inherent in the process of investment recovery in post-war Ukraine. In the context of our research, among the internal risks, touching investment process into Ukrainian economy in the circumstances of the war, the following can be considered:

1. **The lack of a well-established institutional framework** for the functioning of the state under the conditions of a persistent external threat of attack over a long-term period. Despite the ongoing large-scale invasion of Ukraine by the Russian Federation for almost 2 years, such an institutional system has yet to be formed. Even though in 2023, Israel also faced a significant attack by HAMAS militants (Encyclopædia Britannica 2023), the experience of forming national security in that country could be beneficial for Ukraine both during wartime and in the post-war period (given that Israel's comprehensive national security system has proven its effectiveness over an extended period). Notably, the fundamental principles ensuring its institutionalized security include: the use of active military and political efforts to maintain peace; prevention of war and avoidance of confrontations; prioritizing quality over quantity (as Israel has been at a disadvantage compared to its adversaries from the beginning, it compensates with qualitative advantages); conducting operations within enemy territory; minimizing the duration of military actions to reduce harm to its own population and infrastructure; securing the state border; fostering a fighting spirit within the armed forces and population (Eisenkot and Siboni 2019). Moreover, it is essential to consider that the current and future generations of Ukraine will likely live under the prospect of a persistent threat of attack from the Russian Federation (not excluding other potential adversaries).

2. **Insufficient effectiveness of the judicial branch's performance**. According to the Executive Director of the European Business Association, A. Derevyanko, although the overall score of the judicial index in Ukraine slightly improved in 2023 compared to 2021 (by 0.22 points), reaching 2.73 out of five possible (High Council of Justice 2023), there still exist significant challenges within the judicial branch of government that severely complicate the protection of the rights of individuals and legal entities.

Among these issues are: **staff shortages within the judicial system** (as of the end of 2022, in courts that remained operational, there were 1840 vacant judge positions, constituting 28% of the total number of positions, and this number continues to rise. Additionally, according to the Chair of the High Council of Justice, H. Usyk, over 2,000 acting judges have yet to undergo initial qualification assessment, while 361 judges are unable to perform judicial duties due to the completion of their 5-year term of appointment) (Levyy Bereg weekly, 2023); *inefficient court network* (the development of a methodology aimed at optimizing the court network began as early as 2021. It was anticipated to conclude by January 1, 2023; however, the full-scale Russian Federation's invasion altered this timeline. Specifically, due to the occupation of parts of Ukraine's territory, there has been a significant increase in the number of courts unable to administer justice. Consequently, there has been an increased workload on other courts due to the temporary change in territorial jurisdiction. Nevertheless, the necessity of conducting this aspect of reform remains pertinent, and the challenges in funding only amplify its urgency); *the necessity for reforming the Supreme Court* (the introduction of competitive selection for judges of the Constitutional Court of Ukraine was the primary recommendation of the European Commission to maintain Ukraine's candidacy status within the EU. Thus, on December 13, 2022, a law was enacted to comply with this requirement. Presently, the selection of candidates for positions in the Constitutional Court of Ukraine must be conducted through a specially created auxiliary body—the Advisory Group of Experts (AGE), tasked with evaluating candidates' moral qualities and level of professional competence in the field of law. However, a significant drawback of the current law remains the possibility of appointing Constitutional Court judges who, despite the AGE's assessment, do not meet the required level of professional competence in law and high moral standards (The next steps of judicial reform: what are they? 2023). Additionally, several recommendations by the Venice Commission were not adhered to, further complicating Ukraine's integration process into the EU); *insufficiently effective digitalization system in the judiciary*, which diminishes the transparency of the process (although Ukraine is considered a global leader in the digital transformation of public services, the full transition of the national judiciary to digital technologies is still not fully observed. It's important to note that despite the launch of the Unified Judicial Information and Telecommunications System (UJITS) in January 2019) (NAAU and the Judicial Administration began testing the "Electronic Court 2020). Conceptually, it was intended to gradually ensure the automation of a significant portion of processes occurring within the courts: document flow, particularly among case participants, centralized material storage, automated case distribution, video conferencing, data collection, and processing of statistical information, among others. The plan was to commence operation in March of the same year with 8 out of 18 modules in a test mode. However, as of 2023, for various reasons, only three modules have functioned fully: the 'Electronic Cabinet' subsystem, the 'Electronic Court,' and the video conferencing module. Consequently, since the launch of the UJITS, over four years have passed, and during this time, the system has partially become outdated and no longer fully meets current user needs: some modules interact poorly with each other and face challenges with updates. Therefore, there is an urgent need for a comprehensive independent audit of the UJITS system; *necessity to establish a Service of Disciplinary Inspectors within the High Council of Justice; activation of the Higher Intellectual Property Court* (HIPC)

(The Supreme Court on Intellectual Property Issues was established in Ukraine 2017), as stipulated by the judicial reform of 2016.

3. **Debate over the effectiveness of land reform in balancing state and investor interests** (the issue of agricultural land incorporated into companies' charter capital is already noticeable for companies in Ukraine. Despite the lifting of the moratorium in 2021, the Ukrainian agricultural land market remains restricted for commercial use and investment. According to Article 130 of the Ukrainian Land Code (LC), "foreigners, stateless persons, and legal entities are prohibited from acquiring shares in charter capital, stocks, shares, membership in legal entities (…), who are owners of agricultural land" (Land Code of Ukraine 2023b). Violation of this norm could lead to land confiscation. In practice, this new "corporate moratorium" complicates operations not only with land (the intended focus of this restriction) but also with business investments, which unfortunately do not possess agricultural land in ownership. Additionally, based on the results of the sociological survey conducted by AgroPolit.com, Latufundist.com, Kurkul.com, Elevatorist.com, and Zemlak.com, approximately 74% of Ukraine's agricultural market participants support the initiative to postpone the implementation of the second stage of land market reform for legal entities during wartime (The land market in Ukraine—two years: the price of land, land transactions and the impact of the war, 2023) (which was scheduled to commence on January 1, 2024).

4. **Absence of an effective institutional framework for the functioning of the financial market**. Even prior to the war, Ukraine's financial sector was inadequately developed and predominantly reliant on commercial banks. Although commercial banks formed the backbone of Ukraine's financial sector, the total volume of bank lending to the private sector in 2021 accounted for only 28% of the GDP. The country's deposit base also remained low (Fig. 20.1). For many years, the Ukrainian banking system suffered from ineffective risk management, widespread lending to related parties, and regulatory leniency (consequently, the capitalization of Ukraine's stock market in 2021 amounted to only 5% of the GDP, and the capital market infrastructure was significantly fragmented, with loans in the economy largely failing to lead to constructive transformation in the Ukrainian economy compared to other countries), which is partially observed even today.

Therefore, to mitigate financial risks in this sphere, Ukraine needs to:

– *conduct an inventory and explore avenues to minimize the incidence of 'non-performing' loans* (Fig. 20.2).

Currently, auditors are unable to physically access a large number of commercial premises, thus a comprehensive and detailed assessment of asset quality can only be conducted once hostilities cease. Soon after the conflict, a comprehensive Asset Quality Review (AQR) will enable the National Bank of Ukraine (NBU) to determine the recapitalization needs of individual banks.

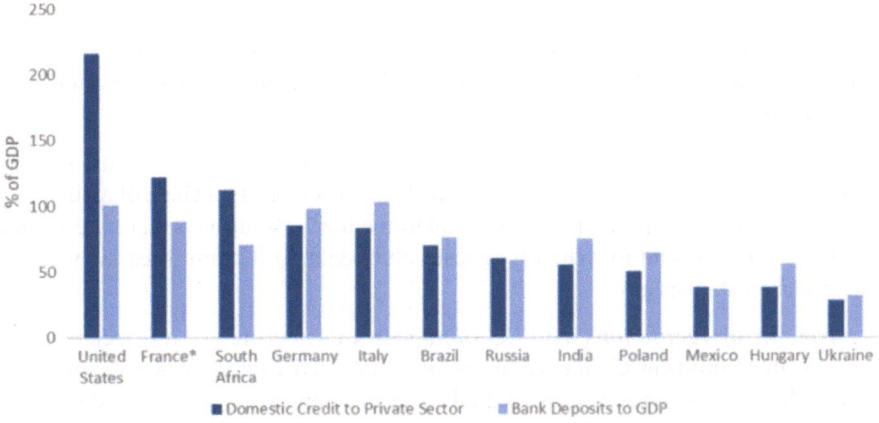

Fig. 20.1 Banking system of Ukraine in comparison to other countries, beginning of 2020. *Source* Haas (2023)

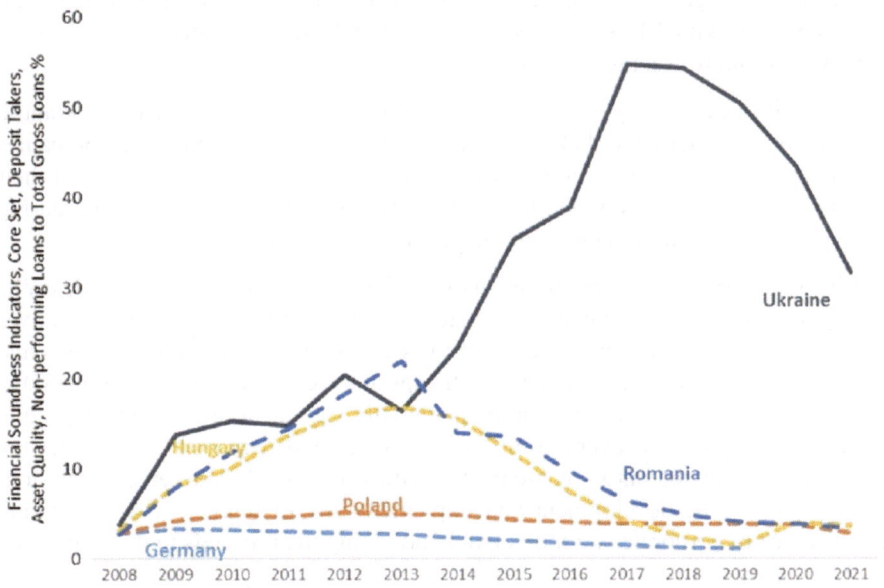

Fig. 20.2 Non-performing' loans in Ukraine and some other countries, 2009–2021. *Source* Haas (2023)

- *the necessity to enhance the level of commercialization in the banking sector.* For an extended period, the Ukrainian banking sector suffered from the detrimental impact of politically motivated lending. Therefore, its post-war recovery will serve as an opportunity for the Ukrainian government and the NBU to optimize not only the balance sheets of commercial banks but also (where necessary) within the shareholders' structure and management (with international support). This will require even stricter due diligence procedures for owners and bank executives to rid themselves of related lending, drawing from the positive experience of 2014–2015.

The improvement of Ukraine's banking sector will require the privatization of the majority of key state-owned banks, which will hold an even larger share of all banking assets post-war. An important issue needing immediate resolution is that state-owned banks are still reluctant to write off or restructure debts if it diminishes the value of any (secured) state assets.

– *harmonizing regulations with EU standards*. Considering that on December 14, 2023, the European Commission officially decided to commence negotiations regarding Ukraine's accession to the EU, in this context, regulatory and supervisory compliance could aid in creating equal conditions for subsidiaries of international banking groups and supporting the long-term sustainable development of cross-border operations in Ukraine.

For instance, alignment of Ukrainian laws concerning professional secrecy and confidentiality with EU standards would enable Ukrainian participation in joint supervisory and restructuring bodies.

– *development of money, bond, and securities markets*. It's critically important to steer the financial sector balance toward capital markets, as Ukrainian companies and entrepreneurs will require a wider spectrum of instruments to support their business growth, considering that many enterprises depleted their capital during the war, thus limiting their ability to take on additional debt obligations.

Further regulatory reforms are necessary to invigorate the securities market. Ukraine still lacks legislation on financial collateral, and essential reforms in derivatives markets are needed to ensure Ukraine receives a clear legal conclusion on netting and close-out netting from the International Swaps and Derivatives Association (ISDA) (International Swaps 2023). Given the potentially significant interest in supporting Ukraine's economic recovery from various social and responsible investors, the securities market regulator (National Securities and Stock Market Commission) must prioritize the development and implementation of regulatory acts enabling the issuance of corporate and municipal bonds using the proceeds to meet specified societal needs.

– *financing of small businesses and financial inclusivity*. Deepening Ukraine's financial sector is intended to benefit broad segments of the country's population, aiding job recovery and means of livelihood. Banks, especially those traditionally focused on large state-owned and/or affiliated companies, will need to adjust their lending practices to become more inclusive and universal lenders. Even before the war, the share of small and medium-sized enterprises with limited access to credit was high and growing rapidly (Fig. 20.3).

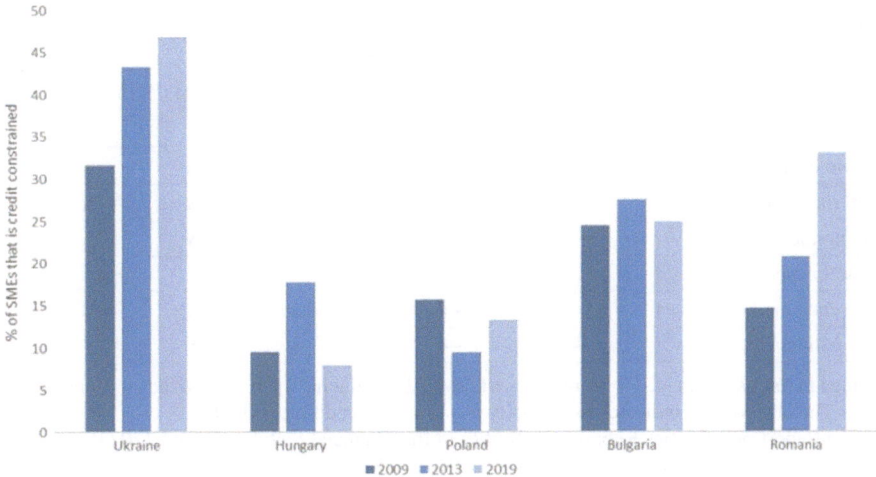

Fig. 20.3 Credit constraints for SMEs in Ukraine and other countries, 2009–2019. *Source* Haas (2023)

- *mobilization of external financing to maintain financial stability.* As international donors assist Ukraine in post-war reconstruction, ensuring predictable donor inflows and gradually relying on commercial decisions to establish an efficient capital market in Ukraine after the reconstruction period is crucial. Allocating substantial resources for risk insurance through specialized agencies such as the Multilateral Investment Guarantee MIGA of the World Bank (MIGA 2023) could be critically important in minimizing political and military risks.

Overall, the Report on Systemic Risks for Ukraine's Financial Sector (as of July 2023), presented by experts from Ukraine's Financial Stability Council, emphasized that the prevailing risks for Ukraine's financial system have been and remain primarily the prolonged full-scale war and Russian terrorist attacks on Ukrainian infrastructure. Although, as highlighted in the mentioned document, the financial sector and the economy as a whole have adapted to operate under force majeure circumstances (Table 20.2).

5. **High level of societal corruption.** The civic engagement program 'Get Involved!' in collaboration with the USAID project 'Supporting Anti-Corruption organizations-leaders in Ukraine, the "Interaction" (SACCI) presented key findings from the 7th stage of the Nationwide Survey 'Corruption in Ukraine: Perceptions, Experiences, Attitudes'. Conducted in winter 2023, this survey marked the first comprehensive study since 2007, encompassing three distinct representative respondent groups to better understand the perceptions and experiences of Ukrainian citizens amidst varying circumstances created by the war: (1) Ukrainians who did not change their place of residence following the full-scale Russian invasion; (2) internally displaced Ukrainians (IDPs); (3) externally displaced Ukrainians (EDPs) (USAID/ENGAGE 2023).

Table 20.2 Evolution of systemic risks for Ukraine's Financial Sector, 2022–2023

№	Change in risk				Risk level			
	2022		2023		08.12.22	03.02.23	04.05.23	19.07.23
	III Q	IV Q	I Q	II Q				
World economy	→	↑	→	↓	x	x	x	x
External conjuncture	↓	→	→	↑	x	x	x	x
Economic conditions	↑	↑	↑	→	•	•	x	x
State finances	↑	↑	→	→	•	•	•	x
Currency market	↑	↑	→	→	•	x	x	x
Geopolitics	↑	→	↑	→	•	•	•	•

Change in risk assessment. ↑—increasing risk level; ↓—reducing the level of risk; → —unchanged level. The level of risk indicates its intensity: •—low; x—medium; ·—high.
Source Report on activities Report on activities of the Financial Stability Counci (2023)

According to the findings of the aforementioned study, corruption is deemed the most serious issue for Ukraine after the full-scale war. This sentiment is shared by 89% of citizens, with political corruption identified as the primary and most serious type of corruption (81%). Despite noticeable improvement in public perception regarding the prevalence of corruption, 94% of respondents still believe that corruption is widespread throughout Ukraine. The percentage of those who are convinced that the level of corruption has increased since the start of the full-scale war exceeds the percentage of those who believe it has decreased, highlighting a highly divided society in assessing this issue. While 53% of Ukrainians consider corruption to be never justifiable (a significant increase compared to 2021), only a small share of respondents have reported instances of corruption to law enforcement agencies. The portion of respondents believing that combating corruption is the responsibility of ordinary citizens has increased from 9% in 2021 to 13% in 2023, yet the majority continue to place the primary responsibility on the President of Ukraine and his Office (44%). Less than 2% believe that this duty should fall on civil society organizations (USAID/ENGAGE 2023).

It's worth noting that trust in specialized anti-corruption bodies has significantly increased compared to 2021. The most substantial increase in public trust occurred towards the Security Service of Ukraine (which had 40% trust in 2023 compared to 12% in 2021) and the National Police (31% compared to 11% respectively). Awareness of anti-corruption measures, reforms, or informational campaigns by state institutions and civil organizations has decreased compared to 2018, but the perception of their effectiveness has significantly increased. Nearly 80% of citizens express readiness to participate in rallies and public protests, although most prefer passive forms of engagement, such as reporting corruption cases in the media and social networks (31%) or initiating and signing electronic petitions (27%). Overall, internally displaced Ukrainians demonstrate the highest level of optimism in perceiving and assessing the level of corruption and anti-corruption efforts among all three respondent groups, while externally displaced Ukrainians are disheartened by the anti-corruption progress in Ukraine (USAID/ENGAGE 2023).

Among the external factors that directly or indirectly impact the efficiency and profitability of the funds being raised for the reconstruction of Ukraine, the following are noteworthy.

1. ***The inability to guarantee the security of capital investments in the absence of a positive and unequivocal decision regarding Ukraine's accession to NATO and the EU*** in the short term (or the creation of an effective alternative to these institutions). While NATO representatives acknowledge Ukraine's security significance to the organization and its members, and the Alliance fully supports Ukraine's inherent right to self-defense and its right to choose its own security systems, since 2014, due to the unlawful annexation of Crimea by the Russian Federation, cooperation has been reinforced in critical directions (NATO 2023). However, even after the full-scale invasion of Ukraine by Russia in 2022, the rapid integration of Ukraine into NATO through a simplified system proposed by Finland and Sweden is highly unlikely. In the short-term perspective, Ukraine's membership in either NATO or the EU is not currently feasible.

2. ***Limited current willingness from international institutions to insure financial and investment risks associated with capital flows to and from Ukraine.*** As of the end of 2023, the Ministry of Economy of Ukraine sought assistance from global financial institutions. The Multilateral Investment Guarantee Agency (MIGA) has already begun providing certain insurance guarantees for war risks in Ukraine from the Reconstruction and Economy Support Trust Fund for Ukraine (SURE). Although the details of all projects are currently undisclosed, it is known that an agreement has been reached to increase guarantees from 17.1 to 40.85 million euros between MIGA and the German banking holding company ProCredit. This decision was announced in London during the Ukraine Recovery Conference 2023 (URC-2023) (MIGA began to provide guarantees for the insurance of war risks from trust fund for the support of Ukraine's reconstruction and economy 2023). However, the increased guarantees from MIGA solely apply to ProCredit's investments in the capital of the Ukrainian subsidiary Joint-stock company 'ProCredit Bank' to support Ukrainian small and medium-sized enterprises. This program will be in effect until December 2025 (MIGA 2023).

Currently, the United Kingdom is also in the process of developing a mechanism for war-related risk insurance to encourage businesses to participate in Ukraine's recovery. This initiative aims to incentivize investment, technology, energy, and defense firms to support reconstruction efforts (Wickham et al. 2023). The Ministry of Economy of Ukraine has proposed risk insurance to the U.S. International Development Finance Corporation (DFC) and all Export Credit Agencies of the G7 countries. Specifically, Ukraine is engaged in discussions with the French Bpifrance (Bpifrance 2023), exploring various collaboration options: insurance through its Export Credit Agency or involvement in the reinsurance trust fund of MIGA for risk mitigation.

Furthermore, the European Union, in collaboration with international partners, is working on implementing military risk insurance to facilitate private investment and business engagement in Ukraine's reconstruction. The National Bank of Ukraine, in conjunction with the World Bank, is also developing a concept for establishing a system to insure military risks in Ukraine. It's worth noting that on December 23, Law No. 3497-IX 'On Amendments to the Law of Ukraine 'On Financial Mechanisms for Stimulating Export Activity' regarding the insurance of investments in Ukraine against military risks' was officially published. This law will come into effect on January 1, 2024 (Verkhovna Rada of Ukraine 2023a, b). However, despite this substantial support for Ukraine in military aspects from international partners, there is currently no discussion on realistic broad-scale opportunities for insuring investment risks for businesses during wartime.

3. *Further unofficial lobbying of interests by Russia and Belarus through a series of countries and their hindrance of necessary decisions regarding Ukraine's support and development*. The U.S. Department of the Treasury released a list of states aiding Russia and Belarus in circumventing sanctions, urging all partners to intensify their oversight of this process. This was highlighted in an official statement from the U.S. government (U.S. Department of the Treasury 2023). It's noted that the U.S. Department of the Treasury's Financial Crimes Enforcement Network (FinCEN) and the Bureau of Industry and Security (BIS) of US Ministry of finance analyzed several countries facilitating the export of sanctioned goods to Russia and Belarus. Notably, this list included Armenia, Brazil, China, Georgia, India, Kyrgyzstan, Mexico, and others. Among the EU countries, whose members generally support Ukraine in various ways, there are specific representatives hindering financial support to Ukraine from the Union. Specifically, Hungary and Slovakia, whose leaders are known for their pro-Russian views, in 2023 blocked the allocation of 50 billion euros in aid from the EU to Ukraine (Politico 2023).

The lack of a new world order system fuels the activation of international criminal and illegal institutions, deepening the global geopolitical crisis. The crises, conflicts, and wars ongoing today vividly demonstrate how deeply the global geopolitical landscape has shifted in recent years, as the exacerbation of unhealthy competition among powerful states has once again become paramount in international relations. According to the Global Peace Index Report as of 2023, there has been a concerning increase in the number and intensity of conflicts that began even before the large-scale Russian invasion of Ukraine, recognized as the most significant conflict since World War II. A significant rise in the number of casualties, both among civilians and military personnel due to conflicts, was observed even before the war between Russia and Ukraine, interrupting a five-year declining trend that started since the peak of the Syrian civil war in 2014 (Izvoshchikova 2023).

In particular, the number of fatalities in conflicts rose by 45% during the period of 2020–2021, primarily in the regions of the Asia–Pacific and Sub-Saharan African regions (Conflict Trends in 2023 2023). After the widespread invasion of Ukraine by Russia on February 24, 2022, according to data published by Statista, the Office of the United Nations High Commissioner for Human Rights (OHCHR) confirmed the deaths of over 9,600 civilians in Ukraine as of September 2023. OHCHR experts note that due to the occupation and inaccessibility of certain territories for monitoring, the actual figures could be significantly higher. Additionally, according to Ukraine's Deputy Minister of Internal Affairs, L. Tymchenko, another 11,000 Ukrainian civilians remain missing. The highest number of fatalities was recorded in March 2022—over 3,900—when Russian forces occupied part of the Kyiv region and advanced towards the capital of Ukraine. OHCHR also notes that, according to preliminary data, over 17,500 Ukrainian civilians suffered various degrees of injuries (Statista 2023). Among them, approximately 1,618 children were affected. As of August 2023, at least 503 children have died, and over 1,115 have sustained injuries of varying severity.

Over the past two decades, the number of 'internationally involved intrastate' conflicts has significantly escalated. Starting from 2004, these conflicts surged nine-fold, reaching a total of 27 instances. Interestingly, these internationally involved intrastate conflicts are now as prevalent as intrastate conflicts, indicating a trend where over 80% of internal conflicts between 1975 and 2017 received external support. Such heightened intervention often aligns with the broader geopolitical ambitions of major global players, evident in cases like increased Russian involvement in Mali and Myanmar (Aung 2023). Additionally, regions like Syria, Libya, and Ukraine have transformed into conflict zones, drawing support from various countries directly or indirectly, thereby exacerbating and perpetuating these conflicts. With the rise of geopolitical rivalries, distinguishing between intrastate and internationally involved intrastate conflicts has become increasingly complex. Recognizing and addressing the nuanced aspects of these evolving conflict dynamics is crucial in fostering global peace and stability.

In the future, the escalation of global challenges will also be linked to destructive climate changes, heightened economic inequality, and political instability, which will undoubtedly impact the redistribution of global financial flows and the profitability of multinational investments.

The combination of the autoregression and Kalman's filter methodologies described above for assessing systemic risks within the context of our research involves analyzing time series of annual data from 1997 to 2022 (where applicable, logarithmic data is used to enhance the statistical significance of modeling): *defining variables* (input parameters—investment flows; resulting variables—GDP per capita, corruption levels, Global Competitiveness Index, sustainable development parameters); *constructing VAR models* (determining model order, variable selection, parameter estimation using the least squares method); *risk assessment using the Kalman filter* (utilizing VAR forecasts to establish initial Kalman filter conditions; using real-time observations to evaluate the system state and adjust forecasts); and *result analysis* (examining systemic risks and their impact on capital investments in wartime conditions).

Employing the aforementioned methods and data concerning the dynamics of Ukraines macroeconomic indicators, which theoretically could be linked with investment flows (whether independent or resulting variables from the period 1997–2022), allows for the following conclusions (Table 20.3). An increase in corruption level significantly reduces the share of investments in GDP (regression coefficient −0.498), the Global Competitiveness Index (−2.286), quality of life index (−2.286), savings as a percentage of GDP (−1.323), and leads to an increase in government debt (2.465). Even considering the pre-war period, investment flows into Ukraine did not yield the theoretically anticipated results for Ukraine's socio-economic development. In particular, the actual growth in investment volumes in Ukraine leads to an increase in the inflation index (−0.586 with a lag of 1), a decrease in GDP per capita (−0.174 with a lag of 3) and the savings level in GDP (−0.592 with a lag of 3), as well as an increase in the national debt to GDP ratio (0.651 with a lag of 2).

However, the unemployment rate decreased (−0.470 starting from a lag of 2) due to the impact of increased investment. Analyzing the influencing factors on the volume of investment in the researched period, we note the following: the corruption perception index (−0.498 with a lag of 2) and the level of external trade security (−0.572 with a lag of 3) are characterized by an inversely proportional impact, while the investment attractiveness index exerts a directly proportional influence (0.436 with a lag of 1). As for other important macroeconomic indicators that theoretically should be closely related to investment processes, they do not exert a statistically significant impact on the respective resulting variables.

Thus, it can be concluded that even before the war, the investment process did not significantly contribute to the socio-economic system of Ukraine due to the substantial political component in this process and the considerable influence of corrupt practices on this economic sector. The war further exacerbated the risks and challenges in this sphere. Therefore, for mutually effective utilization (for Ukraine's economy and potential investors) and a positive synergistic effect from the anticipated capital investment resources for the restoration of Ukraine's economy during and after the war, it is essential to enhance the institutional framework for attracting external and internal investments. Ensuring property rights and improving the investment climate in Ukraine (including minimizing corrupt practices in the process of investment attraction and entrepreneurial activities) is imperative.

Summarizing the analysis results using neural networks (in this case, content analysis of internet sources for 2022–2023), concerning concerns and risks of potential foreign investors and business and government representatives in Ukraine, we note that among the most significant risks identified were the following (Fig. 20.4): political instability, geopolitical tensions, economic instability, currency risks, substantial business losses, significant expenses for securing financial and physical assets, infrastructure destruction, increased poverty rates, labor migration, environmental pollution.

However, as mentioned earlier, besides the risks associated with investors' activities during wartime, there are also significant challenges for the recipient country (in this case, Ukraine). Conflict zones often serve as avenues for "pseudo-investments" for money laundering or terrorism financing due to constrained financial and other monitoring and controlling conditions, both from the recipient country's authorities and international institutions.

Table 20.3 The interplay of investment flows into Ukraine with key macroeconomic indicators, 1997–2022 (vector autoregression, Kalman's filter technics)

№	Dependent variable	Independent variables							Intercept	Std. Err. of b	t(26)	Statistical significance
		Log X1	Log Y1	Log X2	X3	Log Y4	Log Y7	Log X4				
1	Log Y1	-0.498** (-2.814) Lag 2	–	NI	-0.572*** (3.336) Lag 3	NI	NI	0.436*** (2.940) Lag1	1.807	0.100	18.052	p-value = 0,000 R = 0.802; R2 = 595; F(3,22) = 13,237
2	Log Y2	-2.286** (-3.977) Lag 3	NI	-1.285** (-3.577) Lag 3	-1.716*** (4.906) Lag 1	2.067*** (4.037) Lag 2	-0.439** (-2.905) Lag 2	-1.327** (0.934) Lag 2	-2.827	1.226	-2.305	p-value = 0.069 R = 0.916; R2 = 0.712 F(4,5) = 6,556
3	Log Y3	NI	-0.586*** (-3.546) Lag 1	NI	NI	NI	NI	NI	4.089	0.747	5.467	p-value = 0.000 R = 0.586; R2 = 0.316 F(1,24) = 12,580
4	Log Y8	-3.613*** (-1.993) Lag 1	0.045*** (0.224) Lag 3	1.353** (0.867) Lag 2	NI	0.956** (0.445) Lag 1	NI	NI	1.819	0.184	9.861	p-value = 0.000 R == 0.045; R2 = 0.000; F F(1,24) = 0,050

(continued)

Table 20.3 (continued)

№	Dependent variable	Independent variables							Intercept	Std. Err. of b	t(26)	Statistical significance
		Log X1	Log Y1	Log X2	X3	Log Y4	Log Y7	Log X4				
5	Log Y4	NI	−0.174*** (−0.869) Lag 2	NI	NI	–	2.341*** (2.154) Lag 2	NI	1.913	0.231	8.261	p-value = 0.000 R == 0.175; R2 = 0.000; F F(1,24) = 0.755
6	Log Y5	−1.323** (−0.335) Lag 3	−0.592** (−0.765) Lag 3	NI	NI	NI	−2.576** (−1.344) Lag 1	NI	0.281	0.274	1.028	p-value = 0.014 R == 0.592; R2 = 0.323; F F(1,24) = 12,930

(continued)

Table 20.3 (continued)

№	Dependent variable	Independent variables							Intercept	Std. Err. of b	t(26)	Statistical significance
		Log X1	Log Y1	Log X2	X3	Log Y4	Log Y7	Log X4				
7	Log Y6	NI	−0.470*** (−2.613) Lag 2	−0.978* (−0.323) Lag 2	NI	0.632* (0.534) Lag 2	1.243** (0.935) Lag 2	NI	1.595	0.243	6.550	p-value = 0.000 R == 0.471 R2 = 0.189; F F(1,24) = 6,8306
8	Log Y7	2.465*** (1.925) Lag 2	0.651*** (4.210) Lag 2	NI	NI	NI	–	NI	3.317	0.409	8.099	p-value = 0.000 R == 0.651; R2 = 0.401; F F(1,24) = 6,8306

Y1—investment level as a percentage of GDP (%); Y2—global competitiveness index; Y3—inflation index; Y4—GDP per capita (USD); Y5—savings level in GDP (%); Y6—unemployment rate (%); Y7—national debt to GDP ratio (%); Y8—quality of life index; X1—corruption perception index; X2—level of external trade security; X3—balance of current account in the balance of payments; X4—investment attractiveness index

Note *, **, *** statistical significance at the level 90%; 95%, 10% respectively and NI—no impact

Source Calculated by the authors based on data from the International Monetary Fund, World Bank, Transparency International, Ukrstat

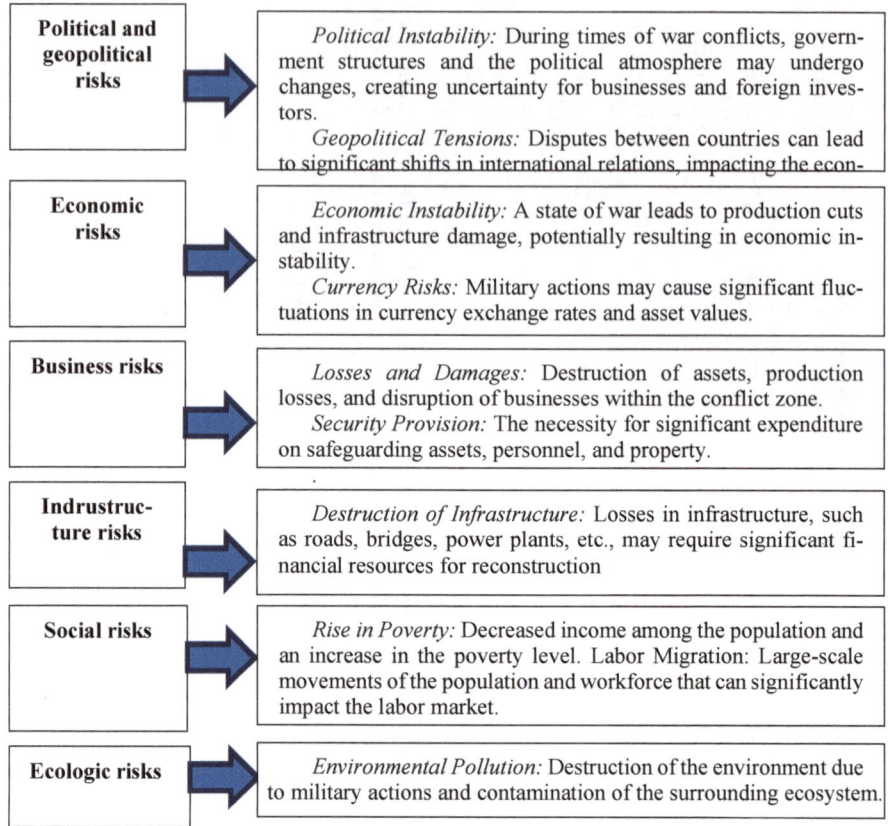

Fig. 20.4 Key factors influencing investment processes in Ukraine during wartime identified using neural network tools (content analysis of internet sources), 2022–2023. *Source* Developed by authors

In our case, as potential investors, we examined countries collaborating within the Ramstein format to support Ukraine in countering Russian aggression (a total of 60 countries). During the 17th meeting within this format, 50 defense ministers from various regions and countries worldwide participated, while other countries indirectly engaged and provided specific support (military, economic, humanitarian) to Ukraine (Nazarenko 2023).

For the purpose of categorizing these countries into groups based on their investment risk indicators concerning Ukraine, we employed cluster analysis (a machine learning method used to group similar objects in a collection in such a way that objects within one cluster are more similar to each other than to those in other clusters). This method was described above.

The basis for grouping was the Corruption Perception Index according to Transparency International 2022 (Corruption Perceptions Index 2022 2022), GDP per capita, and the status (or characteristics) of the country as an offshore jurisdiction as of 2022.

As known, a high level of corruption perception adversely affects the socio-economic development of countries (Fig. 20.5), therefore, attracting investments from such countries may pose a threat to the economic security of Ukraine. Additionally, the influx of investments from offshore jurisdictions (a fictitious variable using a binary approach) may be linked to money laundering and the growth of the shadow economy.

From the figure, it's evident that Ukraine falls within Group 1 of countries with a high level of corruption and low economic development. However, attracting investments from countries in this group is associated with a high level of risk in exacerbating societal corruption. From our perspective, it's prudent to intensify investment activities from Groups 2 and 3, with certain reservations about Group 5, by employing additional control and monitoring tools to counter money laundering, terrorism financing, and the proliferation of weapons of mass destruction in line with FATF (Financing of Terrorism & Proliferation. The FATF Recommendations 2023) and Egmont Group principles (Egmont Group 2023).

Summarizing the results of the cluster analysis, which, as mentioned earlier, divided all potential investor countries among Ramstein group partners into 5 groups, let's provide a brief characterization of them (Fig. 20.6): (1) high corruption, low economic development; (2) moderate corruption, moderate economic development; (3) low corruption, high economic development; (4) highest corruption, lowest economic development; (5) countries with offshore jurisdiction characteristics, high economic development.

Conclusions and Proposals

In light of the information presented and the findings of our econometric analysis, it is imperative to underscore the necessity of adopting the paradigm of 'Concentration of Efforts and Resources for Ukraine's Future Development Security' amidst the persistent and formidable challenges to national security from internal and external factors.

The research highlights several primary directions essential for concentrating actions and resources to address the current imperatives of safeguarding and rebuilding Ukraine, while also ensuring the profitability of both internal and external investors:

1. **Implementation of a war economy:** It is essential to align economic strategies with the contemporary socio-economic landscape of Ukraine and the challenges posed by the global environment. References to existing researches (Mokiy 2022; Ritter 2009) underscore the importance of this alignment.
2. **Overcoming internal system preconditions:** Addressing systemic dysfunctions, notably corruption, requires a concerted effort to integrate judicial and

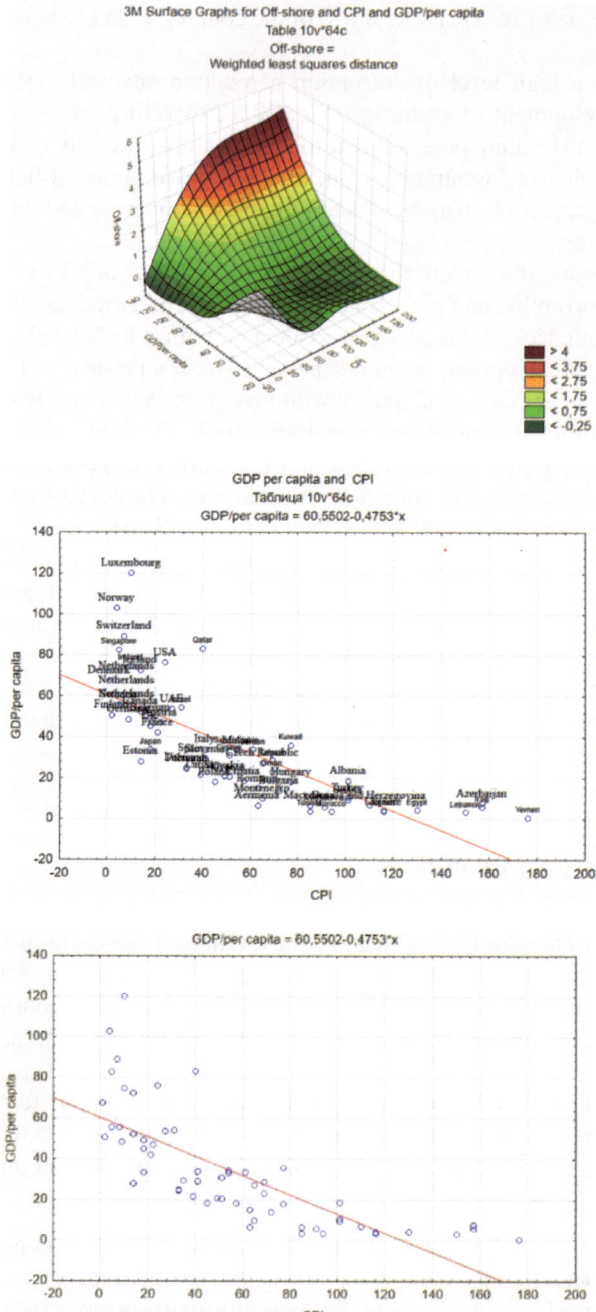

Fig. 20.5 Scatterplot depicting the relationship between the corruption perception level, GDP per capita (USD per capita), and characteristics of offshore jurisdiction in the countries of the 'Ramstein' group (as per the 17th meeting on December 22, 2023), 2023

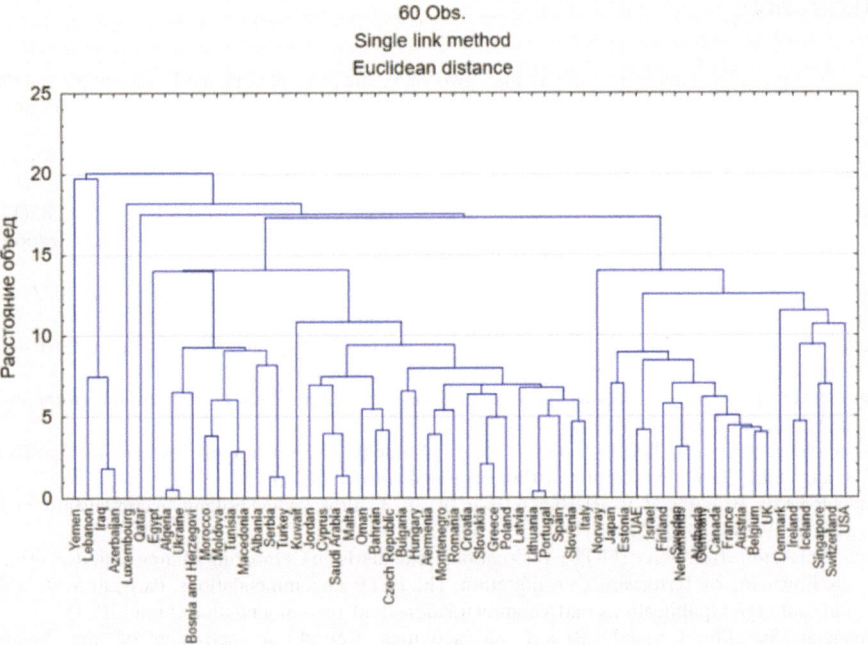

Fig. 20.6 Graph-tree of country-partner connections within the 'Ramstein' group based on indicators of corruption perception, GDP per capita, and offshore jurisdiction characteristics (dummy variable), 2023

law enforcement institutions within a structural model that complements macro-systemic Euro-integration. Mandatory validation of ownership legitimacy by public officials, with repercussions for lack thereof, is crucial in this endeavor.

3. **Identifying factors of productive capability:** Recognizing the strategic importance of factors such as human capital, including intellectual, and associated sectors like microelectronics, is paramount for future security.
4. **Implementing systemic measures for social legitimization of property rights:** Investment of financial resources in priority sectors and spheres is necessary to legitimize property rights socially.
5. **Initiating amendments into principles of stable capital flows and equitable debt restructuring:** The government of Ukraine should spearhead the development and incorporation of amendments into existing principles to ensure stable capital flows and equitable debt restructuring.

By following these outlined directions, Ukraine can navigate its current challenges more effectively, laying the groundwork for sustainable development and enhanced national security in the future.

References

AgroPolit.com: The land market in Ukraine - two years: the price of land, land transactions and the impact of the war. https://agropolit.com/spetsproekty/1018-dva-roki-rinku-zemli-v-ukrayini--dosyagnennya-ta-provali-i-yak-yogo-zminili-viyna (023)

Aung, W.: Conflict Trends in 2023: A Growing Threat to Global Peace. https://www.visionofhumanity.org/conflict-trends-in-2023-a-growing-threat-to-global-peace/ (2023)

Beck, M.: NeuralNetTools: visualization and analysis tools for neural networks. J. Stat. Soft. **85**(11) (2018). https://www.researchgate.net/publication/326703602_NeuralNetTools_Visualization_and_Analysis_Tools_for_Neural_Networks

Bpifrance: Bpifrance, the one-stop shop for entrepreneurs! https://www.bpifrance.com/ (2023)

Corruption Perceptions Index 2022: https://www.transparency.org/en/cpi/2022 (2022)

Eberly College of Science: Vector Autoregressive models VAR(p) models. PennState: Statistics Online Courses. https://online.stat.psu.edu/stat510/lesson/11/11.2 (2023)

Egmont Group: Money Laundering and Terrorist Financing. https://egmontgroup.org/about/money-laundering-and-terrorist-financing/ (2023)

Eisenkot, G., Siboni, G.: Israel's National Security Strategy. The Washington Institute for Near East Policy. https://www.washingtoninstitute.org/media/4613 (2019)

Encyclopædia Britannica: Israel-Hamas War of 2023. https://www.britannica.com/event/Israel-Hamas-War-of-2023 (2023)

Financial Action Task Force (FATF): International Standards on Combating Money Laundering and the Financing of Terrorism & Proliferation. The FATF Recommendations. Paris. https://www.fatf-gafi.org/en/publications/Fatfrecommendations/Fatf-recommendations.html (2023)

Financial Stability Council: Report on activities Report on activities of the Financial Stability Council. https://bank.gov.ua/admin_uploads/article/FSB_Report%202022-2023.pdf?v=4 (2023)

Haas, R.: The future of the financial sector in post-war Ukraine. Ekonomichna Pravda. https://www.epravda.com.ua/columns/2023/01/5/695722/ (2023)

High Council of Justice: The effectiveness of the judicial system in Ukraine improved in 2023—European Business Association. https://hcj.gov.ua/news/efektyvnist-sudovoyi-systemy-v-ukrayini-u-2023-roci-pokrashchylasya-yevropeyska-biznes (2023)

Hong, D.B.: Financial aspects and financial risk disclosure: evidence from Vietnam. Econ. Finance Manage. Rev. **3**(15), 39–48 (2023). https://public.scnchub.com/efmr/index.php/efmr/article/view/252/191

Hryhoriev, H.: Sovereign debt and post-war Ukrainian economic growth—system dynamics approach. Scientific Papers NaUKMA. Economics **8**(1), 32–39 (2023). http://spne.ukma.edu.ua/article/view/289549

International Monetary Fund. https://www.imf.org/en/Home (2023)

International Swaps and Derivatives Association. https://www.isda.org/ (2023)

Izvoshchikova, A.: In Ukraine, the number of children killed and injured as a result of the war has increased. Syspilne Novyny. https://suspilne.media/554565-v-ukraini-zrosla-kilkist-zagiblih-ta-postrazdalih-vnaslidok-vijni-ditej/ (2023)

Lacey, T.: The Kalman filter [review of the Kalman filter]. https://web.mit.edu/kirtley/kirtley/binlustuff/literature/control/Kalman%20filter.pdf (1998)

Lam, M.: Model selection for vector autoregressive processes. Thesis (M.Phil.)—Chinese University of Hong Kong. http://library.cuhk.edu.hk/record=b5890377 (2000)

Levyy Bereg weekly, LB.ua.: The Supreme Court on Intellectual Property Issues was established in Ukraine. https://lb.ua/news/2017/09/30/377991_ukraine_sozdan_visshiy_sud.html (2017)

Levyy Bereg weekly, LB.ua: NAAU and the Judicial Administration began testing the "Electronic Court". https://lb.ua/pravo/2020/07/28/462806_naau_i_sudova_administratsiya_pochali.html (2020)

Levyy Bereg weekly, LB.ua: The next steps of judicial reform: what are they? https://lb.ua/blog/pravo_justice/552223_nastupni_kroki_sudovoi_reformi_yaki.html (2023)

Ministry of Economy of Ukraine: MIGA began to provide guarantees for the insurance of war risks from trust fund for the support of Ukraine's reconstruction and economy. https://me.gov.ua/News/Detail?lang=en-GB&title=MigaBeganToProvideGuaranteesForTheInsuranceOfWar RisksFromTrustFundForTheSupportOfUkrainesReconstructionAndEconomy (2023)

Mokiy, A.I.: Transformation of the methodological foundations of the study and management of the productive capacity of the economy of the regions in the conditions of the challenges of the global environment. Problems of regional development. [Review of Schultz, S. Productive capacity of regional economy: theoretical, methodological and applied aspects: scientific report] (2022)

Multilateral Investment Guarantee Agency (MIGA): MIGA's Ukraine Response. https://www.miga.org/migas-ukraine-response (2023b)

Multilateral Investment Guarantee Agency (MIGA): https://www.miga.org/?gclid=Cj0KCQiAk KqsBhC3ARIsAEEjuJhlpcMNvEKLM6oZ0Dd64OmYUWcd3JYw9akEwpKlN2LYPfyH B4wMF-4aAoT8EALw_wcB (2023a)

Nazarenko, V.: 17th meeting in the Ramstein format: Ground-Based Air Defence coalition for Ukraine was set up. https://war.ukraine.ua/war-news/meeting-ramstein-format-air-defence-ukr aine/ (2023)

Nell, J., Bilan, O., Becker, T., Gorodnichenko, Y., Mylovanov, T., Shapoval, N.: Financing Ukraine's victory. https://kse.ua/wp-content/uploads/2022/09/Financing-Ukraines-Victory_finaldraft.pdf (2022)

North Atlantic Treaty Organization (NATO): Relations with Ukraine. Eurostat Database. Eurostat. https://www.nato.int/cps/en/natohq/topics_37750.htm (2023)

Politico: Hungary and Slovakia oppose Ukraine funding. https://www.politico.eu/article/european-council-summit-eu-leaders-israel-palestine-hamas-ukraine-war-migration/#1293189 (2023)

Ritter, R.: Transnational governance in global finance: the principles for stable capital flows and fair debt restructuring in emerging markets. SSRN Electron. J. (2009). https://doi.org/10.2139/ssrn.1325244

Sau, L.: Annali della Fondazione Luigi Einaudi globalization and its discontents revisited: anti-globalization in the era of trump. Florence 53(1), 193–195 (2019). https://doi.org/10.26331/1077

Sorokin, A.: How much does it cost to rebuild Ukraine after the war. Prohibition. https://zaborona.com/chomu-pislyavoyenna-vidbudova-ukrayiny-skladnisha-nizh-my-dumayemo/ (2023)

State Statistics Service of Ukraine. https://www.ukrstat.gov.ua/ (2023)

Statista: Number of civilian casualties in Ukraine during Russia's invasion verified by OHCHR from February 24, 2022 to September 10, 2023. https://www.statista.com/statistics/1293492/ukraine-war-casualties/ (2023)

Taylor, S.: What is discourse analysis? Bloomsbury Academic Publishing, London. https://www.bloomsburycollections.com/monograph?docid=b-9781472545213 (2013)

Transparency International: Transparency International Global Movement. https://www.transparency.org/en (2023)

U.S. Department of the Treasury: With Over 300 Sanctions, U.S. Targets Russia's Circumvention and Evasion, Military-Industrial Supply Chains, and Future Energy Revenues. https://home.treasury.gov/news/press-releases/jy1494 (2023)

U.S. International Development Finance Corporation (DFC): Investing in Development. https://www.dfc.gov/ (2023)

Ukrinform: The Russian Blown Up Of The Kahavsky Hydropower. All News. https://www.ukrinform.ua/rubric-other_news/3720776-pidriv-rosianami-kahovskoi-ges-usi-novini.html (2023)

USAID/ENGAGE: National Corruption Perceptions and Experience Poll 2023. https://engage.org.ua/eng/national-corruption-perceptions-and-experience-poll-2023/ (2023)

Vasyechko, O.: Counteracting the risks of international investment in the conditions of war. Stat. Ukraine J. **100**(1), 40–50 (2023). https://doi.org/10.31767/su.1(100)2023.01.04

Verkhovna Rada of Ukraine: About Financial Mechanisms for Stimulating Export Activity. The Law of Ukraine. https://zakon.rada.gov.ua/laws/show/1792-19#Text (2023a)

Verkhovna Rada of Ukraine: Land Code of Ukraine. https://zakon.rada.gov.ua/laws/show/2768-14#Text (2023b)

Wickham, A., Nardelli, A., Champion, M.: UK eyes war-risk insurance scheme for Ukraine's reconstruction. Bloomberg. https://www.bloomberg.com/news/articles/2023-06-08/uk-eyes-war-risk-insurance-scheme-for-ukraine-s-reconstruction?srnd=premium-europe (2023)

World Bank: World Bank Group—International development, poverty, & Sustainability. https://www.worldbank.org/en/home (2023)

Conclusion

Vincenzo Pacelli

When you decide to undertake an impervious and slippery journey such as that of studying a complex topic as systemic risk in financial systems, you already realize that the road ahead will be rough although fascinating to travel. You are aware that along the way there will be more new questions which arise than answers that you will be able to reach. But deep down, a researcher doesn't mind this at all, because new questions are necessary, welcome, hoped for to continue the journey. Which is what every passionate traveler (researcher) is really looking for. Therefore, for the mindful and passionate reader, this Volume will have proved full of new doubts and new questions but also the harbinger of some answers.

Through this Volume "Systemic Risk and Complex Networks in Modern Financial Systems", we wanted to provide a broad and varied source of useful essays for understanding, measuring, and mitigating systemic risk within financial systems. Through a multidisciplinary lens, the chapters in this Volume explore various aspects of systemic risk, ranging from theoretical frameworks to cutting-edge methodologies and empirical insights.

Collectively, the chapters underscore the interdisciplinary nature of systemic risk and emphasise the importance of collaborative efforts across diverse fields, including economics, law, mathematics, statistics, physics, and computer science. By adopting a holistic perspective that acknowledges the interconnectedness of financial entities, the Volume provides insights into the complex dependencies within financial systems.

Key themes addressed in the Volume include the role of network science in understanding systemic risk transmission mechanisms, the development of early warning systems for crisis prediction, and the implications of digitalization and cryptocurrencies on systemic risk. Additionally, the Volume examines the effectiveness of macro-prudential policies in mitigating systemic risk, the impact of shocks originating from local banks on broader economic stability, and the relationship between non-performing loan securitization and systemic risk. Network science emerges as a powerful tool, providing deep insights into systemic risk transmission mechanisms.

© The Editor(s) (if applicable) and The Author(s) 2025   411
V. Pacelli (ed.), *Systemic Risk and Complex Networks in Modern Financial Systems*,
New Economic Windows, https://doi.org/10.1007/978-3-031-64916-5

Climate-related financial risks take central stage, highlighting the pivotal role of network models in stress testing and navigating the low-carbon transition. Moreover, digitalization and cryptocurrencies demand heightened attention due to their unique systemic risk profiles. In this context, the Volume explores emerging areas such as cyber systemic risk in the banking sector, risk contagion dynamics in cryptocurrency markets, and the financial challenges associated with the circular economy. Through empirical analysis, theoretical models and case studies, the contributors offer insights into the complexities of systemic risk and provide practical guidance and food for thought for policymakers, regulators, academics and financial practitioners.

By examining the transmission channels, the Volume also offers insights into the complex relationship and interconnection between the banking and insurance sectors. The role of insurance companies in the financial system, particularly in risk transfer services, is indeed paramount, as these firms play a crucial role in mitigating banks' exposure to credit risks.

Looking ahead, the Volume identifies key areas for future research and policy development. These include a deeper understanding of systemic risk transmission mechanisms, integrating environmental, climate and cyber-technological risk factors into risk assessment frameworks, and enhancing international cooperation for crisis prevention and response. Overall, it is hoped that the Volume can help improve our understanding of systemic risk and provide a framework for addressing this critical challenge in modern financial systems. By integrating theoretical insights with methodological approaches and empirical research, the Volume aims to inform decision-making processes and enhance the resilience of financial systems against systemic threats.

Prof. Vincenzo Pacelli

University of Bari Aldo Moro

April, 2024